This book, whose origin was an international conference held in Hawaii, discusses nano- and microfabrication, and how biologists are using such fabricated devices in their research. The goal of the book is to inform materials scientists and engineers about the needs of biologists, and, equally, to inform biologists about how nano- and microscale fabrication may help to shed light on their own particular research problems. The book also aims to stimulate innovative, productive interactions among materials scientists, engineers, and biologists, and to explore ways in which materials scientists and engineers can exploit biological principles and biological assemblies to produce new and ever smaller devices.

In addition to serving as a resource for scientists in the field, the book is intended for any biologist or physical scientist who wishes to understand the current state of the art and what research is currently being done using biological microfabrication.

NANOFABRICATION AND BIOSYSTEMS

NANOFABRICATION AND BIOSYSTEMS

Integrating materials science, engineering, and biology

Edited by

HARVEY C. HOCH
Cornell University

LYNN W. JELINSKI
Cornell University

HAROLD G. CRAIGHEAD
Cornell University

CAMBRIDGE
UNIVERSITY PRESS

Published by the Press Syndicate of the University of Cambridge
The Pitt Building, Trumpington Street, Cambridge CB2 1RP
40 West 20th Street, New York, NY 10011-4211, USA
10 Stamford Road, Oakleigh, Melbourne 3166, Australia

First published 1996

Printed in the United States of America

Library of Congress Cataloging-in-Publication Data

Nanofabrication and biosystems: integrating materials science,
engineering, and biology / edited by Harvey C. Hoch, Lynn W.
Jelinski, Harold G. Craighead.
p.   cm.
ISBN 0-521-46264-9
1. Biotechnology.   2. Nanotechnology.   I. Hoch, Harvey C.
II. Jelinski, Lynn W. III Craighead, Harold G.
TP248.2.N36 1996
660'.6 – dc20   95-10736
                                                   CIP

A catalog record for this book is available from the British Library

ISBN 0-521-46264-9 Hardback

Contents

Contributing Authors

MASUO AIZAWA
Department of Bioengineering, Tokyo Institute of Technology, Nagatsuta, Midori-ku, Yokohama 227, Japan

DAVID L. ALLARA
Department of Chemistry and Department of Materials Science, Pennsylvania State University, University Park, PA 16802

DONALD. M. BRUNETTE
Department of Oral Biology, Faculty of Dentistry, The University of British Columbia, Vancouver, BC V6T 1Z7, Canada

HELEN M. BUETTNER
Department of Chemical and Biochemical Engineering, Rutgers – The State University of New Jersey, Piscataway, NJ 08855-0909

DAVID T. BURKE
Department of Human Genetics, Human Genome Center, University of Michigan, Ann Arbor, MI 48104

CARLOS BUSTAMANTE
Institute of Molecular Biology and Howard Hughes Medical Institute, University of Oregon, Eugene, OR 97403

PETER CLARK
Department of Anatomy and Cell Biology, St. Mary's Hospital Medical School, Imperial College of Science, Technology and Medicine, Norfolk Place, London W2 1PG, England

HAROLD G. CRAIGHEAD
Director, National Nanofabrication Facility, School of Applied and Engineering Physics, Knight Laboratory, Cornell University, Ithaca, NY 14853

MASAYOSHI ESASHI
Faculty of Engineering, Tohoku University, Aza Acoba Aramaki Acoba-ku Sendai, 980-77 Japan

ANDREW EWING
152 Davey Laboratory, Department of Chemistry, Pennsylvania State University, University Park, PA 16802

ALFRED FORCHEL
Lehrstuhl für Technische Physik, Universität Würzburg, Am Hubland, D-97074 Würzburg, Germany

ROLAND GERMANN
IBM Research Division, Zurich Research Laboratory, Säumerstrasse 4, CH-8803 Rüschlikon, Switzerland

FREDERICK GITTES
Departments of Physics, Biophysics Research Division, University of Michigan, Ann Arbor, MI 48109-1055

PAULA BEYER HIETPAS
152 Davey Laboratory, Department of Chemistry, Pennsylvania State University, University Park, PA 16802

HARVEY C. HOCH
Department of Plant Pathology, Cornell University, NYSAES, Geneva, NY 14456

PHILIP E. HOCKBERGER
Departments of Physiology and Biomaterials, Northwestern University Medical School, 303 East Chicago Ave., Chicago, IL 60611

JONATHAN HOWARD
Departments of Physiology and Biophysics, University of Washington, Seattle, WA 98195

LYNN W. JELINSKI
Director, Center for Advanced Technology-Biotechnology, Biotechnology Building, Cornell University, Ithaca, NY 14853

AKIO KAWANA
NTT Basic Research Laboratories, 3-1, Morinosato Wakamiya, Atsugi-shi, Kanagawa 243-01, Japan

MAGNUS MALMQVIST
Pharmacia Biosensor AB, S-751 82 Uppsala, Sweden

HARDEN M. MCCONNELL
Department of Chemistry, Stanford University, Stanford, CA 94305-5080

RUDOLF OLDENBOURG
Marine Biological Laboratory, Woods Hole, MA 02543

KLAUS PLOOG
Paul-Drude-Institut für Festkörperelektroik, Hausvogteiplatz 5-7, D-10117 Berlin, Germany

GERALD H. POLLACK
Center for Bioengineering WD-12, University of Washington, Seattle, WA 98195

ROBERT TAMPÉ
Lehrstuhl für Biophysik, Technische Universität Munchen, D-85748 Garching, Germany

MASAO WASHIZU
Department of Electrical Engineering and Electronics, Seikei University, 3-3-1 Kichijoji-kitamachi, Musashino-shi Tokyo, Japan 180

KENSALL D. WISE
Department of Electrical Engineering and Computer Science, University of Michigan, Ann Arbor, MI 48109-2122

Preface

This book focuses on the exploration of contemporary and emerging approaches of nano- and microfabrication as they apply to biology. In addition to serving as a resource for scientists in this field, this material is intended for novice biologists and physical scientists who wish to access the state of the art and current research in microfabrication as it pertains to biology. The book also can be used for special-topics graduate courses in biophysics or engineering physics, or as resource material for undergraduate courses in biophysics.

The chapters in this book address the materials-science and engineering aspects of nano- and microfabrication, and how biologists have implemented such fabricated devices in their research endeavors. Our goal is to inform materials scientists and engineers about the needs of biologists, and equally to inform biologists about the possibilities of using nano- and microscale fabrication to unravel research problems not heretofore possible; to initiate innovative, productive interactions between materials scientists, engineers, and biologists; and to explore ways in which materials scientists and engineers can exploit biological principles and biological assemblies to produce new and ever smaller devices. This is truly an exciting opportunity to integrate scientists of disparate disciplines and together to explore important biological and engineering problems. This book is an outgrowth of an international conference, "Nanofabrication and Biosystems: Frontiers and Challenges," held May 8–12, 1994, in Kona, Hawaii.

Nano- and microfabrication techniques are widely used in the semiconductor arena, and it is from these approaches that biologists have much to gain. The first two chapters, by Forchel and colleagues and by Germann, introduce the concepts and approaches used by materials scientists. They are in the forefront of contemporary and emerging techniques. These techniques provide devices by which biological systems can be probed, activated, and monitored so that we can learn how they function. The next eight chapters address these important processes through an evaluation of micromechanical devices and microfabricated instruments. The chapters by Esashi ("Microsensors and microactuators for biomedical applications") and Washizu ("The use of micromachined structures for the manipulation of biological objects") are particularly enlightening in providing exciting examples and ideas of how cells can be probed, activated, and monitored. It is from such ideas presented in these chapters that even more useful and creative devices will be devised. As biologists and engineers explore cell function on an ever increasingly microlevel, the need arises for appropriate instrumentation. The chapter by Beyer and colleagues ("Cell constituent analysis: single cell analysis") poignantly explores how femtoliter samples can be analyzed for chemical constituents from single cells grown in picoliter vials, with an eventual goal of selectively monitoring single quantal events.

Cells grow and respond in specific ways to their extracellular environment. Precise control of the chemical and physical domains surrounding cells, particularly of the substratum with which they have intimate contact, is particularly desirable. Nanoscale engineering of substrata bearing specific chemical domains provides an important step forward in controlling the extracellular environment. The approaches described by Allara ("Nanoscale structures engineered by molecular self-assembly of functionalized monolayers") and Tampé and colleagues ("Biofunctionalized membranes on solid surfaces") to create functionalized substrata are both contemporary and suggestive for future directions. Already many researchers have explored cell growth on such biofunctionalized surfaces. Such studies include ordered growth of neuronal cells and their interactions with each other, as in the chapters by Kawana ("Formation of a simple brain on microfabricated electrode arrays"), Hockberger and colleagues ("Cellular engineering: control of cell–substrate interactions"), and Buettner ("Microcontrol of neuronal outgrowth"). Substrata bearing microtopographies are important in discerning how cells grow and respond under various environmental parameters. For example, it was through the use of microfabricated topographies that Hoch and colleagues ("Microfabricated surfaces in signaling for cell growth and differentiation in fungi") were able to discover how important plant-disease–causing fungi sense submicrometer features on host leaves to develop specialized infection structures and invade the host tissues. Similarly, microfabricated topographies have been used by Clark ("Cell and growth cone behavior on micropatterned surfaces") to discern growth patterns of fibroblasts in vitro, and Brunette ("Effects of surface topography of implant materials on cell behavior in vitro and in vivo") has used such surfaces to discover features important to cell and tissue compatibility toward implant materials. The message should be clear that, through the integration of nano- and microfabrication technologies with biology and medicine, important new research directions and discoveries lie ahead.

The last three chapters of this book present several exciting research fronts in biology and medicine, where nanofabrication is waiting to be exploited. Without doubt, there are many biological questions that are waiting to be addressed in all areas of plant, animal, fungal, and microbial biology. And, we need not forget that biological systems could just as well serve as the forging template for nanofabrication. There are important lessons to be gleaned from understanding, and perhaps mimicking, biological materials and the ways that biological materials self-assemble. Biological molecules and systems have continually evolved over geological time scales to acquire properties that are highly attractive in today's drive for new engineering materials. Coupled with their tendency to self-assemble into highly organized two- and three-dimensional structures, biological materials make attractive targets for exploitation by materials scientists and engineers. Thus, there is much to be gained by integrating scientists from research and developmental fields who normally do not think they have much in common. Microfabrication offers new tools for biology and medicine; biology offers new materials and techniques for fabrication. Both of these themes underscore the need to encourage an early dialogue and partnership between biologists and physical scientists. It is our hope that this book will catalyze part of that drive.

We are grateful for the contributions of the many authors of this treatise. Clearly it is their book. We also wish to thank the many other individuals who contributed in many ways toward the fruition of this book, especially Barbara Hickernell of the Engineering Foundation, and the late Gordon Fisher, who initiated the idea of bringing together a group of engineers and biologists on the topic of this book, and to the various granting agencies that supported the conference from which the book was derived. They are the U.S. National Science Foundation, National Institutes of Health, U.S. Army Research Office, and the Engineering Foundation (New York).

Geneva and Ithaca, New York
Harvey C. Hoch
Lynn W. Jelinski
Harold G. Craighead

Introduction – The frontiers and challenges

LYNN W. JELINSKI, HARVEY C. HOCH,
and HAROLD G. CRAIGHEAD

1 Nanofabrication and biosystems: a field that is coming of age

A major thrust in the microelectronics industry has been to produce ever smaller devices. Now devices with features in the order of 0.5 µm can be routinely fabricated in production facilities, and the research frontiers are currently focused in such areas as micromachines (gears, cogs, springs, capacitors, pumps, microforceps), electron beam lithography, and in the production of quantum dots and molecular wires. An enormous amount of research has gone into this field, and materials scientists have a good understanding of important areas such as silicon structure and chemistry, surface treatments, gallium arsenide, and etching rates. In addition to its importance in the microelectronics industry, microfabrication can be used to produce small devices with electrical leads that have the dimensions of cells. This makes it possible to use microelectronics devices to begin to solve, and interface with, biological problems.

Conversely, nanofabrication is approaching the size limits of current technology. It may be possible to use some form of biological self-assembly to produce features that are smaller or have more difficult aspect ratios than is currently possible. Biological molecules self-assemble into highly organized supramolecular structures, and it may be possible to harness some of the self-assembling features of biomolecules to push nanofabrication to even smaller limits.

These two complimentary areas of research appear to be coming together. For example, recent progress in producing microfabricated machines makes it possible to create microforceps that can grasp and position single cells (Amato, 1991). Also possible is the fabrication of submicron pores, designed to mimic pores of the endothelial layer that delimits bone marrow from the blood circulatory system (Waugh, 1991; Waugh and Sassi, 1986). In that study, the question being addressed was whether or not nascent or immature blood cells possess the same ability as do mature blood cells to deform and hence to pass through the pores into the blood stream. It turns out that they do not, which suggests a mechano-elastic mechanism by which they are retained in the bone marrow until they are sufficiently developed to be released. In related studies, others have prepared tiny channels 5 microns in diameter, smaller than the 7-micron diameter of red blood cells. The channels were designed to mimic the transverse section of blood capillaries, with the purpose of studying blood cell deformability under flow conditions (Amato, 1991). These examples highlight ways in which microfabrication is being used to manipulate and study cells.

A number of researchers have used microfabrication to selectively pattern and monitor neural outgrowth. This work has included the fabrication of "diving board electrodes" (Pine et al., 1987; Regehr et al., 1989), in which the neuron was held underneath a cantilevered electrode so that its action potential could be recorded. More recently, miniature three-dimensional "baskets" have been fabricated from which neurons grow new processes. Grids of these baskets could possibly be used as interfaces between living tissue and synthetic material in prosthetic devices. Ultrafine topographies have been used to study oriented cell behavior (see Chapter 20; also Clark et al., 1991), and neurons have been patterned in two-dimensional grids using microfabricated templates and specific surface treatments (Kleinfeld et al., 1988, 1990).

Another example of this emerging field is the use of microfabricated ridges to better understand the mechanism by which plant pathogenic rust fungi infect their hosts through the stomata. Polystyrene replicas of the epidermis of leaves triggered the fungus to produce specialized infection structures, illustrating that topographical features present on the leaf surface serve as the inductive signals. Furthermore, microfabricated ridges of differing heights established the height threshold for this behavior (see Chapter 18; also Hoch, 1987) and also suggested an operative mechanism for mechanosensitive channels (Zhou et al., 1991).

In addition to direct biological applications, the interface between biological and nonbiological substances is important for producing novel sensors and devices. When bound to the gate regions of transistors, biological molecules can influence and/or alter the electronic characteristics, thereby providing switching and sensing capabilities. Similarly, mass changes occur when antibodies bind their antigens. The sensitivity of biological molecules to binding events makes them particularly attractive targets for sensor research. Examples include the binding of monoclonal antibodies to the surfaces of planar waveguides to produce an absorbance-based immunoassay (Choquette et al., 1990) and the commercially available Pharmacia biosensor, based on surface plasmon resonance (see Chapter 7; also Fagerstam et al., 1990).

The light-addressable potentiometric sensor (LAPS) (Chapter 6) is an ion-specific field-effect transistor that uses a pH change generated by the enzymatic reaction of urease with urea to quantify the amount of enzyme present (Bousse et al., 1990). The LAPS device has also been modified to determine the presence of other biological molecules, such as DNA (Hughes et al., 1991).

Most applications of biological molecules in microfabricated devices require the immobilization of proteins on nonporous supports. This requirement presents substantial challenges, particularly when one considers the nonrobustness of biological molecules, the needs for the biological molecule to retain its activity, and the relatively slow response times usually involved (Plant et al., 1991). Special scanning techniques, including near-field scanning optical microscopy, scanning tunneling microscopy, and atomic force microscopy (Chapter 10) have been used to study the nature of the recognition events at chemisorbed surfaces (Haussling and Ringsdorf, 1991).

The preceding selected examples illustrate the diverse and novel science that can arise from research at the boundaries between microfabrication and biology. Such research requires scientists who are experts in their own areas of research but, at the same time, are able to traverse the gap between widely disparate fields. It requires an understanding of each other's language and capabilities. And, it requires a genuine cooperation between biologists and engineers and materials scientists.

The field of nanosystems and biology has reached a critical mass, and it was in recognition of this that a conference was held to provide the groundwork and opportunity for future progress in this emerging and powerful area of research. This book is the result of that workshop, entitled "Nanofabrication and Biosystems: Frontiers and Challenges," and held May 8–12, 1994, in Kona, Hawaii. As with most conferences, much of the information transfer and idea generation occurred in free-ranging conversations during meals and breaks, at late-night brainstorming sessions, and during the last day's wrap-up session. In addition to providing an introduction to the field, this introductory chapter is intended to capture and synthesize the main issues of discussion, to call attention to areas of immediate need, and to set forth challenges for the next generation.

2 Issues

Several interesting questions arose during the course of the conference. One unresolved issue centered around a balance between the perceived needs of biologists and the current research frontiers in nanofabrication. While nanofabricators are currently producing quantum dots, studying molecular wires, and are concerned about surface damage from etching processes (Chapters 1, 2, and 14), biologists are looking for devices and widgets on a much larger size scale. The research frontiers for nanofabrication are on a scale (30 nm) that is small compared to the size of a typical neuron or cell (ca. 10 μm). The question then arises, is nanofabrication as it pertains to biology a technique in search of a question? Related to this is the question, is there enough of an intellectual challenge in micro- and nanofabrication as it pertains to biology to garner excitement and interest from the best fabrication scientists? The answer from the conferees and speakers (Chapters 3–8) was a resounding yes.

Another issue concerns whether GaAs/Si is really useful for biology. Would some other material, perhaps more biocompatible or even a biomade material, be better? What materials should fabricators be working on? It turns out that semiconductor people have a wealth of experience in dealing with clean or deliberately modified surfaces, and this embedded base of knowledge has historically proved to be a boon to the field.

Several societal issues were also raised. Many participants questioned whether adequate fabrication expertise is being focused on biology and vice versa. Would a community knowing more about each others' areas and key scientific questions lead to more productive and innovative research? The conference participants explored ways to enhance these interactions; the generally agreed-upon solution was to have funding specially earmarked for *genuine* collaborations between the disciplines. As the realization surfaced that we are moving closer to a meshing of the disciplines, several attendees remarked that they felt that they were witnessing a historic event at this meeting.

3 Areas of immediate need

Many conferees identified as important a better understanding of the interface between the surface and biomolecules, cells, and membranes (Chapters 9–13, 18). Such an understanding is essential, for example, for improved prosthetic devices (Chapter 19), for sensors (Chapters 6 and 7), and for neural computation devices (Chapters 15–17). This area of immediate need involves developing a *molecular-level* understanding of surface adhesion and surface interactions. Molecular modeling at surfaces is one technique that offers promise.

Another area of immediate need is the Human Genome Project (Chapter 23), which is already well underway (see the next section for a discussion of future challenges for the Human Genome Project). Some bottlenecks were identified, including sample preparation and biochemical steps that require human manipulation. Microfabrication could be used to reduce reagent volumes, saving perhaps as much as $100 million per year. It was noted that carrying out efficient chemical reactions in tiny vials will require knowledge of water structure and diffusion in small chambers. There is also an immediate need for improved analytical detection methods.

Other areas of equal importance are how drug delivery systems could be addressed by nanofabrication. In addition, related to health care are combinatorial libraries and phage display, which could be used to deal with "orphan" receptors. And, despite all the research being performed on microfabricated devices, a need still exists for achieving functional reliability of silicon devices under water and in 0.15 M NaCl.

Effective graduate education for this emerging field was also identified as an area of immediate need. About 20% of the conferees were international graduate students, and each one participated fully in the conference discussions and in the conference poster sessions. The presence of a large number of young scientists involved in the interface between biology and the physical sciences attested to the vitality of the field.

4 Challenges for the next generation

The ability to assemble molecular-scale devices via a combination of biology and nanofabrication was cited as one of the premier challenges for the next generation. Nanofabrication is not a static field, and it is important to keep in touch with its progress. What tool comes after photolithography? Today, photolithography can achieve lines and spaces with dimensions of about 0.5 µm. Conventional wisdom says photolithography can go to 0.2 µm. Below that, what will the production tool be? It may be x-ray lithography or the "projection" of ions and electrons. Electron-beam lithography is viable only on a research scale. What is the ultimate limit in size? It is predicted to be about 30 nm. (Below 30 nm, scaling breaks down and there is no reproducibility because of fluctuations.)

The Human Genome Project also has many challenges for the next generation. Perhaps these challenges were best summarized by one conferee, who, referring to the fact that the DNA must first be chopped up, then sequenced, and then reassembled, asked, "If you have a document to read, do you put it in a shredder first?" The implication here is that we should be looking at alternative strategies for reading the DNA, perhaps mimicking the way that the ribosome moves on DNA and reads it. Perhaps a functionalized atomic force microscope (AFM) (Chapter 10) is an option that could be developed for reading the DNA. Once the genetic information from the Human Genome Project is available, there will be a need to reduce the cost for human genetic screening. Microfabricated banks of tests developed for this purpose could be useful for testing drugs as well.

Another challenge for the next generation is that ultimately we will need the ability to manipulate matter on a molecular scale. We will need to be able to control supramolecular chemistry and molecular architecture. How can biology be used for nanofabrication? Perhaps actin, or microtubule-based motility, could be used for supramolecular assembly (Chapters 21 and 22). Other challenges that were discussed involved distinguishing between self and non-self, determining the principles of neural computation, making tissue engineering a reality, characterizing small amounts of active materials on surfaces, and producing effective prosthetic devices.

The topic of biomolecular materials was one area not specifically covered by the conference speakers, yet it was repeatedly mentioned. Sea urchins, for example, develop spicules with special shapes and intricate structures; as another example, spiders make protein fibers almost as strong as the best humanmade fibers. Many expressed the vision that biomolecular materials is an important area of endeavor and acknowledged that the materials community is beginning to take note of it.

References

Amato, I. (1991) The small wonders of microengineering. *Science* 253:387–388.

Bousse, L., Kork, G., and Sigal, G. (1990) Light-addressable potentiometric sensor. *Sens. Actuators* B1: 555.

Choquette, S. J., Locascio-Brown, L., and Durst, R. A. (1990)Chemical, biochemical, and environmental fiber sensors II. *SPIE* 1368:258–263.

Clark, P., Connolly, P., Gurtis, A. S. G., Dow, J. A. T., and Wilkinson, C. D. W. (1991) Cell guidance by ultrafine topography in vitro. *J. Cell Sci.* 99:73–77.

Fågerstam, L. G., Frostell, Å., Karlsson, R., Kullman, M., Larsson, A., Malmqvist, M., and Butt, H. (1990) Detection of antigen-antibody interactions by surface plasmon resonance. *J. Mol. Recog.* 3:208–214.

Haussling, L., and Ringsdorf, H. 1991 Scanning tunneling microscopy of the specific recognition of biotin by streptavadin at chemisorbed self assembled monolayers on gold surfaces. *Abstracts of the 3rd International conference on Synthetic Microstructures in Biological Research.* September 9–12, 1991, Williamsburg, VA.

Hoch, H. C., Staples, R. C., Whitehead, B., Comeau, J., and Wolf, E. D. (1987) Signaling for growth orientation and cell differentiation by surface topography in *Uromyces*. *Science* 235:1659–1662.

Hughes, R. C., Ricco, A. J., Butler, M. A., and Martin, S. J. (1991) Chemical microsensors. *Science* 254:74–80.

Kleinfeld, D., Kahler, K. H., and Hockberger, P. E. (1988) Controlled outgrowth of dissociated neurons on patterned substrates. *J. Neurosci.* 8:4098–4120.

Kleinfeld, D., Raccuia-Behling, F., and Chiel, H. J. (1990) Circuits constructed from identified Aplysia neurons exhibit multiple patterns of persistent activity. *Biophy. J.* 57:699–716.

Pine, J., Gilbert, J., and Regehr, W. (1987) Microdevices for stimulating and recording from cultured neurons. In *Artificial Organs*, ed. J. D. Andrade, pp 573–582. New York: VCH Publishers, Inc.

Plant, A. L., Locascio-Brown, L., Haller, W., and Durst, R. A. 1991 Immobilization of binding proteins on nonpourous supports comparison of protein loading activity and stability. *Appl. Biochem. and Biotech.* 30:83–98.

Regehr, W. G., Pine, J., Cohan, C. S., Mischke, M. D., Tank, D. W. 1989 Sealing cultured invertebrate neurons to embedded dish electrodes facilitates long-term stimulation and recording. *J. Neurosci. Methods* 30:91–106.

Waugh, R. E. (1991) Reticulucyte ridigity and passage through endothelial-like pores. *Blood* 78:3037–3042.

Waugh, R. E. and Sassi, M. (1986) An in vitro model of erythroid egress in bone marrow. *Blood* 68:250–257.

Zhou, X-L., Stumpf, M. A., Hoch, H. C., and Kung, C. (1991) A mechano-sensitive cation channel in the plasma membrane of the topography sensing fungus, *Uromyces*. *Science* 253: 1415–1417.

1

High-resolution lithographic techniques for semiconductor nanofabrication

A. FORCHEL, P. ILS, R. STEFFEN, and M. BAYER

Technische Physik, Universtät Würzburg
Am Hubland, D 97074 Würzburg
Germany

1.1 Introduction

The study of dimensionality-dependent properties of small semiconductor heterostructures is one of the most active semiconductor research areas at present. During the last decade a vast number of new optical and transport phenomena have been discovered in thin semiconductor heterostructures (quantum wells). These structures can be grown by modern epitaxy techniques as, for example, molecular beam epitaxy with atomically sharp interfaces.

If a suitable material combination (for example, InGaAs/InP) is used, electrons and holes may be confined in the material with the smaller band gap (here InGaAs). If the thickness of the small band gap material is comparable to the de Broglie wavelength of the carriers, significant changes of the physical properties compared to those of bulk crystals are observed. In these quantum well structures the energy dependence of the density of states changes from the well-known square root dependence to a step function, and the energy gap between the lowest allowed states in the valence band and conduction band increases due to the confinement. This induces profound changes of the emission and absorption spectra of materials, which are tunable by the selection of the thickness of the low band gap material.

The possibility of controlling the physical properties of semiconductor layer structures by the thickness of the layers has stimulated a large number of experimental and theoretical studies and has also led to the development of new devices (for example, quantum well lasers, high electron mobility transistors, resonant tunneling diodes). The use of ultrathin layers allows one to optimize the device properties (for example, carrier mobility, quantum efficiency) and to adjust specific device parameters according to system requests. For example, by using InGaAsP/InP quantum well lasers the wavelength ranges of minimum absorption and dispersion of optical fibers can be used for high bit rate optical communication.

If, in addition to the use of thin layers, the lateral dimensions of semiconductor microstructures are reduced into the nanometer range, a further strong change in physical properties is expected (Arakawa and Sakaki, 1982; Asada, 1986; Sakaki, 1992).

$$D(E) \sim E^{1/2}$$

$$D(E) \sim \sum \Theta(E\text{-}E_n)$$

$$D(E) \sim \sum (E\text{-}E_{\bar{n}}E_m)^{-1/2}$$

$$D(E) \sim \sum \delta(E\text{-}E_{\bar{n}}E_{\bar{m}}E_l)$$

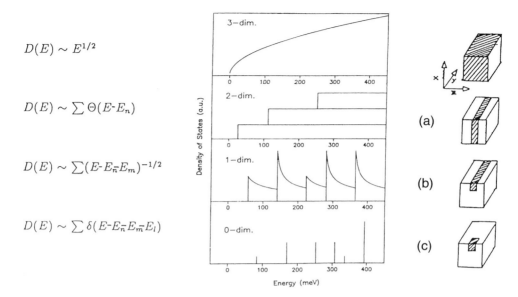

Figure 1.1. Densities of states D(E) as functions of energy for (a) a two-dimensional, (b) a one-dimensional, and (c) a zero-dimensional free electron gas in comparison with the density of states of a three-dimensional system.

Figure 1.1a displays the results of a simple calculation of the conduction band density of states of a two-dimensional GaAs layer (3.5 nm thickness) sandwiched between two AlGaAs layers (Forchel et al., 1988). If the degrees of freedom of the electron motion in the layer plane are restricted to a line of 30 nm width (Figure 1.1b), the density of the states splits into new lateral subbands with resonances at the subband gaps. If the lateral confinement is applied along two edges of the plane, the density of states is composed of a number of narrow peaks (Figure 1.1c). These peaks represent the different lateral levels of a zero-dimensional structure.

Because many optical and transport properties are related to the dimensionality dependence of the density of states, very large changes are expected if a quantum well structure is laterally patterned in the 30-nm range.

1.2 Electron beam lithography for lateral nanometer structures

1.2.1 Basic principles

To date, most of the high-resolution patterning work on semiconductor materials has been based on electron beam lithography (EBL) (Ahmed, 1991; Allee et al., 1991). Electron beams may be generated by using high-brightness point emitters, and they can be focused to spot sizes on the order of 1 nm. By using a suitable deflection system, arbitrary patterns can be defined in electron sensitive resists, which are then transferred into the semiconductors by a suitable etch or implantation step.

Figure 1.2 displays a schematic cross section of the electron optical column of a high-resolution electron beam lithography system (Forchel et al., 1988). The beam is emitted from a high-brightness LaB_6 filament. The electron beam is focused on the substrate by a multi-lens system. The definition of a focus in the nanometer range requires a short working distance between the final lens and the substrate. In order to expose arbitrarily shaped patterns, the electron beam is vector scanned under computer control by electrostatic deflectors. The deflection system is often placed in the final lens to optimize the working distance and to obtain large scanning fields. Deflection fields for high-resolution lithography typically measure 100 μm x 100 μm. For beam currents in the 10-pA range, a focus diameter of 8 nm or less can be maintained throughout this field at an acceleration voltage of 50 kV. The address grid resolution in a high-resolution field is on the order of a few nanometers. In addition to sophisticated and expensive commercial electron beam lithography systems, standard scanning electron microscopes (SEM) or scanning transmission electron microscopes (STEM) may be

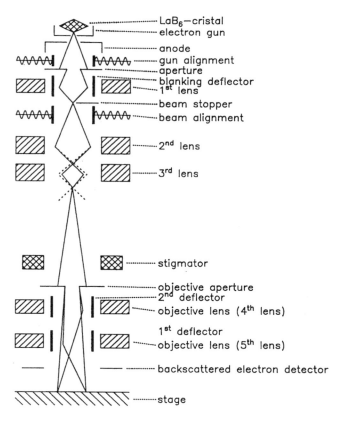

Figure 1.2. A schematic cross section of the electron optical column of a high-resolution electron beam lithography system. The major parts are the electron gun with a LaB_6 cathode, the beam focusing lenses, and the coils for scanning the electron beam.

used for the definition of quantum wire or dot patterns. In this case, the SEM or STEM is interfaced with a pattern generation system, which controls the deflection in the microscopes during the exposures.

In electron beam lithography, the electron beam impinges on a substrate covered by an electron-sensitive material (resist). Depending on the chemical structure of the resist, the electrons either intersect polymer chains or connect smaller chains, thus changing the solubility in organic developers compared to the unexposed areas of the sample. For a positive-tone electron beam resist the exposed material is removed, whereas for a negative-tone resist only the exposed material remains after development. The most commonly used high-resolution electron beam resist is polymethylmethacrylate (PMMA), which acts as a positive-tone resist for typical exposure doses of several hundred $\mu C/cm^2$ at 50 keV.

The basic technological steps for the fabrication of a one-dimensional (or zero-dimensional) structure are illustrated in Figure 1.3 for processes using positive- and negative-tone resists. The process sequence for the definition of a one-dimensional structure using positive electron beam resist can be divided into four major steps (Figure 1.3a). The sample is coated by a typically 100-nm-thick layer of PMMA with different molecular weight. The exposure of the resist with the electron beam followed by a development in a mixture of methylisobutylketone and isopropyl alcohol provides narrow gaps with vertical sidewalls.

The resist mask cannot be used directly as an etch mask for the definition of the quantum wire. Due to the opening in the resist, the wire itself would be etched. A tone reversal is usually obtained by a lift-off process. In this process, a thin layer of a metal (such as Au, Cr, or Ni) or an insulator is evaporated on the sample. With a suitable solvent the unexposed resist with its metal coating is removed. Only the metal that has been deposited directly on the semiconductor surface remains and defines the quantum wires, or quantum dots, for a subsequent etch or implanation step. If etching is used, a free-standing wire or dot structure is

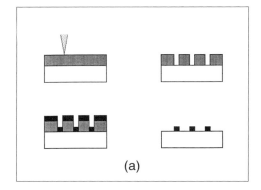 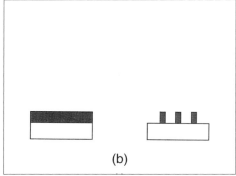

(a)   (b)

Figure 1.3. Basic technological steps for the fabrication of nanostructures using (a) a positive-tone resist and (b) a negative-tone resist.

obtained. In the case of selective ion implantation, buried structures are obtained in a subsequent annealing step.

As shown in Figure 1.3b, the process sequence is somewhat simpler if one uses negative electron beam resist (for example, αM–CMS or HRN). The negative resist is cross-linked in the exposed areas. This reduces the solubility in the developer compared to the unexposed areas. After development of the negative resist, a wire or dot mask – which can be used as an etch or implantation mask – is therefore obtained directly.

In spite of the simplicity of the process, negative resists have been used up to now mainly for low-resolution work. They exhibit poorer contrast and, depending on the chemical composition, may show a swelling of exposed patterns in the development process.

Metal and resist mask patterns are transferred into the semiconductor substrates by dry etching (Maile et al., 1989), wet chemical etching (Ils et al., 1994), or ion implantation (Cibert et al., 1988). By using reactive gases in dry-etch processes to remove the semiconductor, one can optimize the etch rate ratio of the semiconductor and the mask material (selectivity of the etching process). Due to the use of ions with typical energies on the order of 100 eV for the etch process, however, there is a considerable amount of damage in the sidewalls of the etched structures. This leads to a dramatic increase in the nonradiative recombination rate, resulting in the formation of optically inactive ("dead") sidewall layers with extensions of a few tens of nanometers. By using wet chemical etching, surface damage can be avoided completely. These processes therefore are highly suited for the fabrication of ultranarrow lateral structures with typical sizes in the 10-nm range.

1.2.2 High-voltage electron beam lithography

An important criterion for the fabrication of high-quality quantum wire and dot arrays concerns the homogeneity of the wire or dot sizes over the entire array area. In addition to the primary electron dose applied to the resist by the electron beam, backscattering of electrons by the semiconductor substrates contributes significantly to the total electron beam dose in the resist (Allee et al., 1991). Because the range of the backscattered electrons is on the order of a few tenths of a micron to several microns, this implies that the exposure of a particular part of the wire and dot pattern is influenced by backscattered electrons from neighboring structures (the proximity effect). Furthermore, the pattern widths in the resist also include broadening contributions due to forward scattering of the electrons.

In order to obtain ultrasmall mask patterns, high acceleration voltages (100 keV or larger) are particularly attractive (Jones et al., 1987). At these energies, the forward scattering of the electrons by a thin resist layer is negligible and the

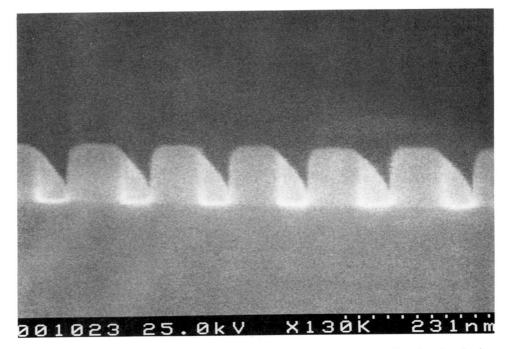

Figure 1.4. Scanning electron micrograph of a PMMA structure after the developing process. The resist was exposed with 200 keV electrons using a commercial transmission electron microscope.

backscattering distribution is broadened strongly compared to the case of exposures at 25 or 50 keV. Hence, the contribution of the backscattered electrons to the exposure dose is significantly diminished compared to the dose of the primary electron beam and is thus uncritical for the exposure. The corresponding reduction of the proximity effects allows the fabrication of ultrasmall wire and dot structures with small periods and with rather large tolerances for the process parameters (for example, electron dose, development conditions).

Figure 1.4 shows a PMMA structure exposed on InP/InGaAs with a commercial scanning transmission electron microscope (TEM) (Hitachi H8000 with a LaB_6 emitter) at an acceleration voltage of 200 kV. A pattern field size of 50 μm x 50 μm was exposed with a positional increment of about 3 nm. The exposure doses range from 600 to 1600 μC/cm² depending on the pattern width and density.

1.2.3 Low-voltage electron beam lithography

Recently, the potential of low-voltage exposure processes for the definition of nanometer structures has been recognized (McCord and Newman, 1992; Stark et al., 1992). At these energies, nearly the total beam energy is transferred into the resist (Peterson et al., 1992). Therefore the fraction of backscattered electrons, which may contribute to the resist exposure, is greatly reduced compared to exposures at higher energies. Low-voltage exposures are therefore expected to be free of the proximity effect. Furthermore, because the electron energy is almost fully used to expose the resist, low-voltage EBL effectively increases the resist sensitivity strongly. On the other hand, the reduction of the beam energy is limited by the resist thickness: If the electrons do not reach the resist–substrate interface, an unexposed resist layer remains at the bottom of the structures. For 100-nm-thick PMMA layers, a minimum electron energy of about 2.2 keV is required for a complete exposure of the layers (Steffen et al., 1994).

For optical investigations on quantum wires, 100 x 100 μm^2–wide arrays of lines with widths down to 55 nm were defined in 100-nm-thick PMMA on InGaAs/GaAs heterostructures at 2.3 keV. For comparison, the same structures were also exposed at 25 keV. For the pattern transfer into the semiconductor an aluminum lift-off was performed. Figure 1.5 displays SEM micrographs of etched structures after exposures without proximity correction at both energies. The micrographs show corners of the wire arrays. The aluminum etch masks are visible on top of the etched wires. In the case of the 25 keV exposure (Figure 1.5a), the width of the etch masks already starts to decrease several microns from the field edges due to the decrease in backscattered exposure near the field edge. This causes a reduction of the wire width, finally resulting in a complete removal of the outermost elements at the corners of the arrays. In contrast, the quantum wire array fabricated by low-voltage exposure (Figure 1.5b) shows no trace of proximity-effect-related width variations. The etch masks keep a constant size out to the corner of the array and they exhibit rectangular mask ends. This is essential for the fabrication of high-quality quantum wires with minimum width fluctuations across the arrays. The energy of quantized electronic levels is to a first approximation inversely proportional to the square of the wire width. Therefore, width fluctuations will lead to different quantization energies in different parts of the wire array. Since for example, photoluminescence spectra are usually obtained from a large number of wires, such variations in quantization would cause a significant broadening of the spectra and would therefore prevent the observation of lateral quantization effects.

1.3 Dimensionality dependence of optical transitions in semiconductor quantum wire structures

1.3.1 Requirements for lateral nanostructures for optical studies and optoelectronic applications

A wide variety of technologies has been used for the fabrication of quantum wire and dot structures. In addition to electron beam lithography and etching (Ahmed, 1991; Izrael et al., 1991; Notomi et al., 1991), focused ion beam techniques (Asahi et al., 1992; Matsui, et al., 1991) as well as scanning probe techniques (Dobisz and Marrian, 1991) are capable of defining lateral patterns in the 10-nm range. In

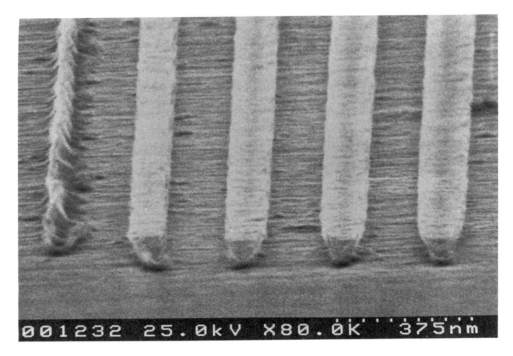

Figure 1.5. SEM micrographs of corners of etched quantum wire arrays; the etch masks are still on top of the wires. (a) Exposure at 25 keV without proximity correction. (b) Low-voltage exposure at 2.3 keV.

these cases, the starting point of the lateral patterning is a two-dimensional heterostructure. Alternatively, there are a number of high-quality approaches based on the in situ definition of wires and dots. These include wires grown by epitaxial techniques into lithographically defined grooves or on top of mesa structures (Arakawa et al., 1993; Kapon et al., 1989; Sogawa et al., 1994; Tsukamoto et al., 1993). Furthermore, wire superlattices have been realized by growth on substrates with steps in the monolayer range (Miller et al., 1992).

In general, techniques used for the fabrication of quantum wires and dots for optical applications must fulfill a number of basic requirements (Sakaki, 1992). In order to observe strong lateral confinement effects, the lateral potential realized by the specific technology should be deep compared to the thermal energy or the quasi-Fermi energy of the experiments. This implies, for example, for room-temperature studies the need for lateral potential depths on the order of 200 meV or more. Due to the electron and hole masses in GaAs and InGaAs lattice matched to InP, the realization of lateral-quantization-induced energy shifts that exceed the thermal energy at room temperature implies lateral wire and dot sizes of 20 nm or less.

In order to observe intense optical transitions in the quantum wire and quantum dot structures, electrons and holes must be confined in the same volume. This excludes the use of electric-field-effect-based approaches, which spatially separate carriers of opposite charge. Furthermore, the patterning process should be performed in such a way that the formation of nonradiative recombination centers is prevented. This is particularly important for the fabrication of wires and dots by dry etching. Due to the incorporation of, for example, ions from the etch gas into the sidewalls of the structures, nonradiative recombination centers are formed, which have been reported to give rise to optically inactive ("dead") sidewall layers (Maile et al., 1989). In addition, the etched surfaces themselves are a source of nonradiative recombination.

In the present work, we have focused on wet-etch processes for pattern transfer because ion-induced damage is completely avoided in these processes. In wet-etch processes, the quantum efficiency degradation in narrow wires and dots is due to surface recombination. In contrast to dry-etched wires, the luminescence energies of wet-etched wires can be modeled quantitatively by using the SEM measured widths of the structures. This implies that wet-etched wires are free of significant dead sidewall layers.

1.3.2 Luminescence studies of InGaAs/InP quantum wires

We have defined $In_{0.53}Ga_{0.47}As/InP$ quantum wires using single quantum well structures, which include a 5-nm $In_{0.53}Ga_{0.47}As$ layer below an 8-nm-thick InP top barrier layer. The high-resolution electron beam exposure was performed on a 100-nm-thick layer of polymethylmethacrylate (PMMA) electron beam resist at an acceleration voltage of 200 kV. By a lift-off process, gold wires are obtained, which serve as a mask for deep wet chemical etching. Details of the fabrication process have been published elsewhere (Ils et al., 1994). The widths of the quantum wires were determined by high-resolution scanning electron microscopy (SEM).

In order to analyze the optical properties of the wet-etched wires, photoluminescence spectroscopy was performed at a temperature of 2 K. The wires were excited with normally incident light of the 514-nm line of a cw argon ion laser. The photoluminescence signal was detected by a liquid-nitrogen-cooled germanium detector using the lock-in technique. The experiments were carried out on wire arrays of about 50 x 50 μm^2 in size. In addition to the wire arrays, mesa structures of the same size were placed on the samples serving as two-dimensional references.

Figure 1.6 shows photoluminescence spectra for different wire widths taken at a fixed excitation density of 350 W/cm². For comparison, a spectrum from a quasi-two-dimensional mesa structure is included at the bottom of the figure.

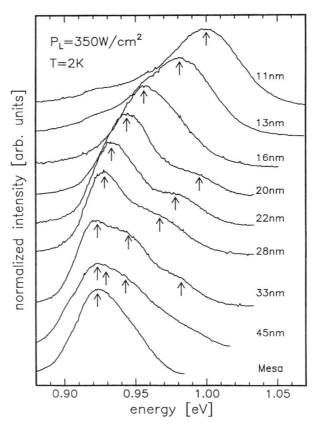

Figure 1.6. Photoluminescence spectra of $In_{0.53}Ga_{0.47}As/InP$ wire structures with different wire widths taken at an excitation density of 350 W/cm². The approximate positions of the different lateral subband transitions are indicated by arrows.

The energy of the photoluminescence emission remains constant for wire widths down to about 60 nm. For smaller wire widths, a significant blue shift is observed, which increases systematically with decreasing wire width and which amounts to about 70 meV for the lowest energy transition of 11-nm-wide wires. For wires with widths between 45 and 20 nm, the electron hole plasma spectra exhibit one or more spectral features on the high-energy edge of the emission; for example, the data for the 33-nm-wide wires in Figure 1.6 clearly show transitions within the first three lateral subbands. With decreasing wire width, the emission features due to higher lateral subband transitions shift strongly to higher energies. Simultaneously, we observe a decrease of the emission intensity. The lateral quantization increases the energetic distance between the e_{11} and hh_{11} ground states and the higher subbands in the wires, resulting in a decrease of the occupation of the higher subbands in the narrow wires. Here the labels e and h denote electron and hole levels in the conduction and valence band; the first index represents the quantum well quantum number, and the second one represents the lateral quantum number due to the size confinement in the wires. The second lateral subband transition e_{12}–hh_{12} vanishes at wire widths smaller than 20 nm. For the smallest wire widths, only the lowest-lying e_{11}–hh_{11} transition is observed.

The arrows in Figure 1.6 display the approximate position of different subband transitions in the wires. The positions have been determined by fitting Gaussian lineshapes to the experimental spectra. The roughness of the wires can be estimated from the systematic increase of the linewidth of the luminescence spectra with decreasing wire width. Based on the calculation of the lateral quantization energies discussed later, the average value for the wire width fluctuations is estimated to about ± 3nm, which is in good agreement with results obtained from SEM micrographs.

Figure 1.7 displays the energy shift of the different photoluminescence transitions of the wires with respect to the two-dimensional reference signal as a function of the SEM-measured geometrical wire widths. The different symbols correspond to the experimental data for the three lateral subband transitions. The energetic distance between the first and second lateral subband transitions increases from a few meV for 45-nm-wide wires to about 40 meV for 20-nm-wide structures. For the energy separation between the second and the third subband transitions, we obtain experimental values increasing from about 15 meV to 45 meV as the wire width is reduced from 45 nm to 30 nm. In conjunction with the large quantization-induced shift of the band edge (up to 80 meV for 8-nm-wide wires) the present wire structures are well suited to investigate properties of the carrier system in the one-dimensional limit. For wire widths smaller than 25 nm, the lateral subband splitting exceeds, for example, the Fermi energy for an electron hole pair density of 1×10^6 cm^{-1} as well as the thermal energy at room temperature and typical longitudinal optical (LO) phonon energies.

The different curves in Figure 1.7 represent the calculated width dependences for the e_{11}–hh_{11} transition (solid line), the e_{12}–hh_{12} transition (dashed line), the e_{13}–hh_{13} transition (dotted dashed line), and the e_{11}–lh_{11} transition (dotted line). The calculation is based on a simple theoretical model using the standard material parameters and assuming a lateral square-well potential of a finite height of 5 eV. This value for the vacuum work function has been found to be typical for many III/V-semiconductor surfaces. For simplicity we used this value for both the electron and the hole barriers. The lateral quantization and the quantization in growth direction are treated to a first approximation as being separable. As shown in Figure 1.7, the measured photoluminescence energy shifts are in good quantitative agreement with the theoretical predictions for all three subband transitions. This suggests that the magnitude of other effects, such as surface states, which might reduce or enhance the lateral confinement, is small.

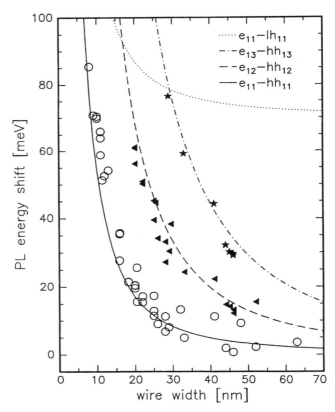

Figure 1.7. Wire width dependence of the lateral subband transition energies observed in the photoluminescence spectra of the In$_{0.53}$Ga$_{0.47}$As/InP wires. The curves represent the calculated energy shift of the first three lateral subband transitions based on a simple theoretical model: e_{11}–hh_{11} (solid line), e_{12}–hh_{12} (dashed line), e_{13}–hh_{13} (dotted dashed line).

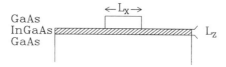

Figure 1.8. Schematic structure of a modulated-barrier InGaAs/GaAs quantum wire with lateral width L_x.

1.3.3 Magnetic field studies of lateral subband transitions in quantum wires

By using modulated-barrier InGaAs/GaAs quantum wires, we have studied the influence of a magnetic field on the transition energies in quantum wires (Bayer et al., 1994). Magnetospectroscopy is particularly interesting because the magnetic field may be used to confine the electron wavefunction in addition to the lateral size confinement (Naganume et al., 1992). For sufficiently large magnetic fields, a transition from size-quantized states to magnetically confined states is therefore expected.

The structure of modulated-barrier InGaAs/GaAs quantum wires is illustrated in Figure 1.8. The quantum wire states lie mainly in the InGaAs well under the GaAs barrier and are confined by a lateral potential barrier that arises because the groundstate energy of the GaAs/InGaAs/GaAs quantum well in the unetched region is lower than that of vacuum/InGaAs/GaAs in the etched region (Gréus et al., 1993, 1994).

The material system used in the experiments is especially attractive for these studies because in the GaAs/InGaAs/GaAs quantum well there is only one confined electron state and one confined heavy-hole state. Further, the strain due to the lattice mismatch between the InGaAs and GaAs causes the light holes to be split from the heavy-hole band by about 60 meV. This implies that features observed within this range of the band edge at zero field can be associated with laterally quantized states.

The wires were fabricated by high-resolution electron beam lithography at 35 keV and selective wet etching on a 5-nm-wide $In_{0.13}Ga_{0.87}As$ quantum well. Details of the fabrication process have been given elsewhere (Gréus et al., 1992). Photoluminescence measurements at a temperature of 2 K were performed using an optical split-coil magneto-cryostat. A cw Ar^+ laser (514.5 nm) was used for excitation. In order to populate several lateral subbands, high-excitation measurements were performed at a power density of 3 kWcm^{-2}. The emitted light was dispersed by a 0.25-m spectrometer and detected by an S1 photomultiplier.

The magnetic field dependences of the energies of the three lateral subband transitions observed in a sample with 29-nm-wide quantum wires are given by the

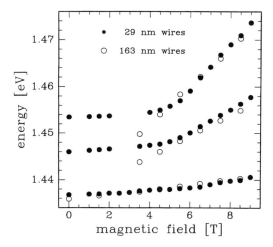

Figure 1.9. The dependence of the experimental transition energy on the normal magnetic field for laterally confined 29-nm wires (solid circles). For comparison, the Landau transitions of wide, quasi-two-dimensional wires (width: 163 nm) are also shown (open circles).

solid circles in Figure 1.9. The magnetic field was oriented parallel to the growth direction and therefore perpendicular to the wire axis. The major qualitative feature to note is a relatively weak increase in the energies with B at low fields and a transition to a stronger dependence at higher fields. To help understand this behavior in more detail, photoluminescence experiments also were performed on effectively two-dimensional (163-nm-wide) GaAs/InGaAs/GaAs wire structures. The energies of the first three Landau transitions in this case are shown by the open circles in Figure 1.9. These results are seen to be similar to those of the quantum wire transitions for high magnetic fields. The similarity of the results for the two cases in this region suggests that for the narrow wires at high fields, the carrier wavefunctions are only slightly influenced by lateral confinement.

We find that for low fields where the magnetic energy is smaller than the confinement energy, the energy shift is given well by perturbation theory in terms of the confined wire states and gives a "diamagnetic shift" proportional to B^2. For high fields where the lateral potential is a small perturbation on the magnetic energy, the energies approach $\hbar(eB)/m_i)(n+1/2)$, where n is the Landau index, and m_i is the electron or hole mass. This high field limit is effectively reached at $B = 10T$. At intermediate fields where the magnetic energy and the lateral confinement energy become comparable, there is a changeover between the high- and low-field behavior.

1.4 Conclusions

As illustrated by the previous examples for InGaAs/InP and InGaAs/GaAs wires, electron beam lithography and wet chemical etching can be used successfully to define lateral nanometer structures with lateral dimensions below 10 nm. The luminescence spectra of the structures display strong lateral-quantization-induced band gap shifts and the formation of higher lateral subbands. The lateral subband structure can be modeled in good agreement with the experimental data by using the measured widths of the structures. This indicates that the present structures are free of significant optically inactive layers. Magnetospectroscopy is particularly suited to study the electronic properties of lateral quantum structures because it allows us to tune the extent of the electron and hole wavefunctions within a given structure. We have used magnetoluminescence to study the transition from a laterally, to a magnetically, quantized system.

Electron beam lithography is well suited to develop lateral nanometer structures. Compared to other techniques (such as nanofabrication based on direct growth), lithography-based approaches are very flexible and permit the realization of an almost unlimited variety of shapes, widths, and distances. The penalty for the enormous flexibility of lithography-based nanofabrication is size fluctuation. Currently, lateral size fluctuations in high-quality wire structures are on the order of ±3 nm. In order to realize wire and dot structures with sizes below 5 nm, all process steps need to be optimized to improve the size control (for example, resist and mask roughness). This will be the main technological task for the next years. These methods, developed largely for electronic devices and physical science research, also can be employed to address new problems in the biological sciences.

Acknowledgments

We are grateful to M. Michel, K. H. Wang, C. Gréus, J. Straka, K. Pieger, and F. Faller for expert experimental support and to T. L. Reinecke, Naval Research Laboratory, Washington, D.C. for helpful discussions. We are grateful to P. Pagnod and L. Goldstein, Alcatel Alsthom Recherche, for the supply of a very high quality InGaAs/InP quantum well wafer. The financial support of the Volkswagen Foundation, the Deutsche Forschungsgemeinschaft, the European Union, and the State of Bavaria is gratefully acknowledged.

References

Ahmed, H. (1991) Nanostructure Fabrication. *Proc. of the IEEE* 79:1140–1148.

Allee, D. R., Broers, A. N. and Pease, F. W. (1991) Limits of nano-gate fabrication. *Proc. of the IEEE* 79:1093–1105.

Arakawa, Y., Naganume, Y., Nishioka, M., and Tsukamoto, S. (1993) Fabrication and optical properties of GaAs quantum wires and dots by MOCVD selective growth. *Semicond. Sci. Technol.* 8:1082–1088.

Arakawa, Y. and Sakaki, H. (1982) Multidimensional quantum well lasers and temperature dependence of its threshold current. *Appl. Phys. Lett.* 40:939–941.

Asada, M., Myamoto, Y., and Suematsu, Y. (1986) Gain and the threshold of three-dimensional quantum-box lasers. *IEEE J. Quant. Elect.* QE-22:1915–1921.

Asahi, H., Yu, S. S., Takizawa, J., Kain, S. G., Okuno, Y., Kaneko, T., Emura, S., Gonda, S., Kubo, H., Hamaguchi, C., and Hirayama, Y. (1992) InGaAs/InP quantum wires fabricated by focussed Ga ion beam implantation. *Surf. Sci.* 267:232–235.

Bayer, M., Forchel, A., Itskevich, I. E., Reinecke, T. L., Knipp, P. A., Gréus, Ch., Spiegel, R., and Faller, F. (1994) Magnetic field induced breakdown of quasi-one-dimensional quantum wire quantization. *Phys. Rev.* B49:14782–14785.

Cibert, J., Petroff, P. M., Dolan, G. J., Dunsmuir, J., Chu, S. N. G., and Panish, M. B. (1988) Optically detected carrier confinement to one and zero dimension in GaAs quantum well wires and boxes. *Appl. Phys. Lett.* 49:1275–1277.

Dobisz, E. A. and Marrian, C. R. K. (1991) Sub-30nm lithography in a negative electron beam resist with a vacuum scanning tunneling microscope. *Appl. Phys. Lett.* 58:2526–2528.

Forchel, A., Leier, H., Maile, B. E., and Germann, R. (1988) Fabrication and optical spectroscopy of ultra small III-V compound semiconductor structures. *Festkörper-probleme* 28:99–119.

Gréus, C., Butov, L., Daiminger, F., Forchel, A., Knipp, P. A., and Reinecke, T. L. (1993) Lateral quantization in the optical emission of barrier-modulated wires. *Phys. Rev.* B47:7626–7629.

Gréus, C., Forchel, A., Straka, J., Pieger, K., and Emmerling, M. (1992) InGaAs/GaAs quantum wires defined by lateral top barrier modulation. *Appl. Phys. Lett.* 61:1199–1201.

Gréus, C., Spiegel, R., Knipp, P. A., Reinecke, T. L., Faller, F., and Forchel, A. (1994) *Phys. Rev.* B49:5753–5756.

Ils, P., Michel, M., Forchel, A., Gyuro, I., Klenk, M., and Zielinski, E. (1994) Room temperature study of strong lateral quantization effects in InGaAs/InP quantum wires. *Appl. Phys. Lett.* 64:496–498.

Izrael, A., Marzin, J. Y., Sermage, B., Birotheau, L., Robein, D., Azoulay, R., Benchimol, J. L., Henry, L., Thierry-Mieg, V., Ladan, F. R., and Taylor, L. (1991) Fabrication and luminescence of narrow reactive ion etched $In_{1-x}Ga_xAs$/InP and GaAs/$Ga_{1-x}Al_xAs$ quantum wires. *Jap. J. Appl. Phys.* 30:3256–3260.

Jones, G. A. C., Blythe, S., and Ahmed, H. (1987) Very high voltage (500kV) electron beam lithography for thick resist and high resolution. *J. Vac. Sci. Technol.* B5:120–123.

Kapon, E., Hwang, D. M. and Bhat, R. (1989) Stimulated emission in semiconductor quantum wire heterostructures. *Phys. Rev. Lett.* 63:430–433.

Maile, B. E., Forchel, A., Germann, R., Grützmacher, D., Meier, H. P., and Reithmaier, J. P. (1989) Fabrication and optical characterization of quantum wires from semi-conductor materials with varying In content. *J. Vac. Sci. Technol.* B7:2030–2033.

Matsui, S., Kojima, Y., Ochiai, Y., and Houda, T. (1991) High-resolution focused ion beam lithography. *J. Vac. Sci. Technol.* B9:2622–2632.

McCord, M. A. and Newman, T .H. (1992) Low-voltage high-resolution studies of electron beam resist exposure and proximity effect. *J. Vac. Sci. Technol.* B10:3083–3087.

Miller, M. S., Weman, H., Pryor, C. E., Krishnamurthy, M., Petroff, P. M., Kroemer, H., and Merz, J. L. (1992) Serpentine superlattice quantum-wire arrays of (Al,Ga)As grown on vicinal GaAs substrates. *Phys. Rev. Lett.* 68:3464–3467.

Nagamune, Y., Arakawa, Y., Tsukamoto, S., Nishioka, M., Sasaki, S., and Miura, N. (1992) Photoluminescence spectra and anisotropic energy shift on GaAs quantum wires in high magnetic fields. *Phys. Rev. Lett.* 69:2963–2966.

Notomi, M., Naganuma, M., Nishida, T., Tamamura, T., Iwamura, H., Nojima, S., and Okamoto, M. (1991) Clear energy shift in ultranarrow InGaAs/InP quantum well wires fabricated by reverse mesa chemical etching. *Appl. Phys. Lett.* 58:720–722.

Peterson, P. A., Radzimski, Z. J., Schwalm, S. A., and Russell, P. E. (1992) Low-voltage electron beam lithography. *J. Vac. Sci. Technol.* B10:3088–3093.

Sakaki, H. (1992) Quantum wires, quantum boxes and related structures: physics, device potentials and structural requirements. *Surf. Sci.* 267:623–629.

Sogawa, T., Ando, S., and Kaube, H. (1994) GaAs/AlAs trench-buried quantum wires with nearly rectangular cross sections grown by metalorganic chemical vapor deposition on V-grooved substrates. *Appl. Phys. Lett.* 64:472–474.

Stark, T. J., Mayer, T. M., Griffis, D. P., and Russell, P. E. (1992) Formation of complex features using electron-beam direct-write decomposition of palladium acetate. *J. Vac. Sci. Technol.* B10:2685–2689.

Steffen, R., Faller,F., and Forchel, A. (1994) Low-voltage electron beam lithography on GaAs substrates for quantum wire fabrication. *J. Vac. Sci. Technol.* B12:3653–3657.

Tsukamoto, S., Naganume, Y., Nishioka, M., and Arakawa, Y. (1993) Fabrication of GaAs quantum wires (~10nm) by metal organic chemical vapor selective deposition growth. *Appl. Phys. Lett.* 63:355–357.

2

Principles of materials etching

R. GERMANN

*IBM Research Division, Zurich Research Laboratory,
Säumerstr. 4, CH-8803 Rüschlikon, Switzerland*

2.1 Introduction

Microfabrication and nanofabrication rely to a great extent on the ability to etch various types of materials. A standard approach for defining the topography on a substrate is shown schematically in Figure 2.1a. After forming a layer on a substrate with epitaxy, evaporation, or sputter deposition, a pattern is defined by lithography in a masking layer, typically in an organic resist. The mask pattern is transferred into the underlying layer by etching away the material in the unmasked regions. The etching can be done by wet chemical etching (WCE) with solutions of acids or bases, or by dry etching with reactive and/or inert gases, using the physical sputtering effect of low-energy ions and/or the chemical assault of reactive gas species.

An alternative to this "subtractive" patterning method is an "additive" approach, which is shown in Figure 2.1b. In the additive approach, a T-shaped masking layer is formed by a combination of directional and isotropic etching techniques, followed by the evaporation of material, such as a metal. As the last step, the mask is removed, which, in the case of an organic underlayer, is done with solvents (lift-off). The pattern created may now also serve as a mask for a subsequent etching process.

Both examples show the importance of etching techniques for micro- and nanofabrication. By repeating the described patterning steps several times, complicated three-dimensional structures can be formed, consisting of layers of semiconductors, metals, and dielectric and organic materials.

The most important commercial applications of etching techniques are in the area of silicon circuit fabrication. Current complementary metal-oxide semiconductor processes for very large scale integration (VLSI) silicon circuits require up to 10 steps of reactive ion etching for different types of materials like silicon, dielectrics (SiO_2, Si_3N_4), silicides, metals (aluminum, tungsten), and organic layers like polyimide (Kaga et al., 1991). Minimum feature sizes of 0.5 μm for 16-Mbit dynamic random access memories are now in production, and feature sizes are expected to be as low as 0.35 μm and 0.25 μm in the near future for 64- and 256-Mbit devices,

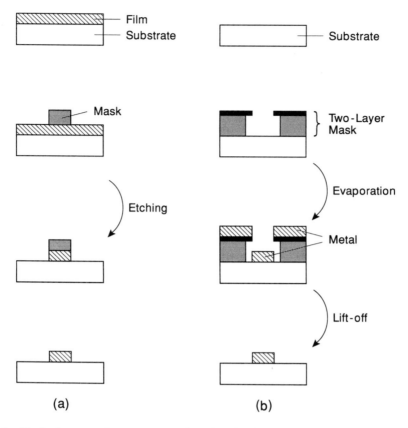

Figure 2.1. Typical processing sequences in microfabrication. (a) Subtractive method. (b) Additive approach (lift-off).

respectively. Such issues as the controllability of the geometrical dimension and the form of the etching profile, the reproducibility of the etching process and the ability to achieve high etch rates, selectivity to underlying layers of various materials, high aspect ratios (ratio between etch depth and feature size), and smooth etched sidewalls and surfaces play an important role in the miniaturization process.

Other important commercial applications are in the field of optoelectronics, where the etching of such III/V compound semiconductors as the GaAs/AlGaAs and the InP/InGaAs(P) material systems is the basis for fabricating devices like detectors, lasers, and light-emitting diodes.

The scientific domain concentrates on finding new etch recipes and methods for new material systems and on optimizing current ones, for example, for higher etch rates, lower material damage, or environmentally safer etch chemistries. The downscaling of VLSI devices and the emergence of new device concepts with mesoscopic and nanoscale dimensions push the exploitation of feature sizes down to the range of only a few nanometers.

Because the need to etch many different materials with very different requirements has led to the development of a variety of etching chemistries, methods, and
tools, it is impossible to give a comprehensive overview within the scope of this
chapter. Instead, I will attempt to describe some basic underlying principles illustrated by examples from the area of optoelectronics.

This chapter is organized in three main sections. The first part gives an
overview of wet chemical etching of silicon and III/V materials. The second part
deals with dry etching for the same material systems and covers etching
chemistries and mechanisms. The etching of ridge waveguide structures with
reactive ion etching (RIE) and the etching of laser facets with chemically assisted
ion beam etching (CAIBE) will be used for illustration. The third part presents a
brief overview of etching-induced damage.

2.2 Wet chemical etching (WCE)

WCE uses solutions of acids or bases. For III/V semiconductors a redox reaction
is the dominant etching mechanism. In some special cases, hydrolysis occurs. A
review of the reaction kinetics of WCE can be found in Loewe et al. (1991), which
also gives an overview of useful etching recipes for Si, GaAs, GaP, and InP.

In the beginning of semiconductor device technology, WCE was the only
method for pattern transfer (before dry etching methods were introduced). WCE
has certain advantages: It is a simple and inexpensive process and, due to its purely chemical nature, can be very selective for different materials, dopants, or doping levels, and for crystallographic planes. Its material and doping sensitivity
allows an etching process to be stopped with high precision, whereas its crystallographic sensitivity makes the fabrication of structural features possible, for
which the sidewalls are exactly defined by low-index crystallographic planes.
Silicon bulk micromachining is a well-known example of this (Petersen, 1982).
WCE is also a low-damage process, because, in contrast to dry etching methods,
no energetic ions are involved.

Nevertheless, there are some severe drawbacks and limitations to WCE. In
general an isotropic sidewall profile with circularly rounded edges is obtained
(see Figure 2.2). Typical for this isotropic etching is the lateral undercutting of the
etch mask. The shape of the etch profile can depend very strongly on the composition of the etchant and on temperature. Mask adhesion is sometimes difficult to
control owing to oxide or contamination layers and can lead to enhanced etching
below the mask. These points together make the exact control of the linewidth and
of the profile form very critical, especially for features with high aspect ratio or
submicron structures. Crystallographic etching as depicted in Figure 2.3 depends
strongly on substrate and pattern orientation (Bassous, 1978). Whereas a defined
profile shape can be obtained in a certain direction, the etching of circular or arbitrarily shaped features with the same sidewall profile on every side is generally

impossible. From a practical point of view, the reproducibility of WCE is often questioned, because the etching result depends strongly on composition, temperature, and sample or bath agitation and can be difficult to control. Moreover, WCE is not very compatible with highly automated wafer handling in VLSI processing.

Despite these disadvantages, WCE has its place in production and research. Three main application areas are the polishing of substrates, revealing of defects, and pattern transfer.

2.2.1 Wet chemical polishing

The preparation of planar, high-quality semiconductor substrates entails sawing and mechanical lapping, as well as polishing steps. To remove crystal damage and to obtain a smooth, mirrorlike surface, chemical or mechanochemical polishing steps are used. Si substrates can be polished with alkaline solutions combined with mild abrasives such as colloidally dispersed silicic acid (Loewe et al., 1990). For GaAs, alkaline NaOCl solutions (Rideout, 1972) or diluted solutions of bromine in methanol (Br_2/CH_3OH) (Sullivan and Kolb, 1963) are used. Br_2/CH_3OH can also be used to polish InP (Aspnes and Studna, 1981).

2.2.2 Revealing of structural defects

Substrates or epitaxially grown films can contain various structural defects such as dislocations or point defects. To characterize their number, density, and local distribution, it is necessary to make them visible. WCE is a suitable method for this, because the etch rate can be locally reduced or enhanced near a cluster of point defects or at the point where a dislocation line penetrates the sample surface. This leads to the formation of so-called etch pits with a diameter of a few micrometers, which are visible under an optical microscope. Typical etchants for this purpose are mixtures of $HNO_3/HF/CH_3COOH$ (Dash, 1956) for silicon substrates; HNO_3 for (111)-GaAs; KOH for (100)-GaAs (Grabmaier and Watson, 1969); and a mixture of H_3PO_4/HBr for InP (Huber and Linh, 1975).

2.2.3 Pattern transfer: Wet chemical etching of microstructures

Advantages and disadvantages of WCE for pattern transfer have already been mentioned. The shape of the etched sidewalls is either isotropic or crystallographic, as depicted in Figures 2.2 and 2.3. Some typical etchants for the fabrication of microstructures in silicon and in the GaAlAs and the InGaAs(P) material systems will be discussed.

Silicon microstructures. WCE is now rarely used for the fabrication of VLSI circuits, mainly due to controllability and reproducibility problems. WCE is more important in the field of micromechanics, for the etching of bulk structures with dimensions of up to a few hundred micrometers or even millimeters (Petersen, 1982). The most commonly used crystallographic etchants for this purpose are KOH in water (Kendall, 1975); mixtures of ethylene diamine, pyrocatechol, and water (EDP) (Finne and Klein, 1967); or mixtures of HF, HNO_3, and CH_3COOH (Schwarz and Robbins, 1976). The etch rates for various crystallographic planes such as (100), (111), or (110) can differ greatly. For a KOH-based etch, the etch rate ratio between the (110) and the (111) plane can reach 400:1 (Kendall, 1975), allowing a very anisotropic etch. On (100) substrates, V-grooves are formed with an inclination angle of 54.7° to the (100) surface, whereas on (110) substrates vertical walls or sidewalls with an inclination angle of 35.26° to the surface occur. Boron-doped layers can be used for a highly selective etch stop (Bogh, 1971). Anisotropically etched structures in silicon are used for a variety of applications as nozzles, gratings, x-ray masks, free-standing cantilevers, and fiber alignment grooves, among other things (Csepregi, 1985; Kaminsky, 1985; Petersen, 1982).

GaAs/GaAlAs material system. The mechanism of WCE of GaAs is a redox reaction in solutions with an oxidizing agent such as H_2O_2 or Cr(VI). The most commonly used etchants for GaAs are acidic or basic solutions containing H_2O_2, such as $H_2SO_4/H_2O_2/H_2O$ (Iida and Ito, 1971), $H_3PO_4/H_2O_2/H_2O$ (Mori and Watanabe, 1978), or NH_4OH $(NaOH)/H_2O_2/H_2O$ (Gannon and Nuese, 1974), and mixtures of bromine in methanol (Br_2/CH_3OH) (Tarui et al., 1971). Cr(VI)-based chemistries such as the $HCl/CH_3COOH/K_2Cr_2O_7$ system have also been used (Adachi and Oe, 1984). Controllable etch rates in the range of 0.01 to 1.0 µm/min have been achieved in this case. More data can also be found in Ashby (1990).

Figure 2.2 shows a scanning electron microscopy (SEM) cross section of a stripe in a GaAs/GaAlAs heterostructure, etched with $H_2SO_4/H_2O_2/H_2$) (8:1:100). Etch rates for GaAs and AlGaAs are comparable in this case, and the etch profile is isotropic. Mask undercutting can also be clearly seen. This etch process has been used for the fabrication of ridge waveguide lasers in the 0.8-µm wavelength region (Harder et al., 1986). WCE has also been applied to etched laser facets (Bouadma et al., 1982), to the etching of grooves and mesa structures for epitaxial growth on patterned substrates (Jaeckel et al., 1989), and to the fabrication of corrugated periodic structures for distributed feedback and distributed Bragg reflector (DBR) lasers (Comerford and Zory, 1974).

InP/InGaAs(P) material system. InP can be etched with halogen acids such as HCl by a hydrolysis mechanism, and with bromine-containing solutions by a redox reaction. Basic solutions cannot be used because of the insolubility of the indium reaction products. The etching of the ternary (InGaAs) or the quaternary (InGaAsP) compounds is more complicated. Pure halogen acids etch selectively

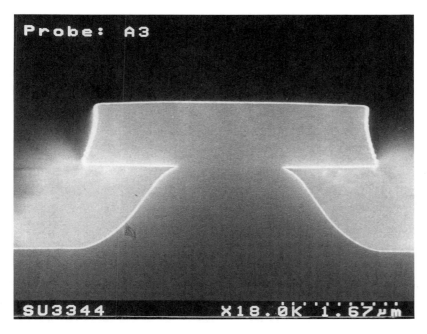

Figure 2.2. Isotropic wet etching. GaAs/AlGaAs heterostructure etched with $H_2SO_4/H_2O_2/H_2O$ (8:1:100). Photoresist mask is still present on top of the etched stripe.

InP and not InGaAs(P), whereas bromine-containing etchants attack all these materials unselectively. The addition of oxidizing agents (H_2O_2, HNO_3) leads to transitions between the two types of reactions (Loewe et al., 1990). For these reasons, many different etchants are used to etch the InP/InGaAs(P) material system.

Pure HCl or mixtures of HCl with CH_3COOH or H_3PO_4 are used for the selective etching of InP with a crystallographic profile. The addition of HNO_3 to HCl makes the unselective etching of InP and InGaAs(P) possible (Adachi et al., 1982). Similar behavior can be found for HBr. Pure HBr or mixtures of HBr with nonoxidizing acids (CH_3COOH, H_3PO_4) etch only InP, whereas mixtures of HBr with oxidizing agents (HCl, H_2O_2, HNO_3, $K_2Cr_2O_7$) also etch InGaAs(P) (Adachi, 1982; Adachi et al., 1982). The solution of bromine in methanol (Br_2/H_3OH) etches InP and InGaAs(P) unselectively with similar etch rates (Turley and Greene, 1982).

Figure 2.3 shows an SEM cross section of an InP sample that has been etched with a mixture of H_3PO_4 and HCl (3:1). The stripes were oriented in the < 011> direction on a (100) substrate. The etch profile is formed by (111) planes in the lower part and by (100) planes in the nearly vertical, upper part. The resist mask and a thin underlying InGaAs layer are still present on the sample. The InGaAs layer has been patterned in a first step before etching the InP. The shape of the profile can be controlled by the ratio between HCl and H_3PO_4. For pure HCl it is given by (111) planes without any vertical part, whereas for a H_3PO_4 proportion greater than 80%, the etch is nearly vertical (Buchmann and Houghton, 1982). For

Figure 2.3. Crystallographic wet etching. SEM cross section of a stripe in InP etched in H$_3$PO$_4$/HCl (3:1).

stripes oriented perpendicular to the shown orientation, the etch profile has an inclination angle of ≈26° in all cases.

WCE of InP/InGaAs(P) materials has been used, for instance, for buried laser structures (Logan et al., 1982), laser mirrors for 1.3-μm lasers (Iga et al., 1980), and 1.5-μm DBR lasers (Tanbun-Ek et al., 1984). These DBR gratings are submicron structures with a grating period of ≈240 nm, etched 80 nm deep. Much finer structures, such as barrier-modulated quantum wires with a width of as little as 25 nm and a very shallow etch depth of ≈20 nm, have also been demonstrated by WCE (Gréus et al., 1993).

2.2.4 Summary

WCE still has important applications for substrate preparation and for pattern transfer with dimensions well above a few microns, such as silicon micromachining and structures for optoelectronics. Advantages of WCE are that it is low-cost, easy to process, has high possible material selectivity, and incurs low damage. It should be pointed out, however, that there are also numerous disadvantages for WCE: for example, the process suffers from poor controllability and reproducibility, it undercuts the mask, and its etch profile depends strongly on the substrate and pattern orientation. These drawbacks restrict the application of WCE for the fabrication of submicron structures and nanostructures to certain special cases, such as a few specific pattern orientations and shallow etch depths. These disadvantages can be overcome with dry etching methods presented in the next section.

2.3 Dry etching

For the reproducible fabrication of submicron structures and nanostructures, it is necessary to find etching processes that allow a mask pattern to be transferred with high geometrical fidelity into the underlying substrate or film. As pointed out in the previous section, WCE general cannot fulfill these requirements. Dry etching methods allow anisotropic etching of structures down to a width of a few nanometers, and they feature a wide range of material selectivity, high aspect ratios, arbitrary pattern orientations and forms, good reproducibility, and automatibility.

2.3.1 Classification of dry etching processes

Plasma-based etching was introduced 25 years ago as a new method for removing photoresists (Irving, 1971). Since then, many different dry etching processes have been developed. With the exception of photochemical etching, all these processes are based on low-pressure plasmas of inert and/or reactive gases and make use of positively charged ions, radicals, and reactive neutrals.

Based on different types of reactors, pressure ranges, gases, electrode configurations, and excitation frequencies, various dry etching techniques have been developed, often unsystematically and inconsistently named. A useful classification is possible through the underlying etching mechanisms to distinguish among four cases (Fonash, 1985).

Physical etching mechanism. In sputter or ion beam etching (IBE) (Lee, 1984), noble gas ions are accelerated with energies of up to a few hundred electron volts toward the sample, and they remove material by physical sputtering. Physical etching is characterized by low etch rates (up to tens of nanometers per minute), poor material selectivity, and secondary effects such as redeposition and trenching that affect the shape of the etched sidewalls. Advantages of physical etching are the anisotropic etching characteristic, which is due to the directionality of the ions, and the ability to control the shape of the etch profile by the inclination and rotation of the sample relative to the ion beam. Physical etching is also used for materials for which a satisfying chemical etch cannot be found, such as certain metals and alloys of magnetic materials.

Chemical etching mechanism. Pure chemical etching occurs in a plasma of reactive gases if the energy of the produced ions is negligible and the etching is dominated by the chemically active gas species. This is true for a high plasma pressure where the mean free path of the ions is too small to attain high energies. The reactive gas species are adsorbed at the sample surface and form volatile etch products that are pumped away. Chemical plasma etching is comparable to WCE and also yields isotropic or crystallographic profiles. Figure 2.4 shows a T-shaped mask consisting of a thin SiO_2 layer on top and a thick underlying layer of polyimide. The undercut in the polyimide has been formed by isotropic dry etching in an oxygen plasma. Chemical etching is characterized by high etch rates, high material selectivity, low ion damage, and poor profile control, and it is therefore complementary to pure physical etching. An important process in semiconductor device fabrication is the stripping of organic resist material with high-pressure oxygen plasmas. More information on chemical etching can be found elsewhere (Smith, 1984).

Chemical–physical etching. For most applications, it is necessary to combine the advantages of physical and chemical etching mechanisms. By finding the proper balance between the physical and chemical components (this is the "art" of dry etching!) it is possible to achieve high etch rates and high material selectivity while at the same time obtaining anisotropic etch profiles with a good control of the geometry. Typical etching chemistries and the mechanisms for the interplay between the chemical and physical components will be given in Sections 2.3.2 and 2.3.3.

The most commonly used process in the semiconductor industry is reactive ion etching (RIE) (Gorowitz and Saia, 1984), which will be discussed in more detail in Section 2.3.4. Another example of a combination of physical and chemical etching is chemically assisted ion beam etching (CAIBE), covered in Section 2.3.5.

Figure 2.4. T-shaped lift-off mask formed by isotropic dry etching in an oxygen plasma.

These two etch methods rely in their original form on the plasma excitation with RF at a frequency of 13.56 MHz. Other plasma excitation methods have also gained importance, such as electron cyclotron resonance excitation at 2.45 GHz, magnetically enhanced RIE (MERIE), inductively coupled plasmas (ICP), and transmission coupled plasmas (Singer, 1991, 1992). All these approaches aim for higher plasma densities at lower pressures to achieve higher etch rates, lower damage, and good uniformity.

Photochemical etching. This etch process uses gases but no plasma for the excitation. Gas phase reactions and chemical reactions between the adsorbed chemical species and the substrate material are driven by photons, normally with laser light (photochemical mechanism). Other mechanisms are based on photogeneration of carriers or on heating effects (Ashby, 1990). This etching process can also be directional and material selective. Typical applications are the etching of deep trenches with high etch rates and maskless etching (Ashby, 1990).

2.3.2 *Plasma–surface interactions in chemical–physical etching*

The interaction of gaseous fragments with the plasma and the sample surface can be described after Oehrlein and Rembetski (1992) and Ibbotson and Flamm (1988) by the following steps:

(a) Etchant formation: The fed-in gas breaks down into ions, radicals, and excited fragments by means of electron impact reactions while neutrals are also still present. The gas species are transported to the sample surface by electric fields or diffusion.
(b) Adsorption: Chemically active species are adsorbed at the sample surface.
(c) Reaction: The etch products are formed by a chemical reaction in the surface layer.
(d) Desorption: The etch products are desorbed from the sample surface and pumped out of the system.

Each of these steps can limit the etch process and the etch rate. This simplified process scheme is complicated by several feedback mechanisms. The etch products can also dissociate in the plasma and influence the plasma chemistry. Organic or organometallic films can be deposited on chamber walls and electrodes and thus can change the electrical characteristics of the system. Material from the chamber walls can be eroded and can influence the plasma chemistry or be redeposited on the sample.

An important point for the etching result is the interplay between the energetic ions and the chemical surface reactions. For certain etch chemistries, such as oxygen plasmas for the etching of organic resist material, volatile etch products are formed and desorbed spontaneously. In general, some "ion enhancement" is necessary to facilitate reaction and desorption (steps [c] and [d]) and to achieve anisotropic etching. There are three models described in the literature of how this ion enhancement works. The first is the surface damage mechanism (Coburn and Winters, 1979), in which ions are believed to damage surface-near regions and enhance the rate of surface reactions, for example, by creating free bonds. In the second model, called chemically enhanced physical sputtering (Mauer et al., 1978), adsorbed radicals change the chemical structure of the surface and make sputtering easier. Third, in the chemical sputtering model (Tu et al., 1981), energetic ions supply energy to the reaction layer, initiating chemical reaction or desorption. This model seems best to describe the etching of silicon with a fluorine-based chemistry.

The anisotropy of the etch is given by the directionality of the ions. Lowering the pressure and increasing the energy of the ions increases the anisotropy of the etch. In many cases, sidewall passivation layers (Oehrlein and Rembetski, 1992) play an important role in achieving anisotropic etching. The growth of these inhibitor films, which in principle occurs on all surfaces, can be controlled by an appropriate plasma chemistry. The role of the ions is to clear the horizontal (etched) surfaces from the film; the remaining film on the sidewalls prevents lateral under-etching. Etching of silicon with $HCl/O_2/BCl_3$ is an example of this mechanism.

2.3.3 *Etching chemistries*

For chemical and chemical–physical etching, it is essential to find etch chemistries from which volatile etch products are formed, otherwise the poor desorption of the etch products will limit the entire etch process.

Silicon. For the etching of silicon, mainly fluorine- and chlorine-based etchants are used. With a fluorine-based etch chemistry, volatile etch products such as SiF_2 and SiF_4 are formed and desorbed spontaneously, leading to isotropic etching and making undercut control difficult (Smith, 1984). Increased ion energies and the sidewall passivation mechanism can facilitate anisotropic etching. CF_4, CF_4/O_2, SF_6, C_2F_6/O_2, and NF_3 have been used etch silicon (Gorowitz and Saia, 1984).

Chlorine-based etchants produce volatile etch products of the form $SiCl_x$ ($x =$ 1–4). Anisotropic etching is achieved with CCl_4, Cl_2, HCl, and BCl_3 in different mixtures and with the addition of O_2, Ar, He, or N_2. Gas mixtures containing fluorine and chlorine (such as CCl_3F, CCl_2F_2, or $C_2F_6Cl_2$) or bromine-based plasmas ($CBrF_3$) have also been reported. References can be found in Gorowitz and Saia (1984) and Smith (1984).

III/V materials. Etchants based on halogens, primarily chlorine, are used to etch III/V materials. The desorption of the group-III halides is the most critical step. Figure 2.5 shows the vapor pressure of various group-III and group-V halides as a function of temperature (Landolt-Boernstein, 1960). For temperatures between

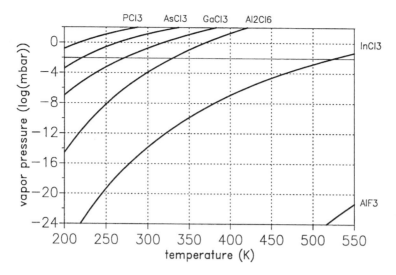

Figure 2.5. Vapor pressure for different III/V-halides as a function of temperature. Calculated using data from Landolt-Boernstein, 1960. Note the logarithmic scale of the *y* axis.

300 and 400 K, as are normally used in RIE, the vapor pressure of the As-, P-, and Ga-chlorides is well above the typical RIE working pressure of $\approx 10^{-2}$ mbar, and these etch products desorb spontaneously. Al_2Cl_6 needs only a small amount of additional energy for desorption, whereas $InCl_3$ has a significantly lower vapor pressure and needs a greater energy supply, such as elevated temperatures or high ion energies.

InP and InGaAsP have been etched in chlorine-containing plasmas with Cl_2, CCl_4, BCl_3, $SiCl_4$, and CCl_2F_2. The poor volatility of the In-chlorides makes it necessary to use substrate temperatures in the range of 520 K or ion energies of up to a few hundred electron volts to achieve useful etch rates and a tolerable surface quality. Bromine- and iodine-based etchants such as CH_3I, Br_2, or $C_2H_4Br_2$ have also been used because of the comparable, or even higher, vapor pressures for the III/V reaction products (Flanders et al., 1989). RIE with CH_4 and H_2 has been reported with anisotropic etching characteristics and smooth surfaces (Niggebruegge et al., 1985). This process can be simply viewed as the reverse process of metal–organic chemical-vapor deposition with metalorganics such as $In(CH_3)$ and hydrides such as PH_3 as etch reaction products. An overview of dry etching recipes and applications for InP and InGaAs(P) can be found in Matsushita and Hartnagel (1991).

GaAs and AlGaAs material is etched primarily in chlorine-based plasmas containing Cl_2, BCl_3, CCl_4, or $SiCl_4$. Etching with CH_4/H_2 is also possible, but with rather low etch rates for Al-containing materials. Addition of chlorine and argon can solve this problem (Vojdani and Parrens, 1987).

Fluorine-based etchants are generally not used because of the very low vapor pressure of the III/V-fluorides, as depicted in Figure 2.5 for AlF_3. An exception is the selective etching of GaAs over AlGaAs. RIE with CCl_2F_2 and He with a selectivity of greater than 200:1 between GaAs and AlGaAs has been reported (Hikosaka et al., 1981). A summary of dry etching recipes and applications for GaAs and AlGaAs can be found in Ashby (1990).

Metal and organic materials. Metals like Ti, Ta, Mo, W, and Nb can be etched in fluorine plasmas, whereas fluorine-based etching is difficult for Cr, Au, or Al because of their involatile fluorides. Etching of Al, Au, and Cr has been reported in chlorine- and bromine-containing plasmas, which also give good results for the etching of Ti (Flamm et al., 1984).

Organic material such as photoresists, polyimides, and other polymers can be etched in pure oxygen plasmas or in mixtures of oxygen and fluorine-containing gases (O_2/CF_4, O_2/SF_6). Photoresists are often removed by isotropic high-pressure plasma etching, but anisotropic etching of organics is also possible (Flamm et al., 1984; Gorowitz and Saia, 1984),

2.3.4 Reactive ion etching (RIE) of InAlAs

The configuration of a typical RIE system is shown schematically in Figure 2.6. A plasma is maintained in a low-pressure (typically 10^{-1}–10^{-3} mbar) glow discharge between two parallel electrodes. The upper electrode is grounded, the lower one is capacitively coupled to an RF generator (normally 13.56 MHz) via an impedance matching network. The asymmetry between the electrodes (the powered one is smaller than the grounded one) leads to a time-averaged potential between the electrodes, as depicted also in Figure 2.6. There is a plasma body

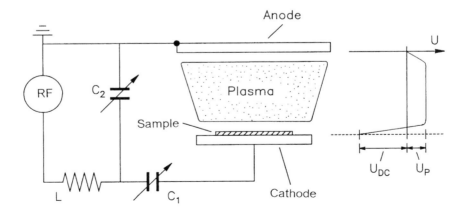

Figure 2.6. Schematic configuration of an RIE system and electrical potential between the electrodes.

with a constant positive potential of a few tens of volts and two dark spaces near the two electrodes. Positive ions produced in the plasma body are accelerated across the dark spaces toward the electrodes. The accelerating voltage for the grounded anode is rather small, whereas the so-called dc-bias for the powered cathode, where the etch samples are placed, can reach up to a few hundred volts. A detailed description of glow discharges can be found in Chapman (1980). Reactive gas species (radicals, neutrals) diffuse from the plasma to the sample and, together with the ions, lead to chemical–physical etching as explained in Sections 2.3.1 an 2.3.2.

The anisotropic etching of InAlAs is taken as an example. Anistropic etching is necessary for the fabrication of In(Ga)AlAs/InGaAs ridge waveguide lasers (Hausser et al., 1993). RIE of InAlAs has been reported with mixtures of CH_4, H_2, and Ar yielding rather low etch rates (Pearton et al., 1991). We have obtained etch rates of up to a few hundred nm/min with Ar/Cl_2 mixtures (Germann et al., 1993). The low volatility of the In-chlorides limits the process. Elevated temperatures of greater than 520 K have been attempted, but lead to isotropic etching with severe mask undercut. Anisotropic etching with nearly vertical sidewalls and a smooth etched surface is obtained with high ion energies at a dc-bias of 750 V. Figure 2.7 depicts a series of InAlAs samples etched under various bias conditions. For a bias of 300 V, the etch rate is low ($\approx$20 nm/min), and the rough surface indicates poor desorption of the etch products. For increasing dc-bias (Figures 2.7b and 2.7c), the etch rate increases, and for a bias of 750 V, the etched surface is as smooth as in the part that was masked during etching. Under these high-bias conditions, all etch products are desorbed from the etched surface.

Low-threshold ridge waveguide lasers on InGaAs/InGaAs multiple-quantum-well, separation-confinement-heterostructure material have been produced with the process (Germann et al., 1993).

2.3.5 Chemically assisted ion beam etching (CAIBE) for etched laser facets

With RIE it is not possible to control the chemical component (density of reactive species) and the physical component (energy and density of the ions) independently, because ions as well as reactive species are generated in the same plasma. With CAIBE, the production of the ions is separated from the injection of the reactive components, which allows better control of the etching process.

Dry etching of semiconductor laser facets is a very challenging process to which CAIBE is well suited. The etching of laser facets requires vertical sidewalls with a deviation of not more than a few degrees, a sidewall roughness in the range of 100 Å, an etch depth of up to 10 μm, and unselective etching for all different materials of the laser structure.

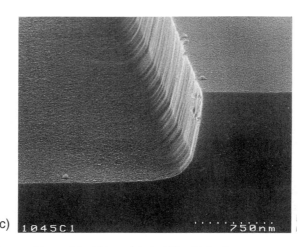

Figure 2.7. Dry etching of InAlAs with Ar/Cl$_2$–RIE (8 sccm Ar, 2 sccm Cl$_2$, 10 μbar). Samples were etched for the same time with three different conditions for the dc-bias: (a) 300 V, (b) 480 V, (c) 750 V. Reproduced with permission from Germann et al. (1993).

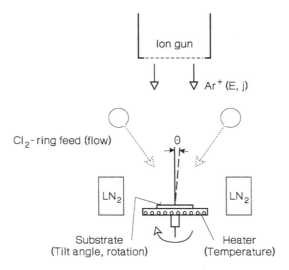

Figure 2.8. Setup of a CAIBE system. Reproduced with permission from Vettiger et al. (1991).

Figure 2.8 shows a CAIBE system. An argon plasma is excited in a Kaufman-type ion source. Ions are extracted from the plasma, formed into a parallel beam, and accelerated with an energy of 500 eV toward the sample by applying appropriate potentials to the grids in front of the ion source. In the CAIBE mode, Cl_2 molecules are injected via a ring feed in close proximity to the sample and are adsorbed on the sample surface. Surface reactions and the desorption of the etch products are enhanced by the argon ions. A low partial water pressure is maintained by a liquid nitrogen (LN_2) trap to achieve the same etch rates for GaAs and GaAlAs. Too much residual water causes the formation of aluminum oxide on the sample, which lowers the etch rate of AlGaAs significantly. The sample holder can be tilted to adjust the angle of the etched sidewalls and rotated to eliminate beam inhomogeneities.

Two other possible modes of this type of etching system are the IBE with noble gas ions (Lee, 1984), and reactive ion beam etching (RIBE), where reactive ions are used through the ion source, resulting in chemical sputtering (Heath and Mayer, 1984). The described CAIBE process as been used to fabricate 5–8-μm-deep laser mirrors with a roughness of less than 200 Å and a deviation from verticality of less than $2°$ in a GaAs/AlGaAs single-quantum-well, graded-index, separate-confinement-heterostructure laser structure (Buchmann et al., 1989). Ridge waveguides with these etched facets have the same performance as lasers with cleaved facets (Vettiger et al., 1991). Etched laser facets for visible semiconductor lasers emitting at a wavelength of 690 nm have also been fabricated with the CAIBE process (Unger et al., 1993, 1994). Aside from the AlGaAs cladding layers, these laser structures contain additional layers of GaInP and AlGaInP. The etched mirrors are slightly rougher than those of the pure GaAs/AlGaAs structure, mainly because of the

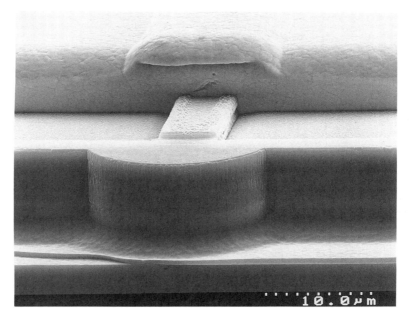

Figure 2.9. Dry-etched laser facet region of a red-emitting visible (Al)GaInP ridge wave-guide laser. Conditions for the Ar/Cl_2–CAIBE are 500 eV Ar ions, current density 0.23 mA/cm^2, 15 sccm Cl_2. Reproduced with permission from Unger et al. (1994).

low volatility of the In-containing etching products. Figure 2.9 shows a convex facet region of a ridge waveguide laser with a dry-etched mirror.

Advantages of lasers with etched facets compared to cleaved ones have been discussed in detail elsewhere (Vettiger et al., 1991).

Concerns exist about the long-term stability of the etched facets because of ion-induced damage and the chemically modified facet surface. Current research addresses these topics by developing etching methods that allow etching with low ion energies without losing anisotropy (Skidmore et al., 1992) and by developing in situ cleaning methods for removing damaged and contaminated surface layers (Asakawa and Sugata, 1986).

2.4 Dry etching–induced damage

Low-energy ions are necessary for anisotropic dry etching. Unfortunately, these ions also induce damage to surface-near regions of the etched samples, degrading the electrical and optical properties. A heavily damaged layer with a thickness of a few nanometers has been observed for dry-etched surfaces by Rutherford backscattering, Auger electron spectroscopy (AES), and x-ray photoelectron spectroscopy (XPS). This layer contains atoms from the etch gas, has numerous structural defects, may even be amorphous, and can, in the case of compound materials, have a strongly disturbed stoichiometry. In addition, the surface can be covered by adsorbed species from the etch gas. Aside from this heavily damaged surface layer, a long-range damaged layer with a depth of up to a few hundred nanometers below the surface has been found, mainly by electrical methods such as current-voltage measurements on Schottky contacts and by optical methods. A brief overview of results for In-based materials can be found in Hayes (1992).

An optical method based on the luminescence of surface-near quantum wells (Wong et al., 1988) is a particularly sensitive tool for the depth-resolved characterization of this long-range damage. The extension of the damaged zone is surprisingly high: The observed damage depths are roughly a factor of 10 higher than the predicted range of implanted ions as calculated by Monte Carlo simulations. Ion channeling can explain this difference as shown experimentally (Germann et al., 1989) and confirmed theoretically (Stoffel, 1992). Dry etching methods are improved by using lower ion energies while still maintaining anisotropy, thus keeping the induced damage at a minimum. Damage can also be annealed in some cases.

In addition to damage on horizontal surfaces, sidewall damage on surfaces parallel to the impinging ions has also been observed by optical methods, showing an increase of surface recombination and an optically "dead" layer (Maile et al., 1989). Surface recombination has several causes. First, low-energy ions hit the sidewall at shallow incidence, creating structural defects that can act as recombination centers. Second, the sidewall can be chemically modified by adsorbates from the etching gas or by oxygen when exposed to air. Third, even an "ideal" undamaged surface has unfilled midgap states. Minimization of sidewall damage is attempted with the following scheme: The third cause requires that "open" surfaces be avoided by burying the sidewall with epitaxial regrowth or by passivating the open surfaces; etching with lowest possible energies or two-step etching with a high-energy anisotropic etch followed by a short, low- or zero-energy chemical etch step is an attempt to avoid or remove damaged layers. The contamination problem is addressed by in situ processing of all steps in a high-vacuum environment. The fabrication of nanostructures and of reliable and stable etched laser facets relies on the successful implementation of this scheme.

2.5 Acknowledgments

The author thanks P. Buchmann for permission to reproduce Figure 2.8, K. Dätwyler for Figures 2.2 and 2.3, and P. Unger for the program for calculating the vapor pressures in Figure 2.5.

References

Adachi, S. (1982) Chemical etching of InP and InGaAsP/InP, *J. Electrochem Soc.* 129:609–613.

Adachi, S., Noguchi, Y., and Kawaguchi, H. (1982) Chemical etching of InGaAsP/InP DH wafer. *J. Electrochem. Soc.* 129:1053–1062.

Adachi, S. and Oe, K. (1984) Chemical etching of GaAs. *J. Electrochem. Soc.* 131:126–130.

Asakawa, K. and Sugata, S. (1986) Damage and contamination-free GaAs and AlGaAs etching with a novel ultrahigh-vacuum reactive ion beam etching system with etched surface monitoring and cleaning method. *J. Vac. Sci. Technol.* A 4:677–680.

Ashby, C. I. H. (1990) GaAs Etching. In *Properties of Galllium Arsenide, 2nd Edition*, EMIS Datareviews Series No. 2, pp. 653–681. London and New York,: INSPEC, The Institution of Electrical Engineers.

Aspnes, D. E. and Studna, A. A. (1981) Chemical etching and cleaning procedures for Si, Ge, and some III-V compound semiconductors. *Appl. Phys. Lett.* 39:316–318.

Bassous, E. (1978) Fabrication of novel three-dimensional microstructures by the anisotropic etching of (100) and (110) silicon. *IEEE Trans. Electron Devices* ED-25:1178–1185.

Bogh, A. (1971) Ethylene diamine-pyrocatechol-water mixture shows etching anomaly in boron-doped silicon. *J. Electrochem. Soc.* 118:401–402.

Bouadma, N., Riou, J., and Bouley, J. C. (1982) Short-cavity GaAlAs laser by wet chemical etching, *Electron. Lett.* 18:879–880.

Buchmann, P., Dietrich, H.P., Sasso, G., Vettiger, P. (1989) Chemically assisted ion beam etching process for high quality laser mirrors. *Microelectronic Engineering,* 9:485–489.

Buchmann, P. and Houghton, A. J. N. (1982) Optical Y-junctions and S-bends formed by preferentially etched single-mode rib waveguides in InP. *Electron. Lett.* 18:850–852.

Chapman, B. (1980) *Glow Discharge Processes, Sputtering and Plasma Etching.* New York: Wiley Interscience.

Coburn, J. W. and Winters, H. F. (1979) Ion- and electron-assisted gas-surface chemistry: -an important effect in plasma etching. *J. Appl. Phys.* 50:3189–3196.

Comerford, L. and Zory, P. (1974) Selectively etched diffraction gratings in GaAs. *Appl. Phys. Lett.* 25:208–210.

Csepregi, L. (1985) Micromechanics: a silicon microfabrication technology. *Microelectronics Engineering.* 3:221–233.

Dash, W. C. (1956) Copper precipitation on dislocations in silicon. *J. Appl. Phys.* 27:1193–1195.

Finne, R. M. and Klein, D. L. (1967) A water-amine-complexing agent system for etching silicon. *J. Electrochem. Soc.* 114:965–970.

Flamm, D. L., Donnelly, V. M., and Ibbotson, D. E. (1984) Basic principles of plasma etching for silicon devices. In *VLSI electronics. Microstructure Science Vol. 8, Plasma Processing for VLSI*, ed. N. G. Einspruch and D.M. Brown, pp. 189–251. Orlando: Academic Press.

Flanders, D. C., Pressman, L. D., and Pinelli, G. (1990, Reactive ion etching of indium compounds using iodine containing plasmas. *J. Vac. Sci. Technol.* B 8:1990–1993.

Fonash, S. J. (1985) Advances in dry etching processes: a review. *Solid State Technol.* 28 (no. 1):150–158.

Gannon, J. J and, Nuese, C. J. (1974) A chemical etchant for the selective removal of GaAs through SiO_2 masks. *J. Electrochem. Soc.* 121:1215–1219.

Germann, R., Forchel, A., Bresch, M., and Meier, H. P. (1989) Energy dependence and depth distribution of dry etching-induced damage in III/V semiconductor heterostructures *J. Vac. Sci. Technol.* B 7:1475–1478.

Germann, R. Hausser, S., and Reithmaier, J. P. (1993) Reactive ion etching of InAlAs with Ar/Cl₂ mixtures for ridge waveguide lasers. *Microelectronic Engineering* 21:345–348.

Gorowitz, B. and Saia, R .J. (1984) Reactive ion etching. In *VLSI electronics. Microstructure Science Vol. 8, Plasma Processing for VLSI*, ed. N. G. Einspruch and D. M. Brown, pp. 298–340. Orlando: Academic Press.

Grabmaier, J. G. and Watson, C. B. (1969) Dislocation etch pits in single crystal GaAs. *Phys. Stat. Sol.* 32:K13–K15.

Gréus, C., Orth, A., Daiminger, F., Butov, L., Straka, J., and Forchel, A. (1993) Optical studies of barrier modulated InGaAs/GaAs quantum wires *Microelectronic Engineering* 21:397–400.

Harder, C., Buchmann, P., and Meier, H. (1986) High-power ridge-waveguide AlGaAs GRIN-SCH laser diode. *Electron. Lett.* 22:1081–1082.

Hausser, S., Meier, H. P., Germann, R., and Harder, C. S. (1993) 1.3 µm multiquantum well decoupled confinement heterostructure (MQW-DCH) laser diodes *IEEE J. Quantum Electron.* 29:1596–1600.

Hayes, T. R. (1992) Dry etching of In-based semiconductors. In *Indium Phosphide and Related Materials: Processing Technology, and Devices*, ed. A. Katz, pp. 277–306. Norwood (MA): Artech House.

Heath, B. A. and Mayer, T. M. (1984) Reactive ion beam etching. In *VLSI Electronics, Microstructure Science, Vol. 8, Plasma Processing for VLSI*, ed. N. G. Einspruch and D. M. Brown, pp. 365–409. Orlando: Academic Press.

Hikosaka, K., Mimura, T., and Joshun, K. (1981) Selective dry etching of AlGaAs-GaAs heterojunction. *Jpn. J. Appl. Phys.* 20:L847–L850.

Huber, A. and Linh, N. T. (1975) Révélation métallographique des défauts cristallins dans InP. *J. Crystal Growth* 29:80–84.

Ibbotson, D. E. and Flamm, D. L. (1988) Plasma etching for III-V compound devices: Part 1. *Solid State Technol.* October 1988:77–79.

Iga, K., Pollack, M. A., Miller, B. I., and Martin, R. J. (1980) GaInAsP/InP DH lasers with a chemically etched facet. *IEEE J. Quantum Electron.* QE-16:1044–1047.

Iida, S. and Ito, K. (1971) Selective etching of gallium arsenide crystals in H₂SO₄-H₂O₂-H₂O System. *J. Electrochem. Soc.* 118:768–771.

Irving (1971) Plasma oxidation process for removing photoresist films. *Solid State Technol.* 14:47.

Jaeckel, H., Meier, H. P., Bona, G. L., Walter, W., Webb, D. J., and Van Gieson, E. (1989) High-power fundamental mode AlGaAs quantum well channeled substrate laser grown by molecular beam epitaxy. *Appl. Phys. Lett.* 55:1059–1061.

Kaga, T., Shinriki, H., Murai, F., Kawamoto, Y., Nakagome, Y., Takeda, E., and Itoh, K. (1991) DRAM manufacturing in the '90s – Part 3: A Case Study. *Semiconductor International* May 1991:98–101.

Kaminsky, G. (1985) Micromachining of silicon mechanical structures. *J. Vac. Sci. Technol. B* 3: 1015–1024.

Kendall, D. L. (1975) On etching very narrow grooves in silicon. *Appl. Phys. Lett.* 26:195-198.

Landolt-Boernstein (1960) *Zahlenwerte und Tabellen, 6. Auflage, Vol. II/2a*, pp. 1–3 and 31–63. Berlin: Springer-Verlag.

Lee, R. E. (1984) Ion-beam etching (milling). In *VLSI Electronics, Microstructure Science, Vol. 8, Plasma Processing for VLSI*, ed. N. G. Einspruch and D. M. Brown, pp. 341–364. Orlando: Academic Press.

Loewe, H., Keppel., P., and Zach, D. (1990) *Halbleiterätzverfahren*. Berlin: Akademie-Verlag.

Logan, R., van der Ziel, J. P., Temkin, H., and Henry, C. H. (1982) InGaAsP/InP (1.3 µm) buried-crescent lasers with separate optical confinement. *Electron. Lett.* 18:895–896.

Maile, B. E., Forchel, A., Germann, R., and Grützmacher, D. (1989) Impact of sidewall recombination on the quantum efficiency of dry etched InGaAs/InP Semiconductor wires. *Appl. Phys. Lett.* 54:1552–1554.

Mauer, J. L., Logan, J. S., Zielinski, L. B., and Schwartz, G. C. (1978) Mechanism of silicon etching by a CF_4 plasma. *J. Vac. Sci. Techno.* 15:1734–1738.

Matsushita, K. and Hartnagel, H. L. (1991) Plasma etching of InP, ion beam milling and sputter etching of InP, reactive ion and ion-beam etching of InP. In *Properties of Indium Phosphide*, EMIS Datareviews Series No. 6, 344–353. London and New York: The Institution of Electrical Engineers.

Mori, Y. and Watanabe, N. (1978) A new etching solution system, H_3PO_4-H_2O_2-H_2O for GaAs and its kinetics. *J. Electrochem. Soc.* 125:1510–1514.

Niggebruegge, U., Klug, M., and Garus, G. (1985) A novel process for reactive ion etching on InP, using CH_4/H_2. *Inst. Phys. Conf. Ser.* 79:367-372 (Proc. Int. Symp. GaAs and Related Compounds, Karuizawa, Japan 1985).

Oehrlein, G. S. and Rembetski, J. F. (1992) Plasma-based dry etching techniques in the silicon integrated circuit technology. *IBM J. Res. Develop.* 36(2):140–157.

Pearton, S .J., Chakrabarti, U. K., Katz, A., Perley, A. P., and Hobson, W. S. (1991) Comparison of CH_4/H_2/Ar reactive ion etching and electron cyclotron resonance plasma etching of In-based III-V alloys. *J. Vac. Sci. Technol. B* 9:1421.

Petersen, K. E. (1982) Silicon as a mechanical material. *Proceedings of the IEEE* 70:420–457.

Rideout, V. L. (1972) An improved polishing technique for GaAs. *J. Electrochem. Soc.* 119:1778–1779.

Schwartz, B. and Robbins, H. (1976) Chemical Etching of Silicon : IV. Etching technology. *J. Electrochem. Soc.* 123:1903–1909.

Singer, P. (1991) ECR: Is the magic gone? *Semiconductor International* July 1991:46–48.

Singer, P. (1992) Trends in plasma sources: the search continues. *Semiconductor International* July 1992:52–56.

Skidmore, J. A., Lishan, D. G., Young, D. B., Hu, E. L., and Coldren, L. A. (1992) HCl, H_2, and Cl_2 radical-beam ion-beam etching of $Al_xGa_{1-x}As$ substrates with varying Al mole fraction. *J. Vac. Sci. Technol. B* 10:2720–2724.

Smith, D. L. (1984) High-pressure etching. In *VLSI Electronics, Microstructure Science, Vol. 8, Plasma Processing for VLSI*, ed. N. G. Einspruch and D. M. Brown, pp. 253–297. Orlando: Academic Press.

Stoffel, N. G. (1992) Molecular dynamics simulations of deep penetration by channeled ions during low-energy ion bombardment of III/V semiconductors. *J. Vac. Sci. Technol. B* 10:651–658.

Sullivan, M. V. and Kolb, G. A. (1963) The chemical polishing of gallium arsenide in bromine-methanol. *J. Electrochem. Soc.* 110:585.

Takimoto, K., Ohnaka, K., and Shibata, J. (1989) Reactive ion etching of InP with Br_2-containing gases to produce smooth, vertical walls: fabrication of etch-faceted lasers. *Appl. Phys. Lett.* 54:1947–1949.

Tanbun-Ek, T., Suzaki, S., Min, W. S., Suematsu, Y., Koyama, F., and Arai, S. (1984) Static characteristics of 1.5–1.6 µm GaInAsP/InP buried heterostructure butt-jointed built-in integrated lasers. *IEEE J. Quantum Electron.* QE-20:131–140.

Tarui, Y., Komiya, Y., and Harada, Y. (1971) Preferential etching and etched profile of GaAs. *J. Electrochem. Soc.* 118:118–122.

Tu, Y.-Y., Chuang, T .J., and Winters, H. F. (1981) Chemical sputtering of fluorinated silicon. *Phys. Rev. B* 23:823–835.

Turley, S. E. H. and Greene, P. D. (1982) LPE growth on structured (100) InP substrates and their fabrication by preferential etching. *J. Crystal Growth* 58:409–416.

Unger, P., Boegli, V., Buchmann, P., and Germann, R. (1993) Fabrication of curved mirrors for visible semiconductor lasers using electron-beam lithography and chemically assisted ion-beam etching. *J. Vac. Sci. Technol. B* 11:2514–2518.

Unger, P., Boegli, V., Buchmann, P., and Germann, R. (1994) High-resolution electron-beam lithography for fabricating visible semiconductor lasers with curved mirrors and integrated holograms. *Microelectronic Engineering* 23:461–464.

Vettiger, P., Benedict, M. K., Bona, G. L., Buchmann, P., Cahoon, E. C., Dätwyler, K., Dietrich, H. P., Moser, A., Seitz, H. K., Voegeli, O., W ebb, D. J., and Wolf, P. (1991) Full-wafer technology: a new approach to large-scale laser fabrication and integration. *IEEE J. Quantum Electron.* 27:1319–1331.

Vodjdani, N. and Parrens, P. (1987) Reactive ion etching of GaAs with high aspect ratios with Cl_2-CH_4-H_2-Ar mixtures. *J. Vac. Sci. Technol. B* 5:1591–1598.

Wong, H. F., Green, D. L., Liu, T. Y., Lishan, D. G., Bellis, M. Hu, E. L., Petroff, P. M., Holtz, P. O., and Merz, J. L. (1988) Investigation of reactive ion etching induced damage in GaAs-AlGaAs quantum well structures. *J. Vac. Sci. Technol. B* 6:1906–1910.

3

The development and application of micromechanical devices in biosystems

KENSALL D. WISE

Center for Integrated Sensors and Circuits
Department of Electrical Engineering and Computer Science
The University of Michigan,
Ann Arbor, Michigan
48109-2122

3.1 Introduction

During the past few years, substantial progress has been made in the development of integrated microelectromechanical devices for use in biological systems. Combining sensors, microactuators, and microelectronics monolithically on a single chip or in highly integrated multichip hybrids, these devices are potentially capable of monitoring a wide range of physical variables with unprecedented accuracy and of controlling events with a spatial resolution extending down to the cellular level. Biosystems represent one of the most important emerging application areas for these devices, and yet the problems in such systems are particularly challenging. This chapter reviews recent progress in representative areas and discusses some of the problems that still remain.

Microelectronics has experienced spectacular growth over the past 30 years to the point where nearly any function that can be defined by a suitable algorithm can be implemented at high speed and at low cost. It has become increasingly clear, however, that in order to use this technology in many applications, especially those in biomedicine, the weak link is in the peripheral portions of such systems, where sensors and actuators must interface to the nonelectronic (biological) world. Work on these devices has also shown dramatic progress recently, however, as sensors, actuators, and interface circuits are increasingly being merged to form microelectromechanical systems (MEMS) (Wise, 1991; Wise and Najafi, 1991). It is this recent progress on micromechanical devices that gives cause for new optimism regarding the impact of microsystems on health care, and especially on implantable devices. This chapter will focus on devices intended for use in vivo and will not treat external monitoring and diagnostic instruments, even though there will certainly be significant impacts in these areas as well, including important advances in the monitoring and control of anesthesia, in the analysis of blood chemistry, and in automated DNA analysis.

Rooted in early research on materials and processes for the emerging field of integrated circuits, efforts at what are now called integrated sensors first began in the late 1960s. After early temperature and pressure sensors, visible imagers were probably the first integrated sensors to find their way into production. Visible imagers have enjoyed steady progress and today are approaching the resolution of photographic film, promising to revolutionize photography at all levels; however, although they represent some of the largest chips made by the semiconductor industry, they have required few processes or packaging techniques beyond those used for integrated circuits themselves. Pressure sensors, in contrast, have required special etching and sealing techniques in order to be practical in silicon. In the 1970s, considerable progress was made in the development of micromachining techniques for such devices with the emergence of impurity-based etch-stops, which took silicon sensors out of the laboratory and into volume production. The automotive industry was the primary driver for these developments. By the late 1980s, surface micromachining had emerged, making possible a variety of new resonant sensors and microactuators, and circuits were being successfully merged with microsensors and microactuators to ease signal-to-noise and packaging problems. Flowmeters, gyros, and accelerometers were joining pressure sensors as high-volume production devices, and many other devices were in development. Today, the integration of sensors, actuators, and electronics into complete microsystems is being pursued on a worldwide basis. Such systems blend sophisticated functions with small size and are particularly attractive for use in biological systems. However, the packaging issues associated with such applications are extremely demanding and not always properly appreciated. Indeed, many of the early applications envisioned for micromechanical devices were not practical due to fundamental problems with power requirements, packaging, leads, or acceptance by the tissue (biocompatibility). Indeed, most problems in applying electronics and micromechanics in biosystems come down to these four issues. The role of technology is to solve these issues.

This chapter first reviews the currently available technologies for forming micromechanical devices and the microstructures on which they are based. Such structures are based on bulk and surface micromachining, wafer-to-wafer bonding, and electroforming processes such as LIGA (after the German "Lithographie Galvanoformung Abformung"). Although a wide range of microstructures, including some with moving parts (such as microvalves and micropumps), has been demonstrated in the laboratory, difficult challenges remain in their application to biosystems, especially in the development of hermetic packages that allow selective access to the biosystem for sensing and that promote acceptance of the microsystem by the tissue. Some recent successes and remaining challenges are illustrated by three emerging devices. First, an active micromachined pressure-sensing catheter is described. This device allows multipoint pressure measurements in the coronary arteries of the heart with a pressure resolution of <2 mm Hg and a width of only 350 μm. The device operates using only two leads and represents a first step toward introducing more sophisticated diagnostics into catheter-based instruments. A second illustration is provided by the development of a family of multichannel recording and stimulating probes for use in studies of information processing in biological neural networks and in neural prostheses. These devices can be formed in two- and three-dimensional arrays having site spacings of typically 100–200 µm; they are facilitating important new studies of neural systems and now permit chronic studies to be performed routinely for periods of many months. Currently, they are also being extended to support thermal and chemical interactions at the cellular level. Finally, mention will be made of work on hermetically sealed microtelemetry systems that should eventually permit such devices to be implanted with no external leads.

3.2 Fabrication technologies for micromechanical devices

The fabrication of integrated sensors and microactuators is primarily based on extensions of integrated circuit technology, which, in turn, is based on photolithography and the successive deposition of thin films of metals and dielectrics. These processes will not be reviewed here; however, they are primarily two-dimensional and permit submicron dimensional control. The fabrication of sensors and actuators is made possible by the addition of several other technologies to this arsenal of processes; the most notable are precision silicon etching (micromachining), wafer-to-wafer bonding, and high-resolution electroforming. These special technologies are briefly reviewed in this section to define the basis for the current generation of devices being developed for use in biosystems.

3.2.1 Bulk micromachining

What we know today as micromachining was born during the 1960s as researchers at Bell Telephone Laboratories struggled to develop precise silicon etching techniques for beam-lead air-isolated integrated circuits (Lepselter, 1966). This work was first based on isotropic etchants (Klein and D'Stefan, 1962), which were joined in the late 1960s by anisotropic etchants (Declercq et al., 1975; Finne and Klein, 1967; Price, 1973), all of which etch the <100> crystallographic direction in silicon much faster than the <111> direction. Today, bulk silicon micromachining is based on the use of potassium hydroxide (KOH), ethylene diamine pyrocatechol (EDP), or hydrazine, all of which attack <100> silicon at about 1–1.5 µm/min, operate at temperatures from about 80 to 115°C, and are easily masked by the silicon dioxide (fused quartz) and silicon nitride films commonly employed in integrated circuit fabrication.

For many sensors, the mechanical microstructure required is in the form of a diaphragm or beam. Figure 3.1 shows typical device cross sections where the structure is formed from the wafer bulk. Using photolithography, an etch mask is defined on the back of the wafer in alignment with patterns on the front surface. The wafer bulk is selectively etched from the back as the final step in wafer pro-

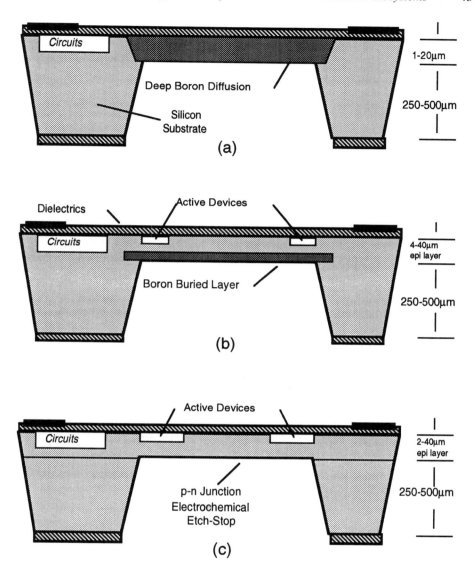

Figure 3.1. Typical cross section of a bulk-micromachined silicon wafer with various methods for etch-stop formation. (a) Diffused boron etch-stop. (b) Boron etch-stop as a buried layer. (c) Electrochemical etch-stop.

cessing to form the microstructure while simultaneously performing die separation. The thicknesses of most beams or diaphragms of interest today, however, are of the order of 1–10 μm, and common wafer thicknesses are in the range of 400–600 μm. Hence, there is a serious problem in achieving the reproducible batch formation of structures controlled to within a fraction of a micron from such thick wafers. The existence of impurity-based etch-stops in silicon (Petersen, 1982) has allowed micromachining to become a high-yield production process during the past decade.

The first and most widely used etch-stop technique is based on the fact that anisotropic etchants, especially EDP, do not attack heavily boron-doped silicon. Thus, a simple boron diffusion introduced from the front of the wafer can be used to create beams and diaphragms, with the etch leaving only the desired material. Layers of p+ silicon having thicknesses from 1 μm to about 20 μm can be formed using this process. Because the boron-doped silicon is in high tensile stress (Cho et al., 1990), the microstructures are flat and do not buckle; however, because the silicon must be heavily doped to achieve a stop, it is not possible to fabricate electronic devices within this material. To overcome this problem, lightly doped epitaxial layers have been used over p+ buried layers as shown in Figure 3.1b; however, epitaxial quality is compromised to some extent by the high buried-layer doping. If instead, a voltage is applied across the sample during the etch, a lightly doped epitaxial layer on a lightly doped substrate can be used with an etch-stop forming at the epi-substrate junction. This electrochemical process (Ozdemir and Smith, 1991), does permit the formation of high-quality devices but requires the added complication of bias during the etch.

Bulk micromachining has been used for most of the integrated sensors produced in the past; however, the long etch-time (several hours) required to etch through the silicon substrate from the back is an increasingly objectionable feature of this process in some applications. Two relatively new approaches overcome this problem: bulk dissolved-wafer processes (Chau and Wise, 1988) and surface micromachining (Howe, 1988). In dissolved-wafer processes, the wafer is processed to form the required etch-stops and active devices and is then bonded to a second wafer of glass or silicon by one of the processes to be mentioned in Section 3.2.3. The lightly doped portions of the first wafer are then etched away to leave the desired microstructures attached to the bottom wafer. Because the wafer dissolution etch can be done in two steps (a fast isotropic etch followed by a slower, anisotropic etch), the processing is all single sided and the need for a slow etch through the entire wafer is avoided. When glass is used as the second wafer, wafer-to-wafer alignment is simplified and stray capacitance is minimized. Active circuitry can be attached to the glass to form a multichip hybrid. When silicon is used as the lower wafer, circuitry can be embedded in it to form the active system.

3.2.2 *Surface micromachining*

The concept of surface micromachining for sensor formation was developed at the University of California at Berkeley (Howe and Muller, 1986) in the early 1980s but had its foundations in early air-bridge crossover work at Bell Laboratories and in early work on electrostatically driven microstructures at Westinghouse (Nathanson et al., 1967). In surface micromachining, a sacrificial layer is deposited on the upper surface of the wafer and is patterned. The material for the beam itself is then deposited over the sacrificial layer and is patterned such that it is anchored to the original wafer surface at the ends of the sacrificial material. As the final step in the wafer process, the sacrificial material is etched away, releasing the beam. This process has been used to produce a variety of microelectromechanical elements (Fan et al., 1988). A related process developed at the University of Wisconsin (Guckel and Burns, 1984; Guckel et al., 1987) has produced vacuum-sealed diaphragms for use as pressure sensors. In most of the work thus far in this area, deposited low-temperature silicon dioxide (LTO) films have been used as the sacrificial layer and *n*-type low-pressure chemical vapor deposited (LPCVD) polysilicon has been used as the beam material. This process has built on the materials available in standard integrated circuit fabrication facilities, but it has not been without problems. The LTO films etch relatively slowly in HF and require careful masking of the remainder of the wafer (usually with silicon nitride). The resulting microstructures also sometimes exhibit a sticking problem following their release (Lober and Howe, 1988). Polysilicon has also had its share of problems, especially those related to its compressive internal stress as normally deposited (Guckel et al., 1988). High-temperature anneals can reduce this stress but introduce additional problems with regard to metallization and other processing steps. Some metals (Kim and Allen, 1991), as well as polysilicon itself (Kong et al., 1990), are promising alternative sacrificial materials, and metals are also being explored for use as the microstructures (Chen and MacDonald, 1991). In addition, other approaches such as merged epitaxial lateral overgrowth (Pak et al., 1991) and the use of undercut bulk etching from the front of the wafer (Petersen, 1982; Sugiyama et al., 1987) offer still further options for the formation of novel microstructures. Figure 3.2 contrasts a bulk-micromachined device with one produced using surface micromachining.

Surface micromachining has allowed a wide variety of microstructures to be produced, including accelerometers, rotary micromotors, laterally driven micropositioners, and resonant sensors. Accelerometers are perhaps the most important current application for this technology and surface-machined structures promise to compete with bulk devices, some of which have been fabricated with built-in self-test capabilities (deBruin et al., 1990). Rotary micromotors (Mehregany et al., 1988, 1990) are usually driven electrostatically but suffer from friction problems at micron dimensions that do not appear to be easily overcome. In addition, electrostatic drive appears difficult to apply in biological applications, and the critical problem of an acceptable microfabricated mechanical feedthrough (for example, for a drive shaft) has yet to be solved. Laterally driven microstructures (Tang et al., 1989) have similar friction problems but are finding wider application due to the relatively simpler problems in coupling to them. However, the mechanical feedthrough problems remain. These microstructures can be made to work within hermetically sealed cavities, but coupling them to the external world remains a missing link. Resonant sensors (Ikeda et al., 1990) are under development for pressure and other variables, where they offer greater batch production capabilities and system compatibility

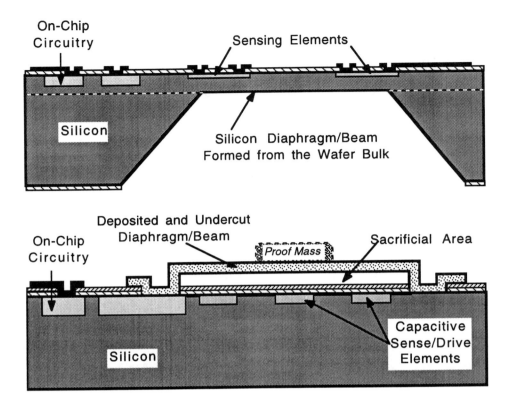

Figure 3.2. Cross sections of bulk-and surface-micromachined devices typical of pressure sensors or accelerometers. By either approach, a thin beam or diaphragm is produced.

compared to quartz resonators formed using alternative technologies. These devices are particularly interesting in that many are being sealed using vacuum chambers fabricated on-chip.

3.2.3 Wafer bonding processes

There are two classes of techniques that are useful as companions or possible alternatives to micromachining. The first group of these can be generally referred to as wafer bonding processes. These have already been referred to in connection with the dissolved-wafer bulk process. Although many options (Hanneborg, 1991) for permanently bonding two wafers together exist, we will describe only two here. The first is electrostatic (anodic) bonding of silicon to glass (Spangler and Wise, 1990). This process was developed in the 1960s, and the first sensor applications took place in the 1970s. In this process, the silicon sensor wafer is placed in contact with a glass wafer whose thermal expansion coefficient closely matches that of silicon. The assembly is raised to a temperature of typically 400–600°C, and a

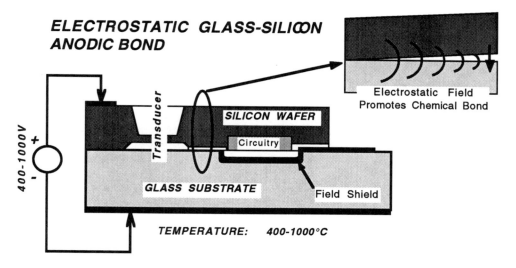

Figure 3.3. Illustration of the electrostatic glass–silicon bonding process used in many micromechanical devices.

voltage of 400–1,000 V is applied (silicon positive). Most of the applied voltage is dropped across the silicon–glass interface, pulling the two materials together and inducing a permanent bond as illustrated in Figure 3.3.

Silicon-to-silicon fusion bonds (Lasky et al., 1985) can also be produced at the wafer level if two silicon wafers are placed in contact with one another and are heated to temperatures in excess of 1,000°C. Such bonds can be used to produce a variety of complex microstructures that are impossible to produce using bulk or surface micromachining alone (Petersen et al., 1988). There are difficulties, however, such as the high bonding temperatures and the need for precision wafer alignment, as well as the need to obtain high coverage of the bonded areas. In the electrostatic bond, the applied field provides a powerful local force drawing the wafers together; with fusion bonding, this external force is absent. With both bonding techniques, the surfaces to be bonded must be highly planar, because the bonds produce very little actual material flow. In the electrostatic bond, planarity must be to within 30–50 nm in order to ensure a hermetic seal.

3.2.4 Electroforming processes (LIGA)

There is one more process capable of producing sophisticated microstructures that must be mentioned. Originally developed in Germany (Ehrfeld et al., 1988) and extended by researchers at the University of Wisconsin (Guckel et al., 1990; 1991), the LIGA process allows the formation of vertical structures using electroplating through a thick resist mask defined with x-ray lithography. Vertical structures

Figure 3.4. A magnetic micromotor produced at the University of Wisconsin using nickel electroforming and the LIGA process. The rotor diameter is 150 μm. (Photo courtesy of H. Guckel)

100 μm or more high and separated by a few microns or less can be formed. Rotary motors, laterally driven microstructures, gears, and a variety of other devices have been realized in batch using this process, which escapes from many of the planar restrictions associated with most wafer-based components. The chief disadvantage is that an appropriate source of x-ray radiation is required. Figure 3.4 shows a view of a magnetically driven micromotor and gear train produced at the University of Wisconsin using the LIGA process.

3.3 Examples of emerging devices

3.3.1 A multiplexed cardiovascular pressure sensor

Ultraminiature pressure sensors are badly needed for studying pressure and flow in the cardiovascular system, particularly within the coronary arteries of the heart. Although catheterization is a major diagnostic tool today worldwide, the sensors for such systems are typically external to the catheter, measuring pressure through the long, fluid-filled lumen and limiting sensing to a single site at a time with a highly restricted signal bandwidth. For measuring pressure gradients across coronary occlusions, sensors are needed that combine high resolution of pressure (≤ 2 mm Hg) with extremely small size (≤ 0.5 mm) so they can be mounted directly on the guide wire or in the catheter lumen near its tip. Two such devices mounted along the catheter would then allow pressure gradients to be resolved across the occlusion. This application offers significant challenges in terms of overall size, pressure resolution, and packaging.

Figure 3.5 shows the structure of a capacitive pressure sensor developed recently for catheter-based applications (Ji et al., 1992). The device consists of a thin silicon pressure diaphragm and a thicker supporting rim, both of which are formed using boron diffusions into a host wafer. This wafer is bonded to a selectively metallized glass substrate at the wafer level, and the silicon wafer is then subsequently etched away to leave the pressure sensors bonded to the glass. Figure 3.6 shows an array of such sensors on glass. The overall width of the sensors is 350 μm. The silicon diaphragms are 220 μm wide x 420 μm long x 1.6 μm thick with a nominal gap between the diaphragm and the glass of 2 μm. The lateral lead tunnels can be vacuum sealed at wafer level using a deposited layer of low-temperature oxide or other suitable insulator. The glass is then separated into individual dies

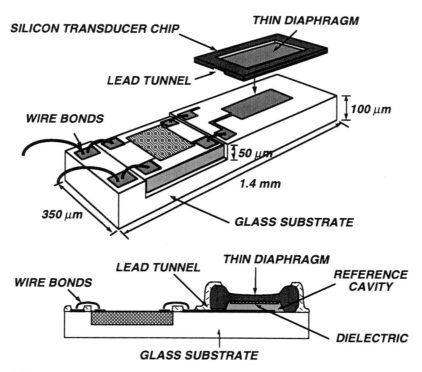

Figure 3.5. Structure of an ultraminiature bulk-micromachined capacitive pressure sensor for a cardiovascular catheter.

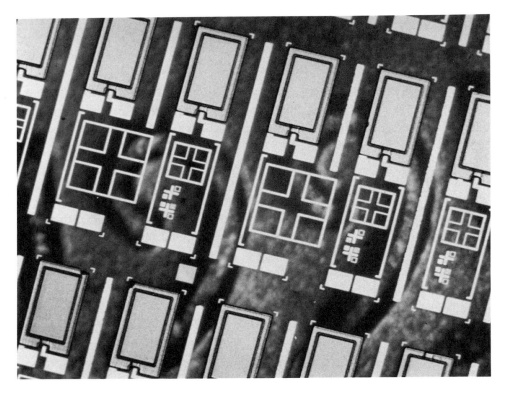

Figure 3.6. Top view of an array of capacitive pressure sensors on glass, being fabricated for cardiovascular catheters. The individual devices are 350 μm wide. The date on a U.S. penny is seen in the background.

using a diamond saw. A recess is sawed in the glass to allow a standard integrated circuit to be mounted in the structure for signal processing. The purpose of this circuit is to limit to two the number of leads on the overall catheter, independent of the number of sensing sites. This is essential for such a device of this size.

The two leads for this sensor are used for four separate functions: supplying power, serial addressing of the various sites and transducers, resetting the chip addresses (on power-up), and signal readout. Each sensing site has four states: pressure measurement, high-reference capacitance, low-reference capacitance, and idle. The high- and low-reference capacitors are used for calibrating the pressure transducer and for measuring the chip temperature for calibration purposes; in the idle state, the leakage current on the overall catheter can be scanned automatically twice each frame for safety purposes. This level is normally <10 μA. If excess leakage is detected, the supply can be interrupted. The nominal supply voltage is 5 V. On this level, an 8-V clock signal is superimposed to increment the site address using an on-chip pulse detector and counter. Pulsing the supply to 11 V resets the counter and reinitializes the device. The signal is read out using the transducer capacitance to set the frequency of a Schmitt oscillator, which inserts spikes on the power supply at a nominal frequency of 2.7 MHz. The frequency sensitivity is 2 kHz/fF with a pressure resolution of <2 mmHg, dropping to <1 mmHg for a 500 μm overall sensor width (Ji et al., 1992).

This presure sensor is an example of a minimally invasive instrument for the acute measurement of pressure in small vessels. It is important in its potential for diagnostic feedback during angioplasty and because it offers an entry point for instrumentation aimed at the measurement of additional variables. For example, such a chip could also serve as a base for integrated transducers measuring lumen diameter, wall compliance, and perhaps blood chemistry. In evolving such devices, however, many challenges remain, and it is important to remember that nearly all the cost is associated with packaging (and testing) in this application. The function of on-board electronics is to make packaging easier, here by limiting the number of required leads. Even with only two leads, the packaging problems for such devices are not yet completely resolved.

3.3.2 *Neural probes for recording and stimulation*

The possibility of in-dwelling microsensors is illustrated by research on advanced neuroelectronic interfaces for the study of signal-processing techniques in biological neural networks and for application in neural prostheses (Hoogerwerf and Wise, 1991; Ji and Wise, 1990; Najafi et al., 1985). Figure 3.7 shows the diagram of a multichannel micromachined recording array intended for such applications,

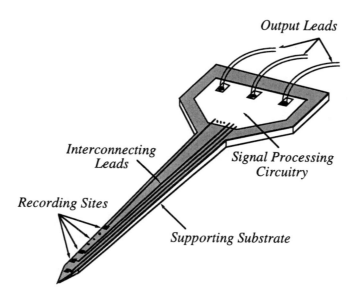

Figure 3.7. Structure of a multichannel intracortical microprobe produced using bulk micromachining.

and Figure 3.8 shows a photograph of one such probe tip. The overall substrate is defined using a boron-diffused etch-stop, with circuitry formed (on an optional basis) in a lightly doped portion of the substrate near the rear of the probe. The bulk-micromachined process is single-sided on wafers of normal thickness; for structures without circuitry the process requires only five masks. Dimensions are controlled to ±1 μm with typical shank widths of 30 μm and thicknesses of 12–15 μm. Using a shallow diffusion at the tip of the probe, the devices can be made very sharp and therefore capable of penetrating structures such as pia arachnoid with little discernible cortical dimpling. Single-unit potentials can be recorded from electrode sites spaced at any desired interval in depth and between shanks typically spaced at 100 μm intervals. The devices can be formed in three-dimensional arrays (Hoogerwerf and Wise, 1991) and are generally well accepted by the tissue. Histology has shown little or no tissue reaction after several months in vivo as long as the implant does not become anchored to the skull due to tissue overgrowth. To prevent such problems, a low probe profile is desirable, and protein

Figure 3.8. View of a multichannel recording probe produced using silicon micromachining technology. The recording sites here are spaced 100 μm apart in depth.

coatings are being explored to enhance and prohibit tissue growth in appropriate areas of such structures.

Leads present a major problem in the development of chronically implantable probes and many other biosensors. The probe should float with the brain as it moves inside the cranial cavity, and the individual recording sites should have impedance levels of more than 1 MΩ. Thus, lead wires must exert minimal tethering on the devices while providing a high level of encapsulation for extended times in vivo. One approach to this problem uses silicon-based ribbon cables (Hetke et al., 1990) as shown in Figure 3.9. A shallow boron diffusion forms the cable substrate. Stress-relieved dielectrics of silicon nitride and silicon dioxide encapsulate thin-film leads of polysilicon or refractory metals that connect the probe to a percutaneous plug or supradural signal-processing platform. A metal or polysilicon shield can be used over the upper surface of the cable if desired, forming a coaxial multilead structure. The overall ribbon thickness is typically 5 μm. These cables have maintained sub-pA leakage currents in saline under ±5-V bias for over three years and appear very promising for these applications. Most important, they can be integrated as a natural extension of the probe substrate so that the probe leads go from the sites to the plug or platform with no intermediate bonds or connections. Used in structures such as that shown in Figure 3.10, they have maintained stable impedance levels and permitted chronic recording from single units for periods approaching one year.

In order to allow large numbers of sites to be integrated in two- and three-dimensional arrays, it is desirable to integrate circuitry into the probe substrates to allow local multiplexing and to limit the number of external leads required. A recent recording probe (Ji and Wise, 1990) incorporates on-chip CMOS (complimentary

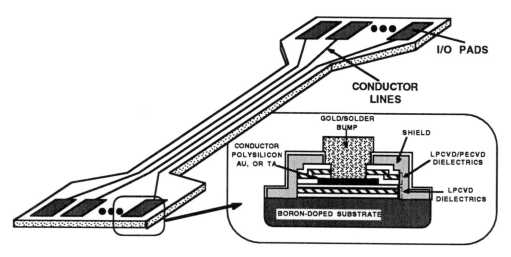

Figure 3.9. Structure of a silicon ribbon cable. Such cables can be integrated with the probes to form a single unit compatible with chronic neural recording.

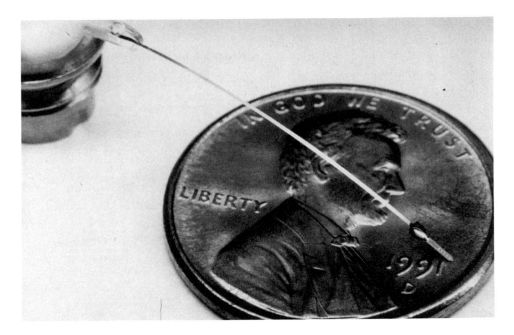

Figure 3.10. Picture of a five-site passive stimulating probe intended for chronic use. The probe is connected to a percutaneous plug using an integral silicon ribbon cable. A drop of silastic is placed over the rear of the probe for handling purposes.

metal oxide semiconductor) circuitry to select 8 of 32 available sites for monitoring and amplification. On-chip amplifiers, each occuping only 0.06 mm^2 in 3 µm CMOS technology, provide a per-channel gain of 300 and bandlimit the recorded signals to 10 Hz to 10 kHz while requiring no off-chip components and allowing a three-lead interface to the external world. A family of companion stimulating probes (Tanghe and Wise, 1992) accepts serial input data to address a particular site and define an 8b-current amplitude. On-chip circuitry then generates the selected current and routes it to the desired site. Figure 3.11 shows a portion of a 16-channel probe that drives current to any of the sites independently over a current range of 0 to ±254 µA with a current resolution of 2 µA. The current pattern from the probe can be changed in 4 µsec using a 4-MHz input clock. The most recent version of the stimulating probes (Kim et al., 1993) allows the user to select 8 of 64 front-end sites for stimulation. All of the stimulating probes operate over five external leads.

The technology is now nearly in place to allow the fabrication of electrode arrays supporting perhaps a thousand sites and capable of instrumenting a considerable volume of tissue. The tissue displaced by these micromachined structures should still be less than 1% of the total, so there is hope that the neural system will continue to function undisturbed by the presence of the array. However, considerably more research will be necessary to ensure that these structures are sufficiently well accepted so that they (and the biosystem) remain functional for years. Today, they are functional for

Figure 3.11. Top view of a micromachined 16-channel stimulating probe. The probe contains sixteen 8b CMOS digital-to-analog converters and associated control circuitry in an area of 11 mm^2 in 3-μm features.

weeks on a rather routine basis, and for months on an occasional basis. They must be viable for many years in order to make next-generation prostheses practical. This will also require additional progress on global leads and on packaging.

There are also possibilities for interfacing with intact neural networks at the cellular level. Highly selective temperature transducers and microheaters can be integrated into the probe structure to allow thermal activation and marking of tissue, and chemical drug delivery channels can be integrated as well (Chen and Wise, 1994). By adding one additional mask to the probe process, multibarrel hollow-core substrates can be formed to allow chemical injection with a resolution of less than 100 μm. Microflow tubes about 10 μm in diameter and separated by as little as a few microns have been integrated with probes and used to stimulate neural activity in acute experiments. Although such structures are currently in their infancy, they have demonstrated the ability to control the chemical environment of tissue on a highly local basis and combine that with the electrical stimulation and recording of cell responses. Though at present such structures are viable only for acute applications, the addition of valves, reservoirs, and perhaps pumps could someday make them viable on a chronic basis. But these are difficult problems that are just beginning to be addressed. Solutions to these problems are perhaps feasible in an engineering sense, but only if important needs can be identified and addressed by close collaboration with biologists and physiologists.

3.3.3 Microtelemetry: toward completely implantable microsystems

Although miniature ribbon cables may solve some of the local lead problems in vivo, communication over distances of more than a few centimeters and with the outside world still poses many unsolved problems. Telemetry systems have slowly evolved over the past several years and, for some systems, may soon offer solutions in this area. The goal is the ability to deliver both power and control signals to the implant and to receive data or status information from it. When the implant itself is large, as in pacemakers and some drug pumps, the power source can be contained within the implant in the form of a battery. Such applications also typically do not have high data rates, making them quite tractable with current technology. For implants that are more size constrained, however, such as the neural probes, this is not possible, and advances are needed.

An example of current trends is illustrated by an implantable microstimulator (Ziaie et al., 1993) for use in functional neuromuscular stimulation (FNS) currently under development at the University of Michigan. It consists of a micromachined substrate that incorporates two stimulating electrodes, a hybrid tantalum chip capacitor for power storage, a receiving coil, a bipolar-CMOS integrated circuit for power regulation and control, and a hermetically sealed glass capsule package. This system is shown in Figure 3.12. The microstimulator measures 2 mm x 2 mm x 10 mm overall and can be implanted through a hypodermic needle. Power and data are transmitted using RF telemetry, and the microstimulator is capable of delivering 10 mA into loads of less than 800 Ω for up to 200 μsec at a rate of up to 40 kHz. The device is addressable, and up to 32 such stimulators can be driven using a single transmitter coil.

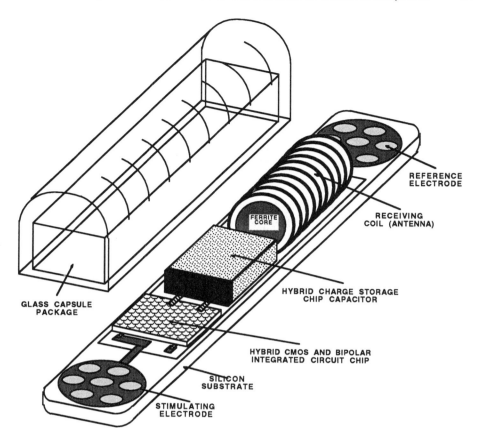

Figure 3.12. Cross section of a microstimulator having a width less than 2 mm being developed for use in functional neuromuscular stimulation. (Courtesy of K. Najafi, University of Michigan)

Additional work on microtelemetry systems is needed to allow still higher performance wireless communication among future implanted sensing nodes, as well as between those nodes and the outside world. Given the continuing progress in microelectronic technology, however, it is probable that very small two-way high-bandwidth systems will be realized during the coming decade as specific system application targets are identified and as the front-end transducers become a reality.

3.4 Conclusions

Athough there has been considerable progress in the development of micromechanical devices for use in biosystems, there are many challenges still ahead. In the engineering area, there is no question that mechanical devices are the most developed of all the various sensor types, followed by thermal (flow), magnetic, and chemical devices. Chemical sensors are among the most needed devices, but the complexity of the in vivo chemistries involved and the packaging difficulties associated with in vivo use have made even acute devices slow to be realized. As we move from sensors to microactuators and MEMS, there are still other problems to be addressed. In the area of microactuators, we badly need better drive mechanisms and ways of delivering the resultant forces to the outside world. Microassembly techniques are also needed but have as yet scarcely been addressed.

As formidable as many of the engineering challenges are, the biological challenges are at least as difficult; they are probably more difficult to address since we are ultimately at the mercy of what the biosystem will tolerate. Insertion and packaging techniques to enhance biocompatibility will be critically important and will require considerable continuing effort. Most important, however, engineers and biologists must continue to work closely together to identify the critical problems and appropriate approaches to the solution of those problems. Micromechanical devices and their technology give us tools to make significant progress during the coming decade if these collaborations between engineering and biology can be established.

Acknowledgments

The author gratefully acknowledges the many efforts of those at the University of Michigan who have contributed to the work illustrated in this paper. In particular, thanks go to Professors K. Najafi and D. J. Anderson, Drs. J. Ji, S. J. Tanghe, and S. T. Cho, and the graduate students and staff of the Solid-State Fabrication Facility. The support and encouragement of Drs. F. Terry Hambrecht and William J. Heetderks of the National Institutes of Health (NINDS) have made our particular activities possible.

References

Chau, H. L. and Wise, K. D. (1988) An ultraminiature solid-state pressure sensor for a cardiovascular catheter. *IEEE Trans. ED* 35:2355–2362.

Chen, L. Y. and. MacDonald, N. C. (1991) A selective CVD tungsten process for micromotors. *Digest IEEE Int. Conf. on Solid-State Sensors and Actuators* June 1991:739–742.

Chen, J. K. and Wise, K. D. (1994) A multichannel neural probe for selective chemical delivery at the cellular level. *Digest Solid-State Sensor and Actuator Workshop, Hilton Head, S.C.* June 1994.

Cho, S. T., Najafi, K., and Wise, K. D. (1990) Scaling and dielectric stress compensation of ultrasensitive boron-doped silicon microstructures. *Proc. IEEE MicroElectro Mechanical Systems, Napa, Ca.* February 1990:50–55.

deBruin, D., Allen, H. V., and Terry, S. C. (1990) Second-order effects in self-testable accelerometers. *Digest IEEE Solid-State Sensor Workshop, Hilton Head, S.C.* June 1990:149–152.

Declercq, M. C., Gerzberg, L., and Meindl, J. D. (1975) Optimization of the hydrazine-water solution for anisotropic etching of silicon in integrated circuit technology. *J. Electrochem. Soc.* 122:545.

Ehrfeld, W., Gotz, F,. Munchmeyer, D., Schelb, W., and Schmidt, D. (1988) LIGA process: sensor construction techniques via X-ray lithography. *IEEE Solid-State Sensor Workshop Hilton Head, S.C.* June 1988:1–4.

Fan, L. S., Tai, Y. C., and Muller, R. S. (1988) Integrated movable micromechanical structures for sensors and actuators. *IEEE Trans. ED* 35:724–730.

Finne, R. M. and Klein, D. L.(1967) A water-amine-complexing-agent system for etching silicon. *J. Electrochem. Soc.* 114:965–970.

Guckel, H. and Burns, D. W. (1984) Planar processed polysilicon sealed cavities for pressure transducer arrays. *Digest IEEE IEDM* December 1984:223–225.

Guckel, H., Burns, D. W., Rutigliano, C. R., Showers, D. K., and Unglow, J. (1987) Fine-grained polysilicon and its application to planar pressure transducers. *Digest IEEE Int. Conf. on Solid-State Sensors and Actuators Tokyo* June 1987:277–282.

Guckel, H., Burns, D. W., Visser, C. C. G., Tilmans, H. A. C., and DeRoo, D. W. (1988) Fine-grained polysilicon films with built-in tensile strain. *IEEE Trans. ED* 35:800–801.

Guckel, H., Christenson, T. R., Skrobis, K. J., Denton, D. D., Choi, B., Lovell, E. G., Lee, J. W., Bajikar, S. S., and Christenson, T. W. (1990) Deep x-ray and UV lithographies for micromechanics. *Digest IEEE Solid-State Sensor Workshop, Hilton Head, S.C.* June 1990:118–122.

Guckel, H., Skrobis, K. J., Christenson, T. R., Klein, J., Han, S., Choi, B., and Lovell, E. G. (1991) Fabrication of assembled micromechanical components via deep x-ray Lithography. *Digest IEEE Workshop on MicroElectroMechanical Systems, Nara, Japan* January 1991:74–79.

Hanneborg, A. (1991) Silicon wafer bonding techniques for assembly of micromechanical elements. *Digest IEEE Workshop on MicroElectroMechanical Systems, Nara, Japan.* January 1991:92–98.

Hetke, J. F., Najafi, K., and Wise, K. D. (1990) Flexible miniature ribbon cables for long-term connection to implantable sensors. *Sensors and Actuators* A23:999–1002.

Hoogerwerf, A. C. and Wise, K. D. (1991) A three-dimensional neural recording array. *Digest IEEE Int. Conf. on Solid-State Sensors and Actuators* June 1991:120–123.

Howe, R. T. (1988) Surface micromachining for microsensors and microactuators. *J. Vac. Sci. Technol. B.* 6:1809–1813.

Howe, R. T. and Muller, R. S. (1986) Resonant-microbridge vapor sensor. *IEEE Trans. ED* 33:499–506.

Ikeda, K. et al. (1990) Silicon pressure sensor integrates resonant strain gauge on diaphragm. *Sensors and Actuators* A21:146–150.

Ji, J., Cho, S. T., Zhang, Y,. Najafi, K., and Wise, K. D.(1992) An ultraminiature CMOS pressure sensor for a multiplexed cardiovascular catheter. *IEEE Trans. Electron Devices* 39:2260–2267.

Ji, J. and Wise K. D. (1990) An implantable CMOS analog signal processor for multiplexed multielectrode recording arrays. *Digest IEEE Solid-State Sensor and Actuator Workshop, Hilton Head* June 1990:107–110.

Kim, Y. W. and Allen, M. G. (1991) Surface micromachined platforms using electroplated sacrificial layers. *Digest IEEE Int. Conf. on Solid-State Sensors and Actuators* June 1991:651–654.

Kim, C., Tanghe, S. J., and Wise, K. D. (1993) Multichannel neural probes with on-chip CMOS circuitry and high-current stimulating sites. *Digest IEEE Int. Conf. on Solid-State Sensors and Actuators (Transducers'93), Yokohama* June 1993:454–457.

Klein, D. L. and D'Stefan, D. J. (1962) Controlled etching of silicon in the HF-HNO$_3$ system. *J. Electrochem. Soc.* 109:37–42.

Kong, L. C., Orr, B. G., and Wise, K. D. (1990) A micromachined silicon scan tip for an atomic force microscope. *Digest 1990 IEEE Solid-State Sensor and Actuator Workshop, Hilton Head, S.C.* June 1990:28–31.

Lasky, J. B., Stiffler, S. R., White, F. R., and Abernathey, J. R. (1985) Silicon-on-insulator (SOI) by bonding and etch-back. *Digest IEEE IEDM* December 1985:684–687.

Lepselter, M. P. (1966) Beam-lead technology. *Bell System Technical Journal* 45:233–254.

Lober, T. A. and Howe, R. T. (1988) Surface-micromachining processes for electrostatic microactuator fabrication. *Digest IEEE Solid-State Sensor and Actuator Workshop, Hilton Head, S.C.* June 1988:59–62.

Mehregany, M., Gabriel, K. J., and Trimmer, W. S. N. (1988) Integrated fabrication of polysilicon mechanisms. *IEEE Trans. ED* 35:719–723.

Mehregany, M., Senturia, S. D., and Lang, J. H. (1990) Friction and wear in microfabricated harmonic side-drive motors. *Digest IEEE Solid-State Sensor Workshop, Hilton Head, S.C.* June 1990:17–22.

Najafi, K., Wise, K. D., and Mochizuki, T. (1985) A high-yield IC-compatible multi-channel recording array. *IEEE Trans. ED*. 32:1206–1211.

Nathanson, H. C., Newell, W. E., Wickstrom, R. A., and Davis, J. R. (1967) The resonant-gate transistor. *IEEE Trans. ED* 14:117–133.

Ozdemir, C. H. and Smith, J. G. (1991) New phenomena observed in electrochemical micromachining of silicon. *Digest IEEE Int. Conf. on Solid-State Sensors and Actuators* June 1991:132–135.

Pak, J. J., Kabir, A. E., Neudeck, G. W., Logsdon, J. H., DeRoo, D. R., and Staller, S. E.(1991) A micromachining technique for a thin silicon membrane using merged epitaxial lateral overgrowth of silicon and SiO_2 for an etch-stop. *Digest IEEE Int. Conf. on Solid-State Sensors and Actuators* June 1991:1028–1031.

Petersen, K. E. (1982) Silicon as a mechanical material. *Proc. IEEE* 70:420–457.

Petersen, K., Barth, P., Poydock, J., Brown, J., Mallon, J., and Bryzek, J. (1988) Silicon fusion bonding for pressure sensors. *IEEE Solid-State Sensor Workshop, Hilton Head, S.C.* June 1988:144–147.

Price, J. B. (1973) Anisotropic etching of silicon with KOH-H_2O-isopropyl alcohol. In *Semiconductor Silicon 1973*, *Electrochem. Soc. Princeton*:339-353.

Spangler, L. J. and Wise, K. D. (1990) A bulk silicon SOI process for active integrated sensors. *Sensors and Actuators* A24:117–122.

Sugiyama, S., Kawahata, K., Abe, M., Funabashi, H., and Igarashi, I. (1987) High-resolution silicon pressure imager with CMOS processing circuits. *Digest Int. Conf. Solid-State Sensors and Actuators, Tokyo* June 1987:444–447.

Tang, W. C., Nguyen, T. C. H., and Howe, R. T. (1989) Laterally-driven polysilicon resonant microstructures. *Sensors and Actuators* 20:25–32.

Tanghe, S. J. and Wise, K. D. (1992) A 16-channel CMOS neural stimulating array. *IEEE Journal of Solid-State Circuits* 27:1819–1825.

Wise, K.D. (1991) Integrated microelectromechanical systems: a perspective on MEMS in the 90s (Invited). *Proc. IEEE MicroElectroMechanical Systems Workshop, Nara* January 1991:31–38.

Wise, K. D. and Najafi, N. (1991) The coming opportunities in microsensor systems. (Invited), *Digest IEEE Int. Conf. on Solid-State Sensors and Actuators* June 1991:2–7.

Ziaie, B., Nardin, M., Von Arx, J., and Najafi, K. (1993) A single-channel implantable microstimulator for functional neuromuscular stimulation (FNS). *Digest IEEE Int. Conf. on Solid-State Sensors and Actuators (Transducers'93), Yokohama* June 1993:450–453.

4

Microsensors and microactuators for biomedical applications

MASAYOSHI ESASHI

Faculty of Engineering, Tohoku University
Aza Aoba Aramaki Aoba-ku Sendai 980-77, Japan

4.1 Introduction

Active and multifunctional catheters are under development for use in instrumentation and treatment inside a blood vessel. The concept of an active catheter, shown in Figure 4.1, involves the integration of very small sensors and actuators of a variety of types. More than one microsensor and microactuator will be necessary in order to allow bending. Communication and control technologies will also be required to minimize the number of lead wires. For this purpose, a common two-lead wire system was developed in which the two lines supplying power also convey signals and select the sensor or the actuator to be activated. Micropackaging technology for assembling, as well as microactuator techniques, are also being studied. Silicon micromachining based on integrated circuit technology and three-dimensional microfabrication technologies, as discussed in Chapters 1–3, are applied to the fabrication of catheters.

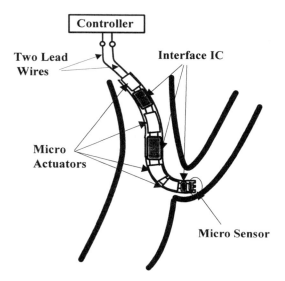

Figure 4.1. Concept of an active catheter.

4.2 Packaged and integrated microsensors

A packaging technology that can integrate a capacitance detection circuit, a capacitive sensor in a parallel electrode structure, or a resonator has been developed. This technology, shown schematically in Figure 4.2, makes small, inexpensive, and reliable sensors available. The electrical feedthrough from the sealed cavity was made through a glass hole. This vertical feedthrough structure can seal the cavity hermetically around the glass hole. The glass holes can be made by electrochemical discharge drilling using spark erosion in NaOH solution. The feedthrough from the silicon was made using a diffused layer. The p–n junction is covered with SiO_2 as shown in Figure 4.2a (Esashi et al., 1990). The feedthrough structure was used for micro pressure sensors for continuous monitoring in an artificial heart (Nitta et al., 1990) and for many other sensors, which will be described later. The feedthrough structures from the electrodes on the glass are shown in Figures 4.2b and 4.2c.

Laser-assisted silicon etching has been developed and applied to accelerometers, resonators, and other devices (Minami et al., 1993b). Resistless dry etching is required to remove a part of the silicon after anodic bonding. Figure 4.3a shows the application of laser-assisted etching to an accelerometer. All parts of the silicon should be monolithically connected with beams before anodic bonding. Fragile parts should also be supported with extra beams. These beams must be removed selectively after bonding the silicon wafer to the bottom glass. This process allows the fabrication of the structure shown in Figure 4.2c. A YAG laser beam is focused on the silicon sample, which is located in an etching chamber as

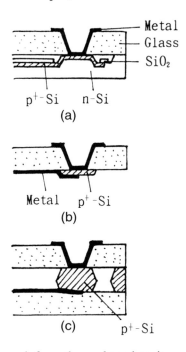

Figure 4.2. Electrical feedthrough from the packaged cavity.

shown in Figure 4.3b. Reactive gases such as NF_3 are used as an etching gas. This process allows fast and dry etching of the silicon without particles.

Large changes in capacitance are available in the capacitive pressure sensor, which detects pressure change by the deformation of a diaphragm. In principle, high sensitivity and low power consumption can be achieved; however, the parasitic capacitance increases with a decrease in the sensor size. Therefore, the capacitance detection circuit must be integrated with the sensing device.

An integrated silicon capacitive pressure sensor has been fabricated (Matsumoto and Esashi, 1993a). The structure is shown in Figure 4.4. A Pyrex glass plate and a silicon wafer are anodically bonded to form an internal cavity for the reference pressure. Integrated in each sensor are a sensing diaphragm and a capacitance detection circuit – a CMOS (complimentary metal oxide

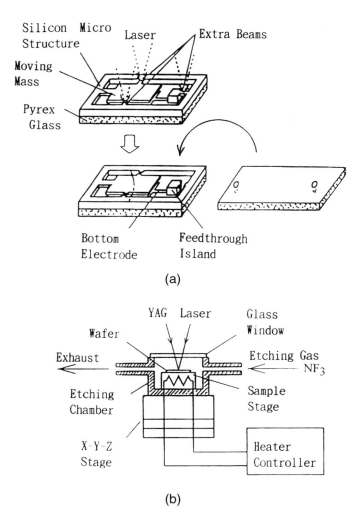

Figure 4.3. Laser-assisted etching.

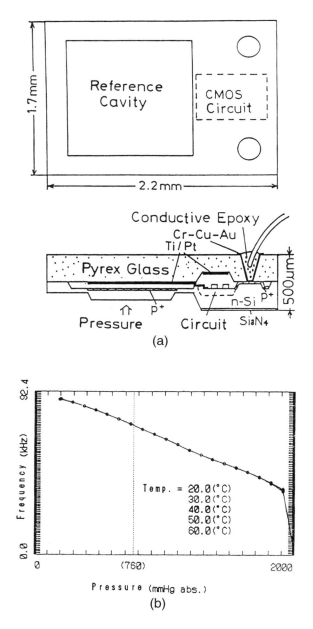

Figure 4.4. Integrated capacitive pressure sensor. (a) Structure. (b) Characteristic.

semiconductor) capacitance-to-frequency converter, described in Section 4.3; the silicon chip acts as the package. The feedthrough from the internal cavity is made through a glass hole as explained previously. This feedthrough structure not only hermetically seals the cavity but also enables batch packaging during wafer process, which gives small, reliable, and low-cost sensors. Figure 4.5 shows the

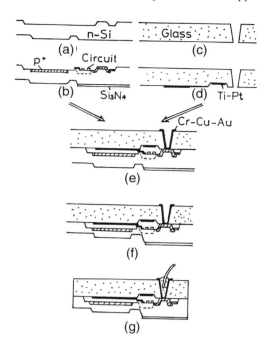

Figure 4.5. Fabrication process of the integrated capacitive pressure sensor.

fabrication process for the sensor. The silicon wafer used as the substrate for the diaphragm and for the CMOS oscillator is anodically bonded with a glass plate. Each packaged sensor chip is obtained after dicing the glass–silicon wafer. Using this method, the packaging is carried out in the wafer stage, and hence low-cost, small-sized devices can be provided.

An integrated capacitive accelerometer that has a capacitive sensor, an integrated circuit, and an electrostatic actuator inside the glass–silicon package is shown in Figure 4.6 (Matsumoto and Esashi, 1993b). A silicon seismic mass is suspended by springs of thin silicon oxynitride beams, and the internal cavity is sealed in a vacuum to avoid viscous damping (known as the squeeze-film effect). The displacement of the mass is detected capacitively, and the output frequency of the integrated circuit is modulated by the capacitance. The output frequency is compared with a reference frequency, and the difference is used as a feedback signal for electrostatic force balancing. Using such a phase-locked loop feedback, one can achieve force balancing. This force balancing mechanism allows the measurement range and the sensitivity of the accelerometer to be varied electrically by the feedback parameter and, in addition, such sensors have the capability of self-diagnosis using the electrostatic force.

A three-axis electrostatic force–balancing accelerometer has also been developed with this packaging technology (Jono et al., 1994). The structure and the principle are shown in Figure 4.7. Tilt of the mass by *x*- or *y*-axis acceleration and

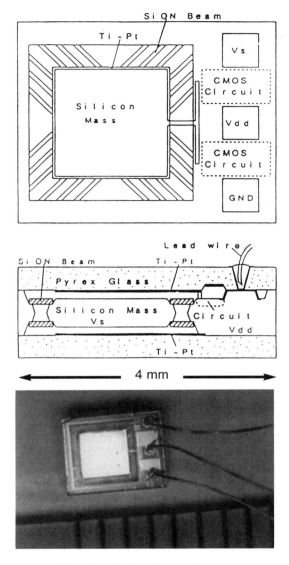

Figure 4.6. Integrated capacitive force balancing accelerometer.

movement by z-axis acceleration are detected capacitively. Differential capacitances are detected separately using a phase-sensitive detector. The feedback voltages for force balancing are superimposed on the signal voltages for capacitive sensing.

Silicon microresonators can have a very high quality factor because of their low internal friction. High quality factors lead to high resolution of resonant sensors. A resonant infrared sensor with the resonance frequency modulated by radi-

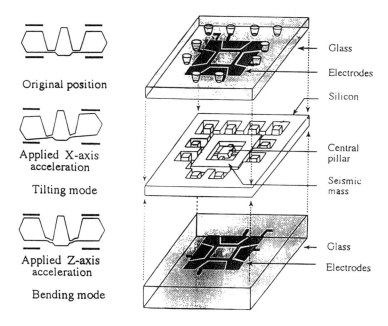

Original position

Applied X-axis
acceleration

Tilting mode

Applied Z-axis
acceleration

Bending mode

Glass

Electrodes

Silicon

Central
pillar

Seismic
mass

Glass

Electrodes

Figure 4.7. Three-axis capacitive force balancing accelerometer.

ation-induced thermal strain variation has been developed (Cabuz et al., 1993). The structure is shown in Figure 4.8. The resonant bridge is clamped to a one-sided silicon frame that is bonded to the glass in order to avoid package stresses. Good thermal isolation and small thermal capacity are required for high sensitivity and fast response, respectively. The resonator bridge consists of a p^+ silicon absorber supported with silicon dioxide beams from the silicon frame. The resonator is driven electrostatically and the movement is detected capacitively. A reference resonator for temperature compensation and an on-chip detection transistor were integrated, as shown in Figure 4.8. The resonant frequency and Q factor were about 100 kHz and 20,000 respectively. The responsivity was about 500 ppm/μW. In order to realize a high Q factor and good thermal isolation, the resonator is placed in a vacuum. For this reason, vacuum packaging (described later) is indispensable.

Resonators and other similar devices require a high vacuum inside a cavity. Even if glass–silicon is bonded in a high vacuum, a good vacuum cannot be obtained in the cavity because oxygen gas is produced from the glass during the bonding process. To solve this problem, nonevaporable getter (NEG) can be added into the cavity (Henmi et al., 1993). The pressure inside the cavity is lower than 1×10^{-5} Torr and is stable over time.

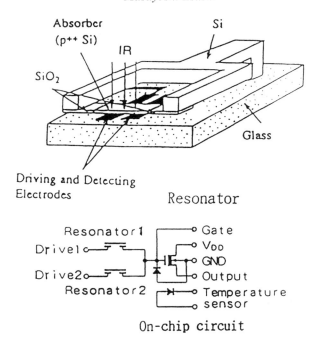

Figure 4.8. Resonant infrared sensor.

4.3 Common two-wire systems

In the integrated capacitive pressure sensor shown in Figure 4.4, the oscillation frequency is changed by the capacitance (Matsumoto and Esashi, 1993a). The Schmitt trigger oscillation circuit used is shown in Figure 4.9a. The output frequency can be detected by monitoring the pulse in the supply current, thus making a two-wire detection feasible with this circuit. Furthermore, the circuit is designed to have an output frequency that is independent of temperature and supply voltage.

Another two-wire detection circuit for differential capacitance sensors has been developed. This was intended to be used as an acceleration sensor similar to that shown in Figure 4.6 (Shirai et al., 1993). The principle and the circuit are shown in Figure 4.10a. Capacitors C1 and C2 are charged with constant current and discharged alternately. The intervals for charging C1 and C2 are proportional to their capacitances. Transient supply current pulses appear when the flip-flop changes state. Figure 4.10b shows the dependence of the current pulse interval of a fabricated chip on the capacitance C1. It is clear that T1 increases with an increasing value of C1. The charging time is unchanged because of the fixed value of C2. To enhance the difference of the amplitude of the two current pulses, capacitor C3 was connected as shown in Figure 4.10a. The charging times T1 and T2 are dependent on temperature, but the effect tends to be canceled by the influence of the ratio of T1 to T2. Because T1 + T2 is temperature dependent, it can

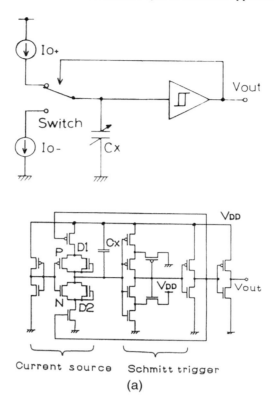

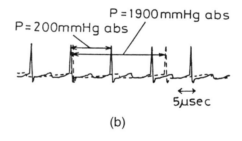

Figure 4.9. Two-wire capacitance-to-frequency converter circuit. (a) Circuit. (b) Supply current waveform.

be used to measure temperature change, thus allowing for compensation of the baseline drift and the sensitivity change.

Complicated systems need advanced communication technology to minimize the number of lead wires. Several sensors can be connected to the same power lines, as shown in Figure 4.11a, and one sensor can be operated selectively. Such a system is called a common two-wire sensing system (Esashi and Matsumoto, 1991). When the power-supply voltage Vp is varied, the clock signal and address signal are detected. The address signal is fed to a shift-register using the clock

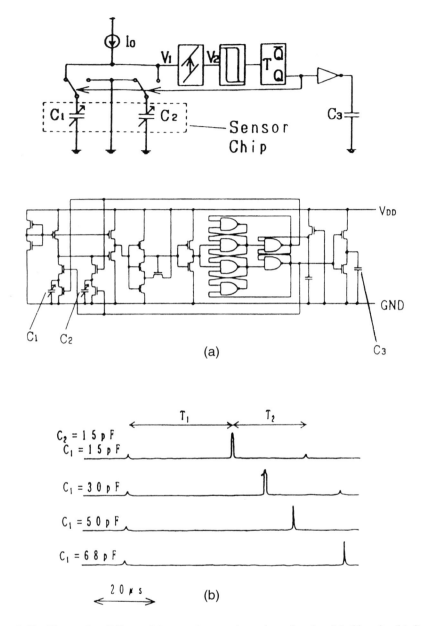

Figure 4.10. Two-wire differential capacitance detection circuit. (a) Circuit. (b) Supply current waveforms.

signal to compare the address of each sensor. When the proper address (in this case, 0011) is chosen, the oscillating sensor circuit becomes active and the oscillation frequency is detected on the power line. Figure 4.11b shows an example of the operation of the fabricated integrated circuit (IC). The extended application of this system to multiactuator systems is being developed.

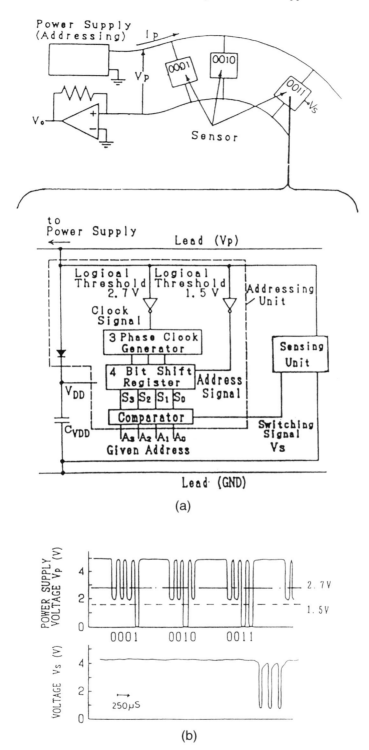

Figure 4.11. Common two-wire sensing system. (a) Circuit. (b) Waveforms.

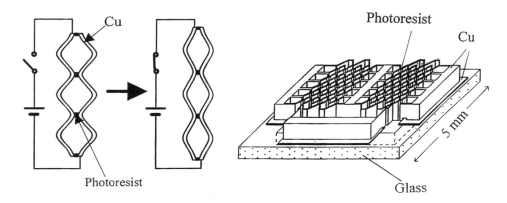

Figure 4.12. Principle of distributed electrostatic microactuator.

4.4 Microactuators

A distributed electrostatic microactuator (DEMA) is proposed, the concept of which is similar to muscles. DEMA has many serially connected driving units, which consist of two wavelike electrodes facing each other. A large electrostatic force and large displacement can be obtained by narrowing the gap and by serially connecting the driving units. The principle and the structure fabricated by photolithography and copper electroplating are shown in Figure 4.12 (Minami et al., 1993a). The structure consists of Cu electrodes and photoresist insulators. The fabrication process is shown in Figure 4.13. Displacement obtained is 28 µm (about 1%) when applied voltage is 160 V. In this actuator, the apparent compliance can be controlled by a feedback control system using capacitive displacement sensing and electrostatic driving (Yamaguchi et al., 1993).

For the active catheter that has multiactuators and interface circuits, a new shape-memory alloy (SMA) actuator has been developed. The resistance of the SMA actuator is too small to be driven by direct current heating. To solve this problem, indirect current heating is used by coating with a thin-film Ni heater. The characteristics are shown in Figure 4.14. The indirectly heated actuator can be driven with only 10 mA, enabling it to be driven by the in-dwelling interface circuit.

4.5 Three-dimensional microfabrication

IC fabrication methods are essential for micromachining, but additional special processes play an important role for three-dimensional fabrication. For example, nonplanar lithography, deep etching, thick film deposition, and interfacial bonding are required for optimized device structure. It is also important to minimize the stress-induced deformation.

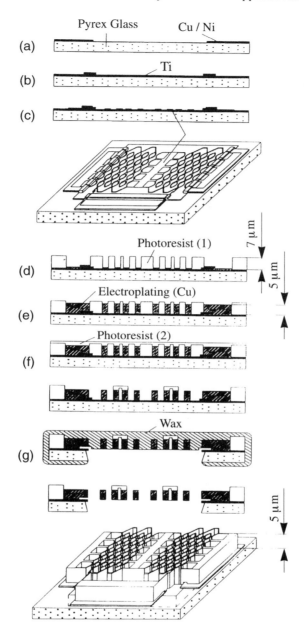

Figure 4.13. Fabrication process of distributed electrostatic microactuator.

High-speed reactive ion etching (RIE) was developed to make silicon and polymer structures that have high aspect ratio (Murakami et al., 1993). The etching apparatus has a cold stage and magnetron RIE system. A small chamber and a large evacuation system to achieve high flow rate of the etching gas are indispensable. Etching through a silicon wafer of 200 μm was performed using a nickel

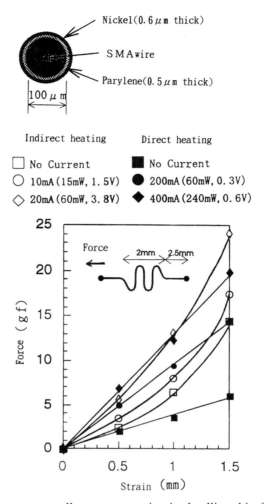

Figure 4.14. Shape-memory alloy actuator using in-dwelling thin-film heater.

mask and SF$_6$ gas. A maximum etching rate of 2.3 μm/min with small side etch ratio of 0.03 was achieved at -120°C. The cross-sectional view of the etched-through silicon wafer, 200 μm thick by 25 μm silicon width, is shown in Figure 4.15. Polyimide also could be etched with high aspect ratio using O$_2$ gas. The etched polyimide shown in Figure 4.15b is 75 μm thick by 5 μm pattern width. The polyimide coated on a silicon substrate that has a metal electroplating base was etched by RIE and was then used as a mold for the electroplating of copper. This technology can realize high aspect ratio microstructures like those realized by the LIGA (after the German, "Lithographie Galvanoformung Abformung") process (Becker et al., 1986), based on deep-etch x-ray lithography, electro-forming, and molding process. This technology can be applied to making a micro coil.

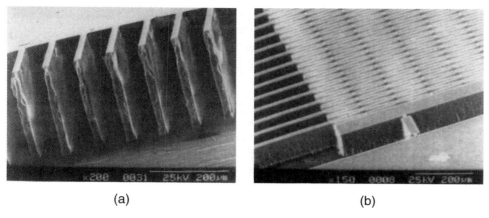

(a)   (b)

Figure 4.15. Deep reactive ion etching. (a) Silicon. (b) Polyimide.

Projection chemical vapor deposition (CVD) using ArF excimer laser has been developed (Takashima et al., 1993). The apparatus is schematically shown in Figure 4.16. By ultraviolet irradiation of a cooled substrate in the organic material gas environment, silicon dioxide, polymethylmethacrylate (PMMA), and tin oxide are deposited selectively with a high deposition rate (about 0.1 μm/min)

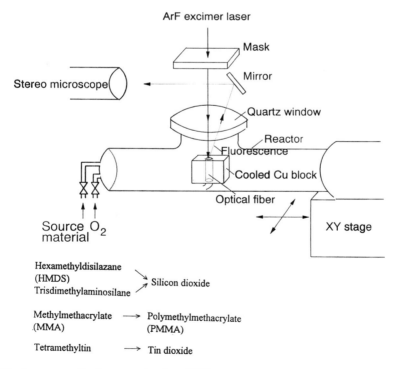

Figure 4.16. Apparatus for laser projection CVD.

and a lateral resolution of 10 μm. This creates microstructures with an insulating film, sacrificial layer film, and conductive film. In order to obtain good film quality, trisdimethylaminosilane is used as source material for silicon dioxide deposition. Thus, CVD can be applied where a uniform resist coating is not possible. A PMMA film 50 μm in diameter can be successfully deposited on the end of a 125-μm diameter optical fiber.

Selective nonplanar metallization was developed for a smart catheter that has a tactile sensor at the end (Maeda et al., 1994). Excimer laser beam–assisted Cr CVD from $Cr(CO)_6$ was carried out using the apparatus shown in Figure 4.17. Chromium film deposited on a polyurethane catheter was contaminated with carbon and oxygen and hence showed low electrical conductivity. To obtain a highly conductive film, deposited chromium film was used as a mask material for patterning the underlying nickel film, as shown in Figure 4.18. The fabricated built-in tactile sensor at the catheter end is shown in Figure 4.19. The sensor structure in Figure 4.19a was shaped by laser ablation, and the metal line in Figure 4.19b was patterned using the laser process shown in Figure 4.18. Another nonplanar metallization process developed for metal microelectrodes is the Parylene laser ablation method (Esashi et al., 1991). Parylene can be coated uniformly on a nonplanar surface and can be patterned by excimer laser ablation. A 10-μm-wide Cr line was patterned by the laser ablation and subsequent lift-off technique.

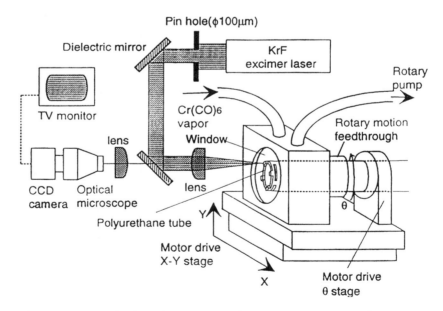

Figure 4.17. Apparatus for laser-induced selective nonplanar metallization.

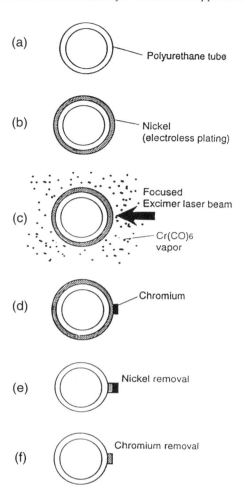

(a) — Polyurethane tube

(b) — Nickel (electroless plating)

(c) — Focused Excimer laser beam
Cr(CO)6 vapor

(d) Chromium

(e) Nickel removal

(f) Chromium removal

Figure 4.18. Fabrication process of metal line on the tube surface.

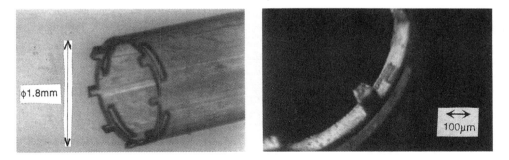

φ1.8mm

100μm

Figure 4.19. Photomicrographs of built-in tactile sensor at the end of catheter. (a) Catheter end shaped by laser ablation process. (b) Metal line patterned using laser process.

4.6 Conclusion

Micromachining technology is very useful in developing high-performance, small-sized microsystems in which multiple sensors, circuits, and actuators are integrated. Advanced microactuators capable of large displacement and large force play an important role in such microsystems, and their development is expected for active catheters.

Acknowledgment

This work was supported by the Japanese Ministry of Education, Science, and Culture under grant-in-aid No. 03102001.

References

Becker, E., Ehrfeld, W., Hagmann, P., Maner, A., and Munchmeyer, D. (1986) Fabrication of microstructures with high aspect ratios and great structural heights by synchrotron radiation lithography, galvanoforming, and plastic moulding (LIGA process). *Microelectronic Engineering* 4:35–56.

Cabuz, C., Shoji, S., Cabuz, E., Minami, K., and Esashi, M. (1993) Highly sensitive resonant infrared sensor. In *Digest of technical papers Transducers '93*, pp. 694–697, IEE of Japan.

Esashi, M., Matsumoto, Y, and Shoji, S. (1990) Absolute pressure sensors by air-tight electrical feedthrough structure. *Sensors & Actuators* A21-A23:1048–1052.

Esashi, M., Minami, K., and Shoji, S. (1991) Optical exposure systems for three-dimensional fabrication of microprobe. In *Proc. IEEE MEMS Workshop*, pp. 39–44, IEEE.

Esashi, M., and Matsumoto, Y. (1991) Common two lead wires sensing system. In *Digest of technical papers Transducers '91,* pp. 330-333, IEEE.

Henmi, H., Yoshimi, K., Shoji, S., and Esashi, M. (1993) Vacuum packaging for microsensors by glass-silicon anodic bonding. In *Digest of technical papers Transducers '93,* pp. 584–587, IEE of Japan.

Jono, K., Hashimoto, M., and Esashi, M. (1994) Electrostatic servo system for multi-axis accelerometers. In *Proc. IEEE MEMS Workshop*, pp. 251–256, IEEE.

Maeda, S., Minami, K., and Esashi, M. (1994) KrF excimer laser induced selective non-planar metallization. In *Proc. IEEE MEMS Workshop*, pp. 75–80, IEEE.

Matsumoto, Y., and Esashi, M. (1993a) An integrated capacitive absolute pressure sensor. *Electronics and Communications in Japan, Part 2* 76:93–106.

Matsumoto, Y., and Esashi, M. (1993b) Integrated silicon capacitive accelerometer with PLL servo technique. *Sensors & Actuator* 39:209–217.

Minami, K., Kawamura, S., and Esashi, M. (1993a) Fabrication of distributed electrostatic microactuator (DEMA). *IEEE J. of Microelectromechanical Systems* 2:121–127.

Minami, K., Wakabayashi, Y., Yoshida, M., Watanabe, K., and Esashi, M. (1993b) YAG laser-assisted etching of silicon for fabricating sensors and actuators. *J. of Micromechanics and Microengineering* 3:81–86.

Murakami, K., Wakabayashi, Y., Minami, K., and Esashi, M.(1993) Cryogenic dry etching for high aspect ratio microstructures. In *Proc. IEEE MEMS Workshop*, pp. 65–70, IEEE.

Nitta, S., Kahira, Y., Yambe, T., Sonobe, T., Hayashi, H., Tanaka, M., Sato, N., Miura, M., Mohri, H., and Esashi, M. (1990) Micro-pressure sensor for continuous monitoring of a ventricular assist device. *Int. J. of Artificial Organs* 13:823–829.

Shirai, T., Esashi, M., and Ura, N. (1993) A two-wire silicon capacitive accelerometer. *Electronics and Communications in Japan, Part 2* 76:73–83.

Takashima, K., Minami, K., Esashi, M., and Nishizawa, J. (1993) Laser projection CVD using low temperature condensation method. In *Abstracts 1st Int.Conf. on Photo-Exited Processes and Applications*, p. 70, The Japan Soc. of Applied Physics.

Yamaguchi, M., Kawamura, S., Minami, K.,and Eshashi, M. (1993) Control of distributed electrostatic microstructures. *J. of Micromechanics and Microengineering* 3:90–95.

5

The use of micromachined structures for the manipulation of biological objects

MASAO WASHIZU

Department of Electrical Engineering and Electronics, Seikei University
3-3-1 Kichijoji-kitamachi, Musashino-shi
Tokyo, Japan 180

5.1 Introduction

Manipulation of biological objects, such as biological cells, organelles, membrane structures, nucleic acid, and protein molecules, constitutes an important unit-operation in biotechnology. Because the dimensions of these objects range from hundreds of micrometers for cells down to nanometers for molecules, their manipulation, especially on a one-by-one basis, requires tools with comparable dimensions. For this purpose, micromachining techniques based on photolithographic processes, which are used for the production of semiconductor integrated circuits, provide the ideal means to manufacture microstructured tools of arbitrary two-dimensional shape.

Specific problems associated with the manipulation of biological objects are (1) they are often fragile and (2) the whole process must take place in aqueous solutions. These factors set a limit on the choice of the method used for actuation. Electrostatic force, which is suitable for handling fine particles, can realize gentle actuation without mechanical contact with the objects; however, special precautions must be taken against chemical reactions at the electrode–water interface. Use of dc voltage causes electrolytic dissociation, so that the electrodes must somehow be separated from the area where the objects are manipulated; thus it is difficult to miniaturize the device. Our solution to this problem is the use of high-frequency voltage, which is less likely to cause electrolytic reactions. High-frequency ac field effects, such as dielectrophoresis (DEP) and electrorotation, are used for translational and rotational actuation of the objects. Devices in which objects are electrostatically manipulated by the microfabricated structure are called fluid integrated circuits (FICs) (Masuda et al., 1989; Washizu, 1990, 1991; Washizu, Kurahashi, et al., 1993; Washizu and Kurosawa, 1990; Washizu, Kurosawa, et al., 1995; Washizu Nanba, and Masuda, 1990; Washizu, Shikida, et al., 1992; Washizu, Suzuki, et al., 1994) in analogy to electronic ICs (integrated circuits), in which a multistaged manipulation process can be miniaturized and integrated onto a substrate. Some applications of FICs for the manipulation of cells, DNA, protein, and membranes are presented here.

5.2 Ac field effects used in FIC

There have been extensive studies on the electrodynamic effects of ac fields (Arnold and Zimmerman, 1988; Chiabrea et al., 1985; Fuhr et al., 1991; Jones, 1979; Miller and Jones, 1987; Pethig, 1991; Pohl, 1978; Saito et al., 1966), some of which are schematically depicted in Figure 5.1

Dielectrophoresis (DEP, Figure 5.1a), developed and named by Pohl (1978), is a nonuniform field effect that yields translational motion of objects. In an electrostatic field, dielectric material polarizes as shown in the figure, and equal amounts of positive and negative polarization charges appear on the downstream and upstream sides of the particle. If the applied field is uniform, Coulombic forces exerted on the polarization charges of both polarities cancel out. However, if the field is nonuniform, there is an imbalance in exerted forces, resulting in a net force.

The "equivalent dipole moment" $\mathbf{p}_{eq}$ induced by the polarization on a spherical particle of radius a in an external field $\mathbf{E}$ is given by Jones (1979):

$$\mathbf{p}_{eq} = 4\pi a^3 \varepsilon_m K^* \mathbf{E} = \alpha \, \mathbf{E} \qquad (1)$$

where

$$K^*(\omega) = \frac{\varepsilon_p^* - \varepsilon_m^*}{\varepsilon_p^* + 2\varepsilon_m^*}, \qquad (2)$$

$$\varepsilon_p^* = \varepsilon_p - j\frac{\sigma_p}{\omega}, \qquad = \varepsilon_m - j\frac{\sigma_m}{\omega}, \qquad (3)$$

α is the polarization factor,
$\varepsilon_p, \varepsilon_m$ are dielectric constants of particle and medium,
σ_p, σ_m are the conductivity of particle and medium, and
ω is the frequency of applied field.

So a mathematical expression for the DEP force becomes

$$\mathbf{F}_d = \overline{(\mathbf{p}_{eq}\cdot\mathbf{V})\,\mathbf{E}} = 2\pi a^3 \varepsilon_m \, \mathrm{Re}\!\left[K^*(\omega)\right] \mathbf{V}(E_{rms}^2) \qquad (4)$$

where — denotes time average.

The direction of the DEP force is either ∇E^2, that is, toward where the field is stronger, or $-\nabla E^2$, that is, toward where the field is weaker, depending upon the polarity of $\mathrm{Re}[K^*(\omega)]$. These directions are termed positive and negative DEP, respectively. Positive DEP is observed more often in biological samples.

Electrostatic orientation (Figure 5.1b) occurs on objects having non-isotropic dielectric properties, such as particles with nonspherical shapes (Miller and Jones

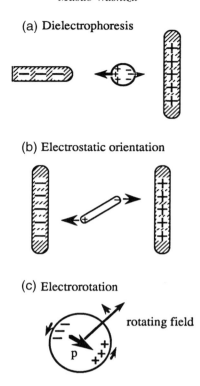

(a) Dielectrophoresis

(b) Electrostatic orientation

(c) Electrorotation

rotating field

p

Figure 5.1. Ac field effects.

1987; Saito et al., 1966). The angle the induced polarization makes with the applied field is the origin of the torque. The equilibrium orientation for a spheroid is known to be with any of three axes parallel to the field; however, the orientation often observed experimentally is with the longest axis parallel to the field, as shown in the figure.

Electrorotation (Figure 5.1c) (Arnold and Zimmerman, 1988) is the rotation of an object in a rotating electric field. Because K^* is a complex constant, Equations (1)–(3) show that there is a phase delay in the induced dipole moment. This yields a rotational torque $\mathbf{T}$ where

$$\mathbf{T} = \overline{\mathbf{p}_{eq} \times \mathbf{E}}. \tag{5}$$

All these effects result from the interaction of the field with the induced dipole moment. In an ac field, the induced moments alternate at the same frequency with the applied field, so the direction of force or torque remains unchanged. If the electrical constants are linear, the magnitude of $\mathbf{p}_{eq}$ is proportional to the applied field $\mathbf{E}$, so that the force and the torque become proportional to $\mathbf{E}^2$.

5.3 Advantages and limitation of FIC

The advantages of FIC can be summarized as follows:

(1) The structure can be made as small as the objects to be manipulated (for example, cells) so that manipulation of individual objects is possible.

(2) Arbitrary two-dimensional structures and electrodes can be fabricated by photographic reduction, resulting in greater flexibility of the design of the field shape.

(3) The device has no mechanical moving parts, which enables easy fabrication. High reliability is expected because the device is free from friction and/or wear.

(4) Micrometer-sized electrode gaps can be fabricated, so that high field-strength, E, and high field gradient, ∇E^2, are obtainable with a practical power supply. Both high E and high ∇E^2 are required for the actuation of micrometer- to nanometer-sized objects. For instance, only 100 V is needed to generate the field intensity of 2×10^6 V/m across a 50-μm gap, whereas 20 kV would be required if the gap were 1 cm.

(5) The high field region is confined to a small volume, which has a large surface-to-volume ratio, so that temperature rise due to Joule heating can be minimized. This enables the continuous application of a very high intensity field. For instance, a 1×10^6 V/m electric field in a 100 kΩ-cm medium yields the Joule loss of 1×10^3 W/cm³. This same field would result in a 240°C rise/sec under adiabatic conditions. On the contrary, the temperature rise will be no more than several degrees centigrade in a FIC with a 100-μm electrode gap.

There is, however, a limitation in the conductivity of the medium, σ_m. The reason is twofold:

(1) For positive DEP, which is usually observed in biological applications, an increase in σ_m reduces polarization, and the effective force becomes smaller; and

(2) A large σ_m often results in excessive Joule heating even in micro electrode gaps.

Therefore, FIC is not a general-purpose device, but if it is properly designed and used, it can be a very powerful tool for specific applications.

5.4 Applications of FIC

5.4.1 Manipulation of cells

Figure 5.2 shows some of the cell manipulation devices. Figure 5.2a is the FIC cell fusion device, which is intended to fuse two types of cells. One cell, say of type A, and another cell, of type B, are fed by micropumps (not shown in the figure) to respective inlets. The inlets lead to fluid channels separated by an insulator wall, except at a small opening made at the center of the electrodes. When a high-frequency voltage is applied between the electrodes, the field lines running through this opening form a constriction, and the field intensity is at its maximum at the opening. Each cell sent from the respective inlet is dielectrophoretically trapped in

a) Cell Fusion Device

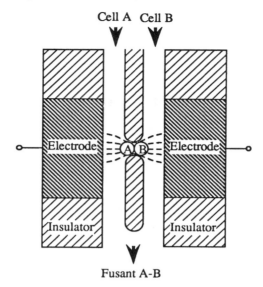

b) Cell Shift Register

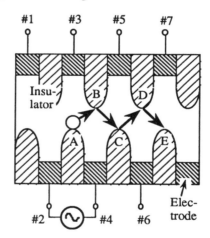

Figure 5.2. Fluid integrated circuit (FIC) cell manipulation devices.

this opening to form a cell pair A–B. Then fusion is initiated by applying a pulsed voltage. This procedure ensures that all fusants are formed from A–B cell pairs.

Figure 5.2b is the FIC device for cell storage, positioning, transport, and delivery. The device is composed of a periodic structure of insulator walls and electrodes. First, a high-frequency voltage is applied between electrodes #2 and #4 in the figure, which creates field maximum at the tip of A. A cell is dielectrophoretically attracted and trapped here. Then the voltage between electrodes #2 and #4 is removed and is applied between #3 and #5 instead. This moves the field maximum to the tip of B, and the cell is attracted toward B and is trapped. Repeating the same procedure, the cell is transported step by step.

5.4.2 Torque-speed measurements of a bacterial motor

Cell manipulation using electrorotation has been applied to the measurement of bacterial motor characteristics.

Some bacterial species, such as *E. coli* or *Salmonella*, are about 2 μm x 1 μm in size, and each bacterium possesses a number of flagella extruding from its body. These bacteria swim in water, rotating their flagella like screws by flagellar motors installed at the root of each flagellum. A flagellar motor is a molecular machine that converts chemical energy (that is, the proton potential difference between interior and exterior of a cell) to mechanical energy. An experimental difficulty in the investigation of the motor mechanism stems from its small size, only 30 nm in diameter. Electrorotation in a microfabricated electrode system has been successfully applied to the direct measurement of the torque-speed (T-ω) characteristics of the flagellar motor (Washizu, Kurahashi, et al., 1993).

When a flagellum is attached to a substrate while the cell body is free, the countertorque of the motor rotates the bacterial body. Such a cell is called a "tethered cell". In a rotating electric field, the tethered cell is subject to electrorotation (Figure 5.3a), and by changing the magnitude of the field, a controllable magnitude of the external torque can be applied to the cell. The direction of the external torque can be chosen to help the spontaneous rotation of the tethered cell, or to reverse it, depending on the rotating direction of the applied field. The frequency at which the cell rotates under the rotating field is optically measured for various external torques, from which the torque-speed curve of the motor is deduced.

Figure 5.3b plots the measured rotation speed against the externally applied electrical torque, which should be proportional to the square of the applied voltage. On the righthand side are plotted the cases when the external torque is in the direction to help the spontaneous rotation, and on the lefthand side are plotted the cases when the bacterial rotation is forced to reverse. The plot is made for one particular bacterium, both when the cell is alive (open squares) and when the cell is dead (solid squares). The discrepancy between the two lines gives the torque generated by the bacterial motor, which appears to be constant in the measured range of rotation speed.

5.4.3 DNA manipulation

DNA is a 2-nm-diameter-long chain molecule, and the genetic information is recorded through its sequence of four bases, A, T, G, and C, at the density of 3.0 kb/μm. In conventional sequencing, cloning, gene analysis, and so forth, DNA is handled in an

a) Principle

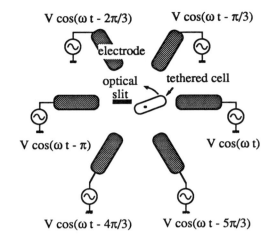

b) Results

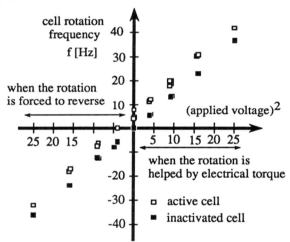

cell : SJW134(polyhook filament, Δche)

Figure 5.3. Measurement of bacterial motor characteristics using electrorotation.

aqueous solution. An inherent limitation of such procedures is that they lack spatial resolution; that is, information about the position is lost during manipulations: Neither a particular DNA strand among others, nor a particular location on a DNA strand, can be specified. For example, consider the case of sequencing large DNA. Because only 1 kb can be sequenced at a time using electrophoresis, the DNA is first enzymatically cut into fragments smaller than this size, subcloned, and the sequence of each fragment is determined. The full sequence of the original DNA is inferred from the sequence of each fragment and from overlapping fragments. The reconstruction is like solving a jigsaw puzzle and becomes increasingly difficult as the size of the DNA increases. If the information about the location of each fragment is preserved, this reconstruction step may be greatly simplified.

Spatial resolution is provided only by a physical manipulation of DNA molecules. DNA can be physically manipulated by electrostatic effects in microfabricated electrode systems.

DNA molecules in water adopt random orientations due to thermal agitation (schematically illustrated in Figure 5.4a). When a high-frequency (ca.1 MHz), high-intensity field in excess of 10 kV/cm is applied, electrostatic orientation occurs and, because all parts of DNA tend to orient parallel to the field applied, DNA is stretched straight (Figure 5.4b). If the electrode has a sharp edge, as is the case with vacuum-evaporated electrodes, a very high intensity field is created around the edge. DNA is

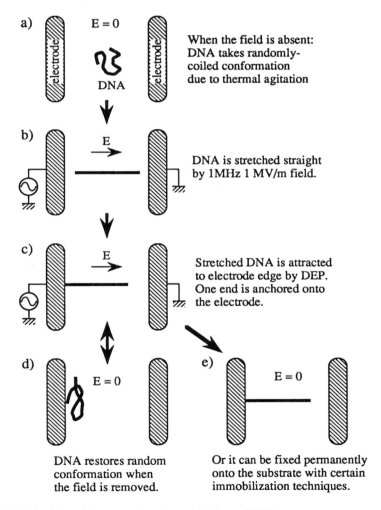

a) E = 0

DNA

When the field is absent: DNA takes randomly-coiled conformation due to thermal agitation

b) E →

DNA is stretched straight by 1MHz 1 MV/m field.

c) E →

Stretched DNA is attracted to electrode edge by DEP. One end is anchored onto the electrode.

d) E = 0

DNA restores random conformation when the field is removed.

e) E = 0

Or it can be fixed permanently onto the substrate with certain immobilization techniques.

Figure 5.4. Principle of the stretch-and-positioning of DNA.

attracted to this region by positive DEP, until one of its molecular ends touches the electrode (Figure 5.4c). The end can be permanently anchored to the electrode, especially when an active metal such as aluminum is used as the electrode material. The stretching is reversible, and when the field is removed, DNA restores itself to its original coiled form, with one end anchored to the electrode (Figure 5.4d). With the use of special techniques, the stretched DNA can be fixed onto the substrate permanently and can maintain a stretched conformation even after removal of the field (Figure 5.4e). We have named this technique "electrostatic stretch-and-positioning." Such a stretch-and-positioning is hardly possible with mechanical micromanipulators, because of the nanometer size of the DNA strand and its low mechanical strength.

An apparent application of this technique is in the measurement of DNA size. DNA is stretched to its full length, so by measuring the stretched length under a microscope, the DNA size is instantaneously determined using the structural constant 3.0 kb/μm within optical resolution (Figure 5.5a). Similar methods can be used for assays of nuclease, because the activity of an exonuclease is measured by the shortening of DNA length as the digestion proceeds (Figure 5.5b). In a similar manner, the cutting sites of a restriction enzyme can be determined by direct observation of the cut position (Figure 5.5c).

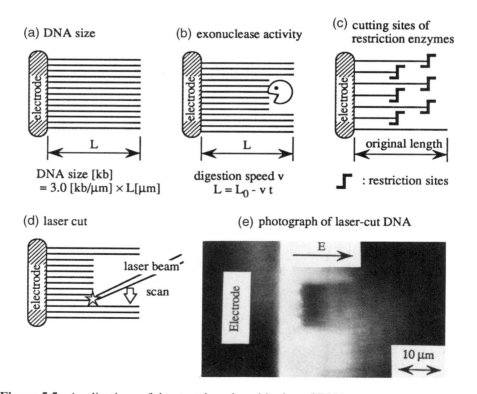

Figure 5.5. Applications of the stretch-and-positioning of DNA.

In scanning microscope (STM and/or AFM) observation of biological macro-molecules such as DNA (Chapter 10), a difficulty is often encountered in locating the object. For this purpose, the stretch-and-positioning method provides a defi-nite location (one end in contact with the electrode edge) and orientation (per-pendicular to the edge).

It has been shown that stretch-and-positioned DNA can be cut at an arbitrary position, again within optical resolution, using a sharp-focused ultraviolet laser beam. Figure 5.5d schematically shows this process: DNA is first aligned onto the electrode edge and then is scanned by a laser beam. Figure 5.5 d' is the photograph taken under a fluorescence microscope, where λ-DNA stained with the fluorescent dye 4',6-diamidino-2-phenyl-indole (DAPI) is scanned by an N_2- laser beam and cut. Our present effort is focused on the application of laser-cut to the successive sequencing of DNA, where stretch-and-positioned DNA is cut into fragments suc-cessively from one of its ends, and each fragment is amplified and sequenced. This method is expected to greatly simplify the reconstruction of the original sequence.

A special precaution must be taken in order that the DNA, which is fixed onto a substrate, can interact with DNA enzymes. If DNA is in contact with the sub-strate over its entire length, steric hindrance occurs, and the polymerase will not be able to bind (Figure 5.6a). The DNA strand must be held at some clearance from the substrate, preferably only at both ends of its molecular contour (Figure 5.6b). We named this "site-specific immobilization." Two methods have been developed to realize such site-specific immobilization:

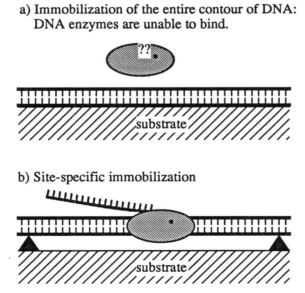

Figure 5.6. Principle of site-specific immobilization.

(1) Insert biotin molecules to DNA ends by a biochemical procedure (Fujioka et al., 1991), and using biotin–avidin binding, anchor the DNA end onto the avidin-coated substrate (Figure 5.7a).

(2) Use floating potential electrodes (Figure 5.7b): The DNA end in contact with an aluminum electrode has a tendency to be fixed permanently. By using electrodes whose gap is slightly smaller than the length of DNA, a DNA chain is stretched, and both of its ends are fixed bridging over two adjacent electrodes. This method allows one to immobilize a stretched DNA at an arbitrary margin, that is, from full tension to loose tension. The tension at which DNA is held may have a crucial effect on the transcription process.

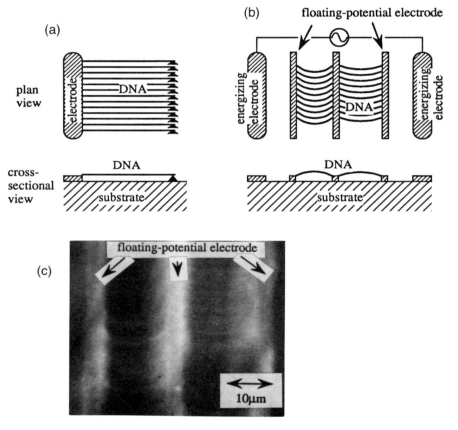

Figure 5.7. Site-specific immobilization methods of DNA. a) Site-specific immobilization using molecular bindings (e.g., biotin–avidin binding). Biotin inserted at the end of DNA molecules is fixed onto the substrate coated with avidin. b) Site-specific immobilization using floating-potential electrodes. c) Photograph of DNA immobilized onto floating-potential electrodes. DNA strands appear somewhat bent due to gentle downward flow of the medium, showing that DNA is held only at its ends. DNA strands fixed at shorter gap (left) are more bent than those at longer gap (right), because they are held loosely.

The site-specific immobilization can be used to investigate the interaction between DNA and DNA-binding enzymes. In fact, a direct observation of fluorescently labeled RNA polymerase sliding along DNA is reported using this method (Kabata et al., 1993).

5.4.4 Manipulation of protein molecules

The size of most protein molecules of biological interest is in the range of ten to several tens of nanometers. A problem associated with the electrostatic manipulation of such nanometer-sized particles is that, because the DEP force in Equation (4) is volumetric, a very high intensity field is required to overcome thermal randomization. It has been reported that the electric field created in micrometer-sized gaps is strong enough to realize molecular DEP of protein (Washizu et al., 1994).

Figure 5.8a shows a DEP chromatography device for molecular separations. On a substrate is deposited an array of interdigitating electrodes having a wavy shape, with a minimum gap of 7 μm. The wavy shape is intended to create field nonuniformity for effective DEP. The electrodes are coated by a thin (about 5 nm) layer of insulating material to prevent electrochemical reactions. On the electrodes are mounted a fluid passage of 25-μm depth, as shown in the figure. Fluorescently labeled protein molecules fed from the inlet encounter 2,000 electrode gaps and are gradually precipitated onto the electrodes by DEP. The protein concentration is monitored at four positions along the fluid passage: the inlet, 1/6 of the way from the inlet, 1/2 of the way from the inlet, and at the outlet. By measuring the fluorescence intensity at each position, the penetration, P, defined as the ratio of the fluorescence intensity between when the field is on and when the field is off, or the collection ratio, $R = 1 - P$, is calculated. Figure 5.9 shows one

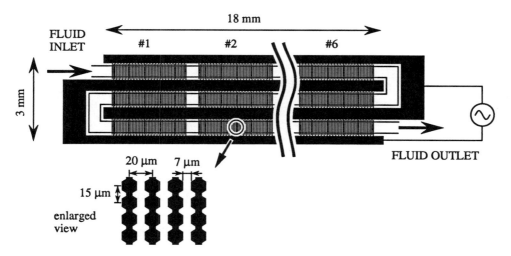

Figure 5.8. Dielectrophoretic (DEP) chromatography.

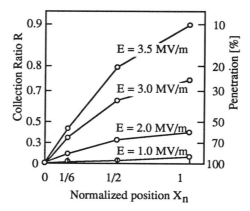

Figure 5.9. Collection ratio *R* as a function of position from the inlet (sample: avidin).

of the results with avidin (68 kDa protein), where the ordinate is plotted on a log-arithmic scale of *P*. The field intensity in the figure is *E* = (peak value in applied 1-MHz field) / (minimum gap). It is seen in the figure that about 90% of the molecules are collected onto the electrode by DEP. We have also achieved separation of different-sized DNA with the same principle (Washizu et al., 1994).

Possible applications of such a device may be in the separation and assay of biomolecules according to the difference in the polarizability of the molecules.

It is also shown that the electrostatic stress can induce conformational changes of proteins (Washizu, Shikida, et al., 1992) just as we have seen with DNA. This phenomenon has implications in that the molecular function of a protein may well be affected by an externally applied high-intensity field. In the future, such an effect may be used for the switching and control of both natural and artificial enzymes (molecular devices).

5.4.5 *Manipulation of membrane structure*

Another area of biological importance is the membrane. The membrane structure in biological systems serves both as the separation wall and as the reaction bed for enzymatic activities. Because the biological membrane consists of a thin layer of lipid bilayer that is fluidic in lateral dimensions, it is essentially deformable and fragile. Large membrane structures are likely to be unstable.

The black membrane technique is the method used to fabricate an artificial phospholipid membrane in water: A small opening of about 1 mm in diameter is drilled in a thin plate made of hydrophobic material such as Teflon. The plate is immersed in water, and phospholipid dissolved in *n*-decane is applied onto the opening (Figure 5.10a). The solution flows along the hydrophobic surface (Figure

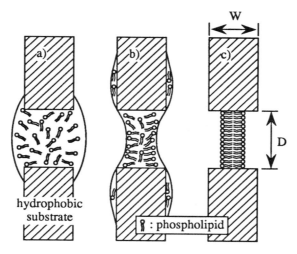

Figure 5.10. Black membrane method. (a) Apply solution of phospholipid in *n*-decane. (b) *n*-decane spreads along hydrophobic substrate. (c) Phospholipid bilayer is formed at the orifice.

5.10b), until a phospholipid bilayer is formed at the opening (Figure 5.10c). Because the thickness of the bilayer is smaller than the wavelength of visible light, it appears as a "black" membrane when it is observed with back-scattered light.

A problem that has prevented the wide use of the black membrane method is its low reproducibility and low stability. As is clear from Figure 5.10, the conditions to obtain good membranes are the following:

(1) The aspect ratio of the orifice, D/W in Figure 5.10c, must be high enough, otherwise the phospholipid solution cannot be thin enough to form a bilayer.
(2) Flatness of the surface of the orifice plate must be high, otherwise the formed bilayer easily breaks.

We have applied microfabrication techniques for the machining of the orifice. Isotropic etching of a glass plate from its back side makes a sharp-edged orifice, whose cross section is shown in Figure 5.11. This fabrication method ensures the flatness of the plate and an equivalently high aspect ratio of the orifice. The hydrophobicity is brought about by the surface treatment of the glass orifice plate with fluorosilane. It is experimentally shown that an orifice fabricated by this method enables reproducible and stable formation of black membranes.

One of the advantages of the photolithographic technique is its ability to fabricate regularly arranged periodic structures. This can be used to manufacture uniform-sized liposomes (Washizu and Hashimoto, 1990). The size of liposomes is defined by microstructures. Regularly arranged circular pits are patterned on a

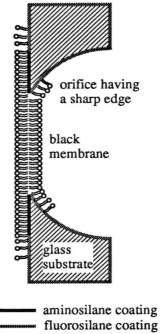

orifice having
a sharp edge

black
membrane

glass
substrate

━━━━ aminosilane coating
╍╍╍╍ fluorosilane coating

Figure 5.11. Formation of black membrane on a microfabricated orifice.

surface-oxidized Si wafer. This exposes lipophilic Si at the pits, surrounded by a hydrophilic SiO_2 surface. When a phospholipid dissolved in organic solvent such as chloroform is poured on the device and is allowed to evaporate slowly, the solvent tends to remain at the hydrophobic Si surface, and when evaporation is complete, the lipid is left preferentially at the pits. When the wafer is immersed in water, liposomes grow from each pit, and because all pits are identical, uniform growth of liposomes occurs, and thus uniform-sized liposomes are obtained.

It should be added that the electrostatic force is essentially what is called "surface force" and is highly suitable for the manipulation of these soft and fragile sheet-like objects. The electrostatic force can be used to deform cells or liposomes, or it can be used to enhance the growth rate of liposomes by pulling up the lipid layer (Washizu, 1990).

5.5 Conclusions

It has been shown that microfabrication techniques, when combined with electrostatic effects, possess a high adaptability for the manipulation of biological objects, such as cells, DNA, protein molecules, and membranes. The spatial resolution provided by such physical manipulations is a great advantage over conventional chemical methods. Automated cell processing, bioassay and separation of biological molecules such as DNA or protein, and novel methods of DNA sequencing are expected applications. In the future, the combination of microfabrication and electrostatic methods might even be applied to the construction of molecular devices using molecules that are not "self-assembling."

Acknowledgments

The author would like to thank Professor Senichi Masuda of Tokyo University for valuable discussions. The work related to the bacterial motor and the DNA immobilization are the result of the collaboration with Dr. Shin-ichi Aizawa of Teikyo University and Dr. Nobuo Shimamoto of the National Institute of Genetics, respectively. The discussions with, and the help of, Mr. Osamu Kurosawa of Advance Company and Dr. Seiichi Suzuki of Seikei University are also appreciated.

References

Arnold, W .M. and Zimmermann, U. (1988) Electro-rotation: developments of a technique for dielectric measurements on individual cells and particles. *J. Electrostatics* 21:151–191.

Chiabrea, A., Nicolini, C., and Schwan, H.P. (eds.) (1985) *Interactions Between Electromagnetic Fields and Cells.* NATO ASI Series Vol. 97. Plenum Press, New York.

Fuhr, G., Hagedorn, R., Müller, T., Wagner, B., and Benecke, W. (1991) Linear motion of dielectric particles and living cells in microfabricated structures induced by traveling electric field. *Proc. '91 IEEE Micro Electro Mechanical Systems*, 259–264.

Fujioka, M., Hirata, T., and Shimamoto, N. (1991) Requirement for the β-γ pyrophosphate bond of ATP in a stage between transcription initiation and elongation by *E. Coli* RNA Polymerase. *Biochemistry* 30:1801–1807.

Jones, T.B. (1979) Dielectrophoretic force calculation. *J. Electrostat.* 6:69–82.

Kabata, H., Kurosawa, O., Arai, I., Washizu, M., Margason, S. A., Glass, R. E., and Shimamoto, N. (1993) Visualization of single molecules of RNA polymerase sliding along DNA. *Science* 262:1561–1563.

Masuda, S., Washizu, M., and Nanba, T. (1989) Novel method of cell fusion in field constriction area in fluid integrated cCircuit. *IEEE Trans. IA* 25(4): 732–737.

Miller, R. D. and Jones, T. B. (1987) Frequency-dependent orientation of ellipsoidal particles in AC electric fields. *Conf. Rec. IEEE 9th ann. EMBS meet*, 710–711.

Pethig, R. (1991) Application of AC. electrical fields to the manipulation and characterization of cells. In *Automation in Biotechnology,* I. Karube, ed., pp. 159–185. Amsterdam: Elsevier.

Pohl, H. A. (1978) *Dielectrophoresis.* Cambridge University Press, New York.

Saito, M., Schwartz, G., and Schwan, H. P. (1966) Response of nonspherical biological particles to alternating electric fields. *Biophys. J.* 6:313–327.

Washizu, M. (1990) Electrostatic manipulation of biological objects in microfabricated structures. In *Integrated Micro-Motion Systems*, F. Harashima, ed., pp.417–432. Amsterdam: Elsevier.

Washizu, M. (1991) Handling of biological molecules and membranes in microfabricated structures. In *Automation in Biotechnology,* I.Karube, ed., pp.113–125. Amsterdam: Elsevier.

Washizu, M. and Hashimoto, T. (1990) Fabrication of uniform-sized liposomes using microfabricated structures. *Rep. of Toyota Phys. Chem. Res. Inst.* 43:17–25 (in Japanese).

Washizu, M., Kurahashi, Y., Iochi, H., Kurosawa, O., Aizawa, S., Kudo, S., Magariyama, Y., and Hotani, H. (1995) Dielectrophoretic measurement of bacterial motor characteristics. *IEEE Transaction IA* 29(2):286–294.

Washizu, M. and Kurosawa, O. (1990) Electrostatic manipulation of DNA in microfabricated structures. *IEEE Transaction IA* 26(6):1165–1172.

Washizu, M., Kurosawa, O., Arai, I., Suzuki, S., and Shimamoto, N. (1995) Applications of electrostatic stretch-and-positioning of DNA. *IEEE Transaction IA* 31(3):447–456.

Washizu, M., Nanba, T., and Masuda, S. (1990) Handling of biological cells using fluid integrated circuit. *IEEE Transaction IA* .25(4):352–358.

Washizu, M., Shikida, M., Aizawa, S., and Hotani, H. (1992) Orientation and transformation of flagella in electrostatic field. *IEEE Transaction IA* 28(5):1194–1202.

Washizu, M., Suzuki, S., Kurosawa, O., Nishizaka, T., and Shinohara, T. (1994) Molecular dielectrophoresis of bio-polymers. *IEEE Transaction IA* 30(4):835-843.

6

Light-addressable potentiometric sensor: applications to drug discovery

HARDEN M. McCONNELL

Department of Chemistry
Stanford University
Stanford, CA 94305-5080

6.1 Introduction

Ongoing discoveries in molecular biology, cell biology, and genetics create the basis for a true revolution in biology and medicine. In part, this revolution has arisen from the development of new biochemical and biophysical techniques. The ongoing rapid sequencing of the human genome is one outcome of the advances in some of these techniques. It is of interest to consider the rate-limiting obstacles that are involved in applying these advances to medicine, specifically the development of new therapeutic drugs. It is safe to say that the rate of progress in fundamental biology and medicine is limited by capital and by the pool of dedicated and talented research workers. Unfortunately, there is little hope at present that either of these rate-limiting factors will be ameliorated. Consequently there is a strong incentive to use the available but limited capital more efficiently, as well as to optimize the research tools available to productive research workers. The purpose of this chapter is to describe a new methodology, LAPS (light-addressable potentiometric sensor) microphysiometry, that has already shown significant utility for modern research in cell biology and pharmacology and that holds promise for applications to the discovery of new therapeutic drugs and hormones. Here LAPS (Hafeman et al., 1988; McConnell et al., 1992; Owicki et al., 1994; Parce et al., 1989) refers to a "light-addressable potentiometric silicon sensor." A commercial product utilizing this sensor is the Cytosensor™, manufactured by Molecular Devices Corporation (MDC) and introduced in 1993.

6.2 New problems in therapeutic drug discovery

Rapid advances in DNA biochemistry and macromolecular structure are presenting the pharmaceutical industry with new opportunities and challenges. For example,

Microphysiometer Cell Capsule and Plunger

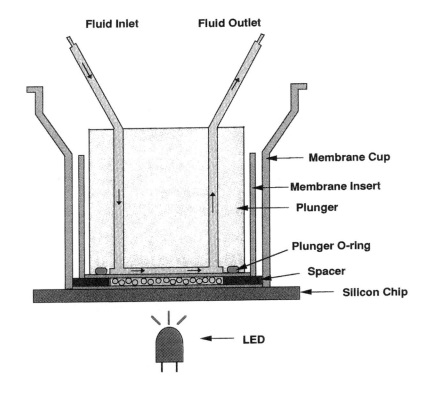

Fluid Inlet Fluid Outlet

Membrane Cup

Membrane Insert

Plunger

Plunger O-ring

Spacer

Silicon Chip

LED

(a)

Schematic Diagram of Microphysiometer System

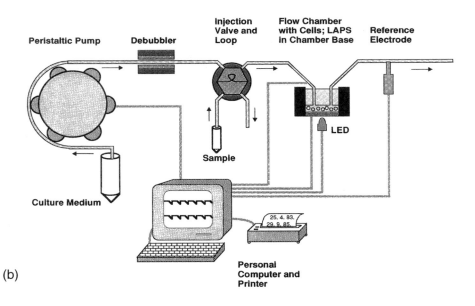

Peristaltic Pump Debubbler Injection Valve and Loop Flow Chamber with Cells; LAPS in Chamber Base Reference Electrode

LED

Sample

Culture Medium

25. 4. 83.
29. 9. 85.

Personal Computer and Printer

(b)

the ongoing sequencing of the human genome is yielding many new genes, some of whose functions are unknown. (The current rate is over one thousand genes discovered each day.) It is likely that dozens, if not hundreds, of these genes are potential targets for new therapeutic drugs. A particularly interesting class of genes is those that code for "orphan receptors." Such genes correspond to DNA sequences that code for putative plasma membrane receptors. The putative receptor is termed an "orphan" if its natural ligand is unknown and if its function is likewise unknown. The discovery of both natural as well as synthetic ligands for orphan receptors thus represents both a challenge and an opportunity for fundamental cell biology as well as the pharmaceutical industry. (A significant portion of current pharmaceuticals serve as ligands for plasma membrane receptors on biological cells.)

The problem of discovering orphan receptors' ligands thus provides a major challenge and seems to be a promising application of LAPS microphysiometry. The difficulty of using classical techniques to discover the ligands of orphan receptors is easily understood. The classical competitive ligand binding technique used by the pharmaceutical industry has little value when *no* ligand is known. That is, one cannot discover one ligand by its competition with a second radioactive ligand, if no ligand is known. One can screen for biological activity in cell supernatants, tissue extracts, natural product mixtures, or synthetic chemical libraries by looking for receptor-specific biochemical triggering of a population of target cells. The receptor-specific triggering of a cell can result in a change of transmembrane potential, release of intracellular calcium, activation of adenyl cyclase, hydrolysis of inositol phosphate, release of cytokines, cell proliferation, or cell death. This wide variety of responses to receptor-ligand–mediated cell triggering could potentially create a rate-limiting obstacle to screening broadly for ligands of orphan receptors. This is because the assays for each of the aforementioned cell responses are specialized and time consuming. A major advantage of the LAPS microphysiometer technique for screening for ligands of orphan receptors is that this technique can use a single type of measurement system to detect ligand-specific receptor activation of cells. This result is discussed in Section 6.3.

Figure 6.1. Schematic representation of the LAPS microphysiometer instrument called Cytosensor™. (a) Biological cells are maintained in a narrow space (50 microns) between the silicon chip and a permeable membrane that is in contact with aqueous medium that undergoes a flow-on, flow-off cycle. The pH of the medium surrounding the cells is sensed by the photoresponse of the silicon chip. During the flow-on period (~ 30 sec) the pH of the solution next to the chip is the same as that of the medium that flows past the permeable membrane. During the flow-off period (~ 20 sec), the medium is acidified by the biochemical events associated with cell metabolism. The acidification rate is then determined from the photoresponse. (b) Schematic diagram of Cytosensor™ showing fluid paths. The commercial instrument can monitor up to eight cell samples simultaneously.

6.3 Enhanced acidification rates as a consequence of ligand-mediated receptor activation

LAPS microphysiometry as embodied in the Cytosensor™ has been described extensively elsewhere (McConnell et al., 1992) and is summarized here in Figure 6.1. Biological cells are placed on a permeable membrane at the bottom of a small plastic cup, in close proximity to a pH-sensitive reverse-biased silicon chip. Cellular metabolism gives rise to acidification of the surrounding medium, and the acidification rate is accurately measured using the ac photoresponse of the chip. Extensive research at MDC has led to the following significant discovery. It has been found in many experiments that specific ligand-mediated receptor triggering leads to an enhanced acidification of the medium surrounding the cells. Furthermore, this enhanced acidification is very general, largely independent of the cell type and the specific biochemical pathways that are triggered within the cell (second messenger pathways). This therefore provides a generic technique to screen for ligands of orphan receptors. For such screens it will often be desirable to transfect an orphan receptor gene into cells in which this gene is not expressed, so that there are two populations of cells: one, receptor positive; and one, receptor negative. The receptor-negative cells thus serve as a control on putative ligand-mediated acidification signals generated by the receptor-positive cells.

6.4 Silicon microfabrication

The Cytosensor™ is being used in a number of laboratories to screen for receptor ligands. However, there is considerable interest in massive screening that is beyond the capacity of the Cytosensor™. To this end, MDC is experimenting with the fabrication of silicon chips that have greatly enhanced throughput capacity. Figure 6.2 shows one version of a higher-throughput chip (Bousse et al., 1993). Whereas the Cytosensor™ uses a single illumination site per chip and uses one fluid flow path over the chip, the chip in Figure 6.2 has eight channels and eight sites for illumination. The flow channels are etched in the silicon. Other chips currently under study have even larger capacities.

6.5 Conclusion

LAPS microphysiometry offers a novel methodology to screen for new receptor ligands, including the ligands of orphan receptors that are being discovered by sequencing the human genome. In order to keep pace with the discoveries of new genes and with the development of large chemical libraries of potential receptor ligands, MDC is exploring silicon chip microfabrication appropriate to LAPS microphysiometry.

Micromachined Sensor Array

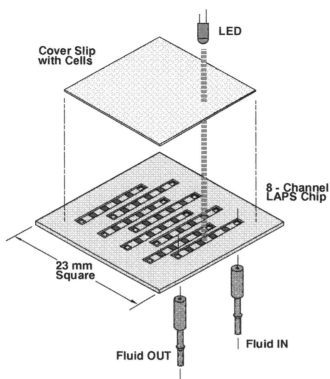

Figure 6.2. LAPS chip with eight parallel fluid channels. Adherent cells are maintained on the glass cover slip that is pressed down upon the silicon chip. Experiments have shown good signal/noise ratio in the system (Bousse et al., 1993).

Acknowledgments

The results described here have been obtained at Molecular Devices Corporation. The microfabrication of silicon has been supported in part by ARPA Contract #MDA 972-92-C-0005.

References

Bousse, L. J., McReynolds, R. J., Dawes, T., Lam, P., Bemiss, W. R., and Parce, J. W. (1994) Micromachined multichannel systems for the measurement of cellular metabolism. *Sensors and Actuators B* 20:145–150.

Hafeman, D. G., Parce, J. W., and McConnell, H. M. (1988) Light-Addressable Potentiometric Sensor for Biochemical Systems. *Science* 250:1182–1185.

McConnell, H. M., Owicki, J. C., Parce, J.W., Miller, D. L., Baxter, G. T., Wada, H. G., and Pitchford, S. (1992) The Cytosensor Microphysiometer: Biological Applications of Silicon Technology. *Science* 257:1906–1912.

Owicki, J. C., Bousse, L. J., Hafeman, D. G., Kirk, G. L., Olson, J. D., Wada, H. G., and
    Parce, J. W. (1994) The Light-Addressable Potentiometric Sensor: Principles and
    Biological Applications. *Ann. Rev. Biophys. Biomol. Struct.* 23:87–113.
Parce, J. W., Owicki, J. C., Kersco, K. M., Sigal, G. B., Wada, H.G., Muir, V. C.,
    Blousse, L. J., Ross, K. L., Sikic, B. I., and McConnell, H. M. (1989) Detection of
    Cell Affecting Agents with a Silicon Biosensor. *Science* 246:253–247.

7

A surface plasmon resonance biosensor for characterization of biospecific interactions

MAGNUS MALMQVIST

Pharmacia Biosensor AB, S-751 82 Uppsala

7.1 Introduction

Biotechnology is in a phase where technology plays a major role and where new ways to describe molecules by methods based on new technology have an impact on findings. In this chapter the discussion will be centered on the tradition of developing methods for biological research, the expected dramatic changes in biotechnology, and a description of a new technology – biomolecular interaction analysis (BIA) – based on biosensor development.

Interactions are the basis for life, regardless of the size of the interacting partners. This holds true for molecules, and we all are dependent on the molecular interactions in the network of balanced operations (enzymes) and control and management (signal transduction). As in the real world, structure is the prerequisite for function.

Today's science describes the central biological questions on the molecular level, and a protein is always one of the components in a biospecific pair. So, for example, are the differentiation and proliferation problems analyzed as to which molecules interact with each other in order to transfer a message from "headquarters" to the operating cell. How is this message transferred over the cell membrane of water-insoluble lipid bilayer?

With BIA, the *function* of biomolecules is described in *physical–chemical terms* with the interaction described by kinetic and affinity constants. BIA technology has made it possible to measure both the strength and the kinetics of the interaction. It has now been found that nature uses different kinetic properties for different purposes, and BIA is crucial for providing a new way of looking at the function of protein interactions.

The structure of molecules is the description of the biomolecules in *organic–chemical terms*. With this language we can get three-dimensional pictures of how all the atoms are located in relation to each other. It is basic knowledge to understand what part of one molecule interacts with another molecule. In situations of chemical catalysis, the organic–chemical language can describe the chemical steps and atoms that are involved in forming the product from the substrate. Binding of two molecules to each other can be described in terms of hydrogen bonds, charge interactions, hydrophobic areas, and so forth. This information is still hard to obtain, and NMR and x-ray crystallography are required to define a detailed three-dimensional structure of a protein.

As always, reaching this detailed information requires effort, time, and appropriate techniques. There is a difference between the characterization of a biomolecule (the description of the molecule in a Linnaean way, which is done only once) and the repeated analysis of characteristic properties describing the molecule or the future use of it. There are many technologies used for characterization of biomolecules; for example, mass (mass spectrometry, MS), sequences of amino acids or nucleotides, and fluorescence to measure relative position of fluorescent amino acids, and circular dicroism (CD) for analysis of α-helix content in a protein. There are also many combined methods to perform this repeated analysis of biomolecules in the search for the detailed characterization of a pure substance. In instances where one wants to use the function of a biomolecule, the analysis of a substance is often repeated in order to answer the traditional analytical questions: what, where, when, and how much? In these situations, these methods or slightly modified versions are used to describe the molecule.

The miniaturization and automation of all these technologies are important steps toward making laboratory work more productive and more industrialized. Most of the important general technologies for use in biotechnology are probably already identified, and the main technology work can be used to reduce the last handicraft – lab work – into a more effective intellectual task. However, there are many challenges in the development of technology for the analysis of individual molecules and for the characterization of individual protein function. What is the variability of function of individual molecules, and what is partial denaturation of proteins when described for individual molecules? Such questions can be important for the future development of biotechnology.

BIA is a technological answer to functional questions related to central biological problems. Today's research has come to a point where the important areas – differentiation, proliferation, defense, metabolism, reproduction, and central control – are analyzed and described in molecular terms. The biological problems have been the same over long periods of time, but new technology has made it possible to describe biological functions on a highly sophisticated level.

What will happen when the scientific community has characterized the major biological pathways for signaling and control after another 10 years of research? Characterization of enzymatic metabolic pathways was done in the period 1950–1970, but few have used these important findings for further work. However, spectroscopy and absorbance measurements are used in all types of routine analysis, such as ELISA (enzyme-linked immunosorbent assay), cell work, and so on. Will we be able to use the technical and analytical knowledge for development of routine analysis based on the structure–function relationships of biospecific pairs?

In this perspective the molecular diagnostic efforts based on DNA-sequence information are important. In principle, a disease is coupled to the primary structure of DNA that codes for a mutated protein. This contradicts all traditional diagnostics that are coupled to the concentration of one substance.

The requirements for higher throughput in the analysis of biomolecules must be met by miniaturization of the present technologies and by new combinations. Multianalysis is a way to increase throughput and precision in measurements; it needs to be addressed in the near future because the technical questions have been discussed for some time. This means that enough people are focusing on the problem and will find the practical solutions in the coming years.

Detectors for multianalysis are one key component in the development of technology. For biological molecules that easily can be labeled, such as DNA, fluorescence is a very versatile and sensitive probe around which to build technology. Other molecules cannot be analyzed with labels and thus they must be labeled at a later stage or analyzed with label-free technologies. Proteins often belong to the latter category, and more rapid methods for separation and specific identification of proteins of interest need to be developed.

The system approach to miniaturized format should be followed, in analogy to the macrosystems for separation and analysis. Thus pumps, valves, separation in miniaturized format, detectors for uv, fluorescence, and specific activity have to be combined for separation and analysis with rapid throughput.

7.2 The BIA concept

With real-time BIA, label-free biospecific interactions are monitored as they occur. This technology can be used to promote a better understanding of the relationship between function and structure. One way to study this is the systematic variation of biomolecular recognition, for example, by producing structural variants of proteins and analyzing the effects on the parameters describing function.

The information obtained by real-time BIA can be summarized as six components:

(1) specificity/identification
(2) concentration
(3) affinity
(4) kinetics
(5) cooperativity
(6) relative binding pattern

There is an obvious link between life-science research and routine applications in medical, veterinary, environmental, food, and defense areas. When a diagnostically significant or therapeutically effective relationship between concentration of biomolecules and a disease has been established, the application method can become routine.

Established techniques such as ELISA and RIA (radio immunoassay) can solve most analytical demands for concentration determinations of proteins. New technology must therefore be able to meet diagnostic demands better than established techniques and compete with an organization specifically generated for established technologies.

Biosensor techniques also have a role to play, because they can fulfill demands for multianalysis, freedom from labels, robustness, and simplicity. Still, most diagnostic criteria rely on the concentration or activity of one or several analytes, which are correlated to specific diseases established in clinical trials. The use of more sophisticated analysis (such as structural analysis) or functional description (such as kinetics) is still in its infancy. However, with the possibility of measuring kinetics and affinity, label-free analysis can open the door for clinical research based on these parameters.

7.3 Surface plasmon resonance for the label-free detection of biomolecular interactions

The basic purpose of any surface-sensitive biospecific detector is to measure the adsorption of the analyte to an immobilized ligand both in the same place and at the same time that absorption occurs (Jönsson et al., 1985b). The adsorption can be followed in real time without labels; the first use of surface plasmon resonance detection (SPR) for label-free analysis of antigen–antibody interaction was by Liedberg et al. (1983).

An instrument system for BIA based on a continuous flow of analyte samples or buffers over a surface with immobilized ligand and detection of the adsorbed analyte by surface plasmon resonance has been described elsewhere (Jönsson et al., 1991; Jönsson and Malmqvist, 1992). The practical use of SPR was not possible until the system approach was integrated with a detector, a sensor chip for immobilization of ligands with low nonspecific binding, and integrated liquid handling fluidics for transport of samples and reagents to the place for interaction. One such device has been made commercially available under the name BIAcore (Pharmacia Biosensor, Uppsala, Sweden). Most work on kinetic and affinity measurements based on surface-sensitive techniques has been done with this instrument (Figure 7.1).

The interface between the liquid phase and the solid surface must fulfill several criteria, such as low nonspecific adsorption, efficient immobilization procedures, and stability. A surface based on a gold film covered by a layer of carboxylated dextran has been described by Löfås and Johnsson (1990), (Figure 7.2). Large and small molecules can be covalently coupled to carboxyl groups on

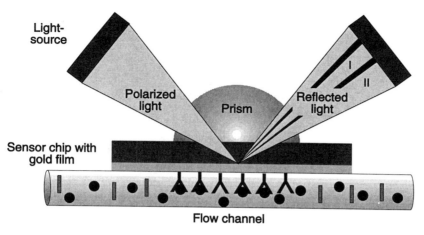

Figure 7.1. Configuration of the surface plasmon resonance detector, the sensor chip with the gold film and carboxylated dextran micromatrix, and the flow cell. The adsorption of biomolecules to the immobilized ligand is measured at the same place and at the same time as it is absorbed. The polarized light from the light-emitting diode of 760 nm is focused through a prism onto the interface between the glass and the liquid in the flow cell. Due to the surface plasmon resonance, extinction of light in the reflected fan-shaped beam is present for one particular angle. This is marked with I in the figure and represents the starting point for the resonance signal, which is the position of the extinct light on the diode array detector. By adsorption of substance from the liquid flow onto the immobilized ligand, the refractive index on the surface increases, and due to the SPR sensitivity to refractive index on and in the vicinity of the surface, the extinction of light has changed position from I to II with time during adsorption.

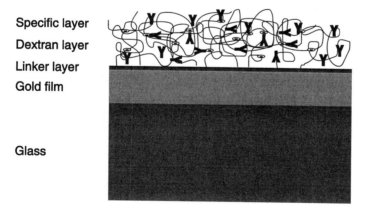

Figure 7.2. Sensor chip with glass substrate and gold film. On top of the gold film, a linker layer is adsorbed through the affinity of thiols to gold, and dextran is coupled to it. As the dextran matrix is bound without crosslinking, the dextran chains are very flexible to about 100 nm out from the surface. The dextran matrix has carboxy groups for activation and coupling of proteins and other biomolecules. (Published with permission.)

the sensor surface through amine, thiol, or aldehyde functional groups (Johnsson et al., 1991; O'Shannessy et al., 1992) or via streptavidin-biotin biospecific binding (Figure 7.3).

The matrix with 2–3% dextran is hydrophilic and coupled to the thin gold film through a linker layer of a long-chain hydrocarbon thiol adsorbed as a self-assembled layer on the gold film (Löfås and Johnsson, 1990). Because there is no crosslinking of the polymer, the matrix is a flexible, hydrophilic, and open structure with a high capacity for protein adsorption (Stenberg et al., 1991; Liedberg et al., 1992). Both tissue culture media and bacterial broth give extremely low nonspecific background on immobilized ligands (Brigham-Burke et al., 1992).

To define the liquid flow characteristics in the thin-layer channels and to direct the liquid to the proper place, a new type of flow injection system was developed (Sjölander and Urbaniczky, 1991). The integrated microfluidics is composed of about 20 valves on an area of about 1 cm² (Figure 7.4).

Dependent on dissociation properties of the analyzed biospecific pair, regeneration conditions for elution of bound material and for saving the activity of the immobilized ligand need to be established. Most antibody systems need regeneration, but there are exceptions (Griffiths et al., 1993). Strong acids (such as hydrochloric and phosphoric acid) are useful reagents (Brigham-Burke et al., 1992). The conditions must be checked for every new immobilized ligand, analyte, and regeneration agent, because the binding forces are not known for the particular interaction.

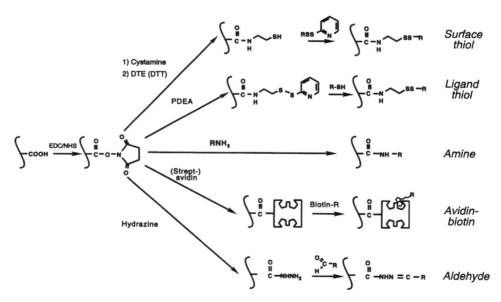

Figure 7.3. Activation of carboxyl groups to active esters. From the esters, it is possible to use different routes for coupling ligands to the sensor chip.

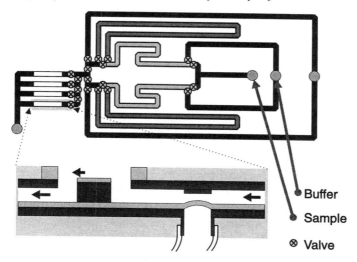

Figure 7.4. Schematic drawing of the integrated fluidics for transport of sample to the flow channels with the sensor chip. In the upper part, the microfluidics is seen from the top with the 5- and 50-µl loops, the valves, and the four flow channels. The lower part of the figure shows the pneumatic valve and the open flow channels without the sensor chip.

Surface sensitive biosensing can be described in the following terms:

Real-time characterization. Qualitative and quantitative analysis of biomolecular interactions based on the principle of real-time monitoring. This is the basis for kinetic analysis of interactions.

Label-free analysis. The interaction can be measured without the use of labels, which makes analysis possible in crude samples and without changing the structure of the target molecules by chemical modification. The potential for identification of biospecific partners and unknown function is a consequence of label-free analysis.

Multianalysis. The same analytical system could, without any changes in hardware, perform analysis of specificity, identity, concentration, kinetics, affinity constants, and relative binding pattern.

Another important aspect would be the self-diagnostic capability inherent in the technique, since all steps in the analysis are followed. In a conventional solid-phase assay only the last step in the reaction process is monitored, for example, in an enzyme-catalyzed reaction.

Miniaturization. A detection principle that monitors surface concentration is independent of the surface size. Thus, miniaturization with multispot analyses can be performed for several analytes from the same sample without complex instrumentation.

7.3.1 Surface plasmon resonance

At an interface between two optically transparent media of different refractive indices, light is reflected back into the medium of higher refractive index if the angle of incidence exceeds a certain critical angle. Although the incident light is totally internally reflected, an electromagnetic field component of the light (called the evanescent wave) penetrates a short distance, in the order of one wavelength, into the medium of lower refractive index. If the light is monochromatic and *p*-polarized, and if the interface between the media is coated with a thin layer of a so-called free-electron metal, a reflectance minimum appears in the reflected light at a specific angle of incidence. This phenomenon is called surface plasmon resonance (SPR) (Kretschman and Raether, 1968).

The angle position of the reflected light at which SPR is observed – called the resonance angle – is sensitive to changes in the refractive index of a thin layer adjacent to the metal surface. A volume, defined by the size of the illuminated area and the penetration depth, is probed. For example, when biomolecules are adsorbed, or interact with already immobilized molecules in this volume, an increase of the surface concentration occurs and the resonance angle shifts.

7.3.2 Optointerface

The traditional method for creating optical coupling between the prism and a separate glass slide is to use an optically matched liquid. Instead, an optically matched soft polymer was developed that could replace the liquid. The interface

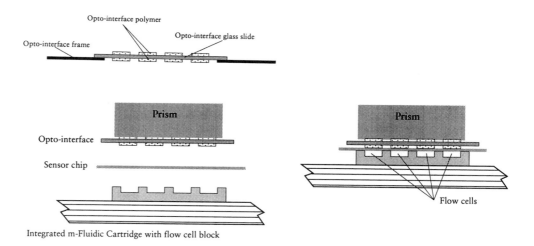

Figure 7.5. Optointerface for reversible optical coupling of the prism and the sensor chip. Use of an optically matched polymer allows the sensor chip to dock to the prism repeatedly and with the integrated fluidics from the opposite side, thus forming a unit where detection and interaction between the ligand and the analyte can take place (Jönsson and Malmqvist, 1992).

consists of a glass slide in a holder; on each side of the glass an optically matched silicon soft polymer is precision cast in a defined pattern (Figure 7.5). The mold for the micropattern originates from an etched silicon wafer, thereby ensuring the geometrical reproducibility of microstructures in large-scale production.

7.3.3 Sensor surface

The sensor surface is the interface between the detector and the biological liquids. Thus, the surface is both a part of the detector with the metal film and a part of the analyzed biospecific system with the immobilized ligand. The commonly used SPR metals in the visible to near-infrared wavelength range are aluminium, gold, and silver in the 20–60-nm thickness range. For chemical stability reasons, gold was chosen.

The matrix was constructed from a composite of linker layer and a covalent-ly bound flexible carboxymethyl modified dextran matrix (Löfås and Johnsson, 1990). The linker layer serves partly as a barrier to prevent proteins and other mol-ecules from coming into contact with the metal, and partly as a functionalized structure for further modifications. The linker is a long-chain hydrocarbon mole-cule with a thiol group at one end for adsorption to the gold and a hydroxyl group in the other end for activation and covalent coupling of dextran.

Dextran is a linear polymer of glucose residues that exhibits very low nonspecific adsorption of biomolecules when used as matrix for chromatographic media. The use of an extended coupling matrix enables exploitation of the volume probed by the evanescent wave. The dextran on the sensor chip is carboxymethy-lated to a level of approximately one carboxyl group per glucose residue. This modification provides (1) a chemical handle for covalent immobilization of the lig-and, and (2) a net negative charge on the dextran at physiological pH values, so that positively charged molecules adsorb electrostatically to the matrix under condi-tions of low ionic strength.

This electrostatic adsorption of molecules on the surface before covalent cou-pling can be compared with the hydrophobic adsorption of substances on plastic surfaces used in ELISA, where fragile molecules sometimes denature due to adsorption and expose new structures.

The dextran matrix swells in aqueous media, providing an extensively solvated matrix. Attachment of biomolecules to the flexible dextran preserves a hydrophilic environment and gives a high degree of accessibility to the molecules.

A matrix simply allows more molecules to interact per unit area than on an unmodified surface. This increases the dynamic range of measurements partly by increasing surface binding capacity from typical monolayer coverage of 1–10 ng/mm^2 to more than 50 ng/mm^2 (Stenberg et al., 1991) and partly by extending the surface layer to cover a greater range of the evanescent wave (Liedberg et al., 1992). Use of radioactively labeled proteins enables direct measurement of absolute surface con-

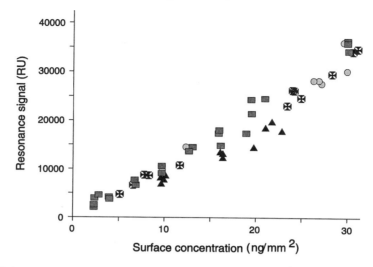

Figure 7.6. Correlation of the SPR resonance signal with the surface concentration of MAb, transferrin, and chymotrypsinogen. (Data from Stenberg et al., 1991.)

centration (Jönsson et al., 1985a). Close correlation between the SPR response and the amount of matrix-bound proteins was demonstrated for monoclonal antibodies (MAb), transferrin, and chymotrypsinogen, and the response was linear over a range of surface concentrations (2–30 ng/mm^2) (Figure 7.6). The measurements gave a specific response of close to 1,000 RU/ng protein/mm^2 for all three proteins (Stenberg et al., 1991). Furthermore, the refractive index increment is similar for a wide range of proteins, independent of amino acid composition (Armstrong et al., 1947). Glycoproteins and lipoproteins will give a slightly lower specific SPR response.

It was therefore possible to increase considerably the binding capacity and sensitivity by using a matrix, provided that the molecules could diffuse and bind freely in the matrix. Furthermore, well-known coupling chemistry can be used to immobilize the biospecific ligand (Figure 7.3).

7.3.4 Integrated microfluidics

Solid-phase technology is time limited by the mass transport of molecules to the surface. In a thin-layer cell, mass transport of analyte to a surface acting as a sink is described by the following equation (Matsuda, 1967):

$$J = kcD^{2/3} u^{1/3}, \qquad (1)$$

where J = mass transfer of analyte to the surface,
k = a constant determined by the dimensions of the flow cell,
c = bulk concentration of analyte,

D = diffusion coefficient of the analyte,
u = linear flow of solution over the surface.

By adjusting the flow rate to complement the flow cell dimensions, the mass transport conditions could be optimized to enable interaction analysis to be performed in minutes rather than hours. Furthermore, the equation is only valid in the laminar flow regime. Turbulent flow conditions would interfere with the interpretation of kinetic measurements and cause uncontrolled mixing of the sample and the continuous flow buffer. One problem often encountered in flow injection analysis was sample plug dispersion due to interdiffusion of the sample and the continuous flow buffer during the transport of the sample from the inlet to the reaction site. This problem was minimized with the design of an integrated fluidic system that placed the sample loops and valves near the reaction site on the sensor chip (Figure 7.4) (Sjölander and Urbaniczky, 1991).

7.3.5 Immobilization of biospecific ligand

Several coupling chemistries can be used in conjunction with the carboxymethylated dextran matrix. The most general is the amine coupling whereby a fraction of the carboxyl groups on the dextran are converted in an activation procedure to reactive N-hydroxy-succinimide esters. These esters react readily with uncharged primary amine groups in proteins and other ligands (Figure 7.3). After coupling of biomolecules to the matrix, remaining esters are inactivated with ethanolamine. The chemistry also makes it suitable for disulphide immobilization of thiol-containing biomolecules (Carlsson et al., 1978) or for aldehyde coupling of carbohydrate-conjugated molecules (O'Shannessy and Quarles, 1987). Through the coupling of streptavidin, the biospecific interaction with biotin-modified substances can be used for the binding of ligands to the sensor surface.

Activation–coupling–deactivation protocols were developed for optimal performance in the automated BIAcore. Typical protein concentrations required for immobilization are 10–100 µg/ml, with a sample volume consumption of 70 µl. The total time for the immobilization sequence is less than 30 min.

Because proteins in general have several amino groups, multipoint attachment of protein molecules to the matrix is possible. This can cause crosslinking of the dextran matrix, with the risk of decreased analyte binding capacity of the immobilized ligand. For this reason, it was important to optimize the amount of ligand attached to the sensor chip for each particular application. The amount of ligand was conveniently controlled by varying the activation time and the ligand concentration. By using a thiol–disulphide exchange reaction, controlled coupling can be performed by using engineered cysteins in protein molecules for defined attachment points (Cunningham and Wells, 1993).

The amine-coupling reaction reduces the possible loss of biological activity in the immobilized ligand, because most biomolecules contain a large amount of potential coupling sites. With real-time BIA, the sensorgram recorded during immobilization gives a direct check on the amount of ligand attached to the matrix. With substances of known molecular weight, the apparent stoichiometry of the surface complex may be calculated from the saturating binding capacity of the surface, because the SPR response is proportional to mass on the sensor surface (Löfås et al., 1993).

7.4 Application areas

By measuring refractive index changes on the sensor chip surface, a general parameter is measured, which gives the technology a general application. The general methods developed for the BIA technology are (1) specificity, (2) concentration, (3) kinetics and affinity, (4) in situ modification, and (5) epitope mapping. All of these methods relate to the information generated by adsorption of the analyte to the immobilized ligand. The measured change in resonance signal with time is presented as a sensorgram (Figure 7.7). The continuous flow of buffer over the sensor surface defines the baseline in the sensorgram from which the signal increases by adsorption of substance. From the asso-

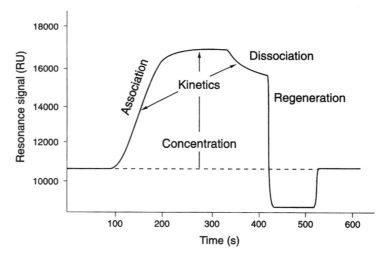

Figure 7.7. Sensorgram showing the resonance signal change with time for the adsorption of an analyte to the immobilized ligand when the sample is introduced in the cell. The association of analyte is diffusion limited at first, with constant slope until the amount of ligand starts to limit the interaction before the equilibrium level is reached. The constant slope can be used for concentration determinations (Karlsson et al., 1993); the bent part, for association rate constant determinations (Karlsson et al., 1991); and the equilibrium level, for affinity and concentration measurements. After the sample pulse has passed, the complex starts to dissociate and the dissociation rate constant can be determined. With suitable regeneration conditions, the immobilized ligand can be used for several experiments.

ciation part of the sensorgram, representing the part when the sample plug passes over the surface, the association rate constant can be calculated. After the sample pulse has passed, the biospecific complex dissociate and the rate constant can be determined.

These general methods can then be used for molecular pairs such as (1) anti-body–antigen, (2) receptor–ligand, (3) enzyme–inhibitor, (4) protein–nucleic acid, and (5) nucleic acid–nucleic acid. The methods were developed mainly with monoclonal antibodies and their interaction with different antigens.

7.4.1 Affinity and kinetic measurements

Kinetic analysis has been the dominant use for BIAcore because it fills the gap in methods for protein–protein interaction analysis. The affinity constant (K_a), the association rate constant (k_a, k_{ass}, or on-rate constant) and the dissociation rate constant (k_d, k_{diss}, or off-rate constant) are all defined by the law of mass action:

$$K_a = [\text{Complex}] \ / \ [A] \times [B], \text{ where } K_a = k_a/k_d. \tag{2}$$

In a biological system, where protein binds the partner, the result is a series of molecular and cellular responses. In the technical use of antibodies and other binding molecules, the binding is detected by other means and usually results in a change of optical or electrical signal. The molecular properties relating to bind-ing are described by the affinity and kinetic rate constants of the reaction.

The on- and off-rate constants describe the kinetic properties of a reaction, in con-trast to the affinity constant, which is a measure of the interaction at equilibrium. Kinetic studies on antigen–antibody reactions are not found frequently; however, stop-flow fluorescent quench techniques (Foote and Milstein, 1991) and Farr tech-niques with radio-labeled haptens (Hoebeke et al., 1987) are methods currently in use.

7.4.2 Kinetics: theory

The association phase of the so-called sensorgram is a function of kinetic properties of the reaction under such experimental conditions that the mass transport is not a limiting factor in the reaction. The calculation of kinetic and affinity constants from the sensorgram has been described elsewhere (Karlsson et al., 1991, 1992).

The law of mass action describing the reaction between analyte and ligand can be rewritten in terms of response signals derived from BIAcore:

$$dR_A/dt = k_aCR_{max} - (k_aC + k_d)R_A, \tag{3}$$

where R_A = response due to analyte adsorption,
C = analyte concentration,
R_{max} = the maximum amount of bound material.

The concentration of analyte is constant because the flow of analyte over the surface gives rise to pseudo-first-order kinetics.

For a series of concentrations, the slope value k_s from each plot of dR_A/dt vs. R_A can be introduced into a new plot vs. analyte concentration (Figure 7.8a–c):

$$-k_s = k_a C + k_d \qquad (4)$$

From this equation, the association rate constant can readily be obtained.

A more favorable experimental situation for determination of k_d is obtained by measuring the dissociation of bound analyte in buffer flow. No analyte is pre-

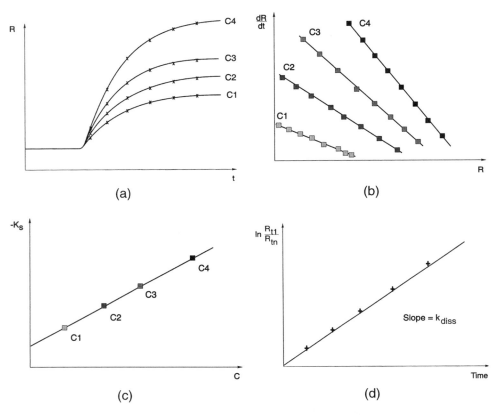

Figure 7.8. Schematic presentation of the calculation of kinetic constants. (a) The associations of the analyte run at different concentrations. (b) From the association sensorgrams, the plot of dR/dt vs. R is created and the slope k_s is determined. In (c), the slopes of the curves for each concentration are plotted against concentration of analyte. The surface concentration is the same over the different experiments. From this plot of slopes vs. concentration, the k_a can be calculated as the concentration-dependent slope. From the dissociation of the complex, the plot of lnR_{t1}/R_{tn} vs. time directly gives the dissociation rate constant. To avoid rebinding of analyte and diffusion limitation, low surface concentration of ligand should be used.

sent and the rate equation can be simplified:

$$dR_A/dt = -k_d R_A \qquad (5)$$

With the dissociation start at time t_1, the k_d can be obtained:

$$\ln(R_{t1}/R_{tn}) = k_d(t_n - t_1), \qquad (6)$$

where R_{tn} and t_n are values obtained from one and the same dissociation curve.

7.4.3 Binding of monoclonal antibodies

Detailed kinetic analysis of monoclonal antibodies has been performed by Karlsson et al. (1991), Altschuh et al. (1992), Zeder-Lutz, Altschuh, et al. (1993), and Zeder-Lutz, Wenger, et al. (1993). In these papers, the theoretical base for the kinetic determinations is made; the use of nonlinear least-squares analysis methods has been described by O'Shannessy et al.(1993). Epitope mapping of the binding site for monoclonal antibodies has been described by Fägerstam et al. (1990). Analysis of active complex formed on the sensor surface has also been performed by Schuster et al. (1993).

7.4.4 Concentration analysis

Concentration determinations in immunoassays are normally based on equilibrium measurements of formed immunocomplexes. In a flow system, other parameters of interest are the diffusion of substance to the sensor chip and the kinetic constants of the immobilized antibody. For further sensitivity and specificity, a sandwich assay can be performed (Fägerstam et al., 1992; Löfås et al., 1991), as well as inhibition assays for haptens.

In a study by VanCott et al. (1992) on antibody reactivity to peptides from the envelope glycoprotein gp160 of HIV-1, BIAcore gave highly reproducible results and, under favorable conditions, similar sensitivity to ELISA.

BIAcore can operate under diffusion-controlled conditions for analyte. This potential use of the flow system has given rise to concentration determination for molecules of the same type based on the on-rate (Karlsson et al., 1993) (Figure 7.9). The same calibration curve can be used for antibodies independent of antigen specificity. The other parameters – such as flow and the flow-cell geometry – are instrument dependent. By calibrating the instrument with one antibody or antibody fragment, the concentration of other antibodies in serum, tissue culture media, or broth, can be determined directly. The analysis must be performed with high surface concentrations of antigen. This contradicts the conditions used for kinetic measurements in the same instrument.

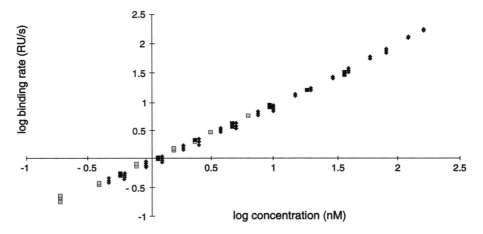

Figure 7.9. Use of log on-rate vs. log concentration analyzed under diffusion-limiting conditions in BIAcore. The figure shows the results from four different antibodies. The results indicate that for a given type of molecule (such as antibodies, antibody fragments, or mutated proteins), the functional concentration can be determined independently of specificity (Karlsson et al., 1993).

7.4.5 *Characterization of recombinant antibodies*

Phage-display technologies and selection methods have dramatically changed the way in which a variety of molecules can be generated. With bacteria and phages it is possible to obtain human antibodies without immunization (Hoogenboom et al., 1992), and peptides of different sizes can be selected for binding to different targets. Single-chain antibodies, scFv, were found to have both high on- and off-rate constants (Griffiths et al., 1993; Marks et al., 1992). Recent results of kinetic analysis of engineered antibody–antigen interactions are discussed in a review by Malmqvist (1994). The biosensor technology was also used for the on-line analysis of specific components in gel filtration (Griffiths et al., 1993). In a paper by Holliger et al. (1993), bifunctional scFv antibodies were characterized for affinity as well as analyzed on-line with gel filtration.

7.4.5 *Dissection of binding sites*

Kinetic analysis has made it possible to determine the affinity in two-component association and dissociation rate constants. By combining the kinetic analysis with site-specific mutations, the importance of individual amino acids for the association or dissociation of the biospecific complex can be determined. In this way the antigen–antibody binding site (Kelley and O'Conell, 1993) and the growth hormone and the extracellular domain of the receptor (Cunningham and Wells, 1993) have been analyzed by alanine-scan mutagenesis. The general finding was that relatively few amino acids that interact in the structural analysis contribute considerably to the affinity. It was also found that the dominant change in affinity was an effect of the dissociation rate constant, caused by relatively few hydrophobic amino acids centrally positioned in the binding site. A series of charged amino acids on the outer edge of the binding site had an impact on the association rate constant.

7.4.6 Receptor–ligand interactions

Receptor–ligand interaction is now the dominant application area, and it is one of the most intensively studied fields of today's bioscience. The affinity for the complex formed by intracellular vesicular proteins and plasma membrane proteins has been studied by Calakos et al. (1994), who found the affinity to be in the μM range, with the dissociation rate constant to be around 10^{-2} sec^{-1}. On the other hand, the cell-adhesion molecules CD2 and CD48 have approximately the same affinity but a dissociation rate constant larger than 1, indicating the half-life for the complex to be less than 1 sec (van der Merwe et al., 1993).

Other studies of the function of receptor–ligand interaction concern intracellular signal transduction by the interaction of SH2 domain and tyrosine-phosphorylated platelet-derived growth factor ß-receptor sequences (Panayotou et al., 1993) and the function of the p85α subunit of the phosphatidylinositol 3-kinase complex (End et al., 1993).

7.4.7 Nucleic acid interactions

Interactions between proteins and DNA such as lactose repressor have been studied for the interaction with DNA sequences (Bondeson et al., 1993) as well as for ETS1 recombinant oncoprotein interactions (Fisher et al., 1994). Surprisingly few analyses of DNA–DNA interactions have been published (Wood, 1993).

7.5 Conclusion

From these few described examples of BIA analysis, it is clear that kinetic analysis has dominated the scene in different application areas. The identification of unknown functions by using the label-free potential has not yet been applied, but in the long run it will be very important in relation to the HUGO program's demand to identify the unknown biological function of specific sequences. The technical solution of surface plasmon resonance detection of biospecific interactions will be followed by new techniques that may have higher sensitivity or other features that can be beneficial for some applications. However, with the surface plasmon resonance detection in BIAcore, a new technology for biological research seems to have been established. It is part of the biological research laboratory's long-term shift to automated information generation, so that more time can be used for creative thinking and formulating applied questions in biotechnology.

References

Altschuh, D., Dubs, M.-C., Weiss, E., Zeder-Lutz, G., and Van Regenmortal, M.,H.,V. (1992) Determination of kinetic constants for the interaction between a monoclonal antibody and peptides using surface plasmon resonance. *Biochemistry* 31 (27):6298–6304.

Armstrong, S. H., Budka, M. J. E., Morrison, K. C., and Hasson, M. (1947) Preparation and properties of serum and plasma proteins: he refractive properties of proteins of human plasma and certain purified fractions. *J. Am. Chem. Soc.* 69:1747–1753.

Bondeson, K., Frostell, Å., Fägerstam, L., and Magnusson, G. (1993) Lactose represperator DNA interactions: kinetic analysis by a surface plasmon resonance biosensor. *Analytical Biochemistry* 214:245–251.

Brigham-Burke, M., Edwards, J. R., and O'Shannessy, D. J. (1992) Detection of receptor-ligand interaction using surface plasmon resonance: model studies employing the HIV-1 gp120/CD4 interaction. *Analytical Biochemistry* 205:125–131.

Calakos, N., Bennett, M .K., Peterson, K. E., and Scheller, R. H. (1994) Protein–protein interaction contributing to the specificity of intracellular vesicular trafficking. *Science* 263:1146–1149.

Carlsson, J., Drevin, H., and Axén, R. (1978) Protein thiolation and reversible protein-protein conjugation. *Biochemical J.* 173:723–727.

Cunningham, B. and Wells, J. A. (1993) Comparison of a structural and a functional epitope. *J. Mol. Biol.* 234:554–563.

End, P., Gout, I., Fry, M. J., Panayotou, G., Dhand, R., Yonezawa, K., Kasuga, M., and and Waterfield, M. D. (1993) A biosensor approach to probe the structure and function of the p85α subunit of the phosphatidylinositol 3-kinase complex. *Journal of Biological Chemistry* 268 (14):10066–10075.

Fisher, R. J., Fivash, M., Casas-Finet, J., Erickson, J. W., Kondoh, A., Bladen, S. V., Fisher, C., Watson, D. K., and Papas, T. (1994) Real-time DNA binding measurements of the ETS1 recombinant oncoproteins reveal significant differences between the p42 and p51 isoforms. *Protein Science* 3:257–266.

Foote, J. and Milstein, C. (1991) Kinetic maturation of an immune response. *Nature* 352:530–532.

Fägerstam, L. G., Frostell, Å, Karlsson, R., Larsson, A., Malmqvist, M., and Butt, H. (1990) Detection of antigen–antibody interactions by surface plasmon resonance. *J. Molecular Recognition* 3:208–214.

Griffiths, A. D., Malmqvist, M., Marks, J. D., Bye, J. M., Embleton, M. J., McCafferty, J., Gorick, B. D., Hughes-Jones, N. C., Hoogenboom, H. R., and Winter, G. (1993) Human anti-self antibodies with high specificity from phage display libraries. *EMBO J.* 12(2):725–734.

Hoebeke, J., Engelsborghs, Y., Chamat, S., and Strosberg, A.D. (1987) The immune response towards beta-adrenergic ligands and their receptors: VII Equilibrium and kinetic binding studies of l-alprenolol to a monclonal anti-alprenolol antibody. *Mol. Immunology* 24(6):621–629.

Holliger, P., Prospero, T., and Winter, G. (1993) "Diabodies": small bivalent and bispecific antibody fragments. *Proc. Natl. Acad. Sci.* 90:6444–6448.

Hoogenboom, H. R., Marks, J. D., Griffiths, A. D., and Winter, G. (1992) Building antibodies from their genes. *Immunological Reviews* 130:41–68.

Johnsson, B., Löfås, S., and Lindqvist, G. (1991) Immobilization of proteins to a carboxymethyldextran modified gold surface for biospecific interaction analysis in surface plasmon resonance. *Analytical Biochemistry* 198:268–277.

Jönsson, U. Fägerstam, L., Ivarsson, B., Lundh, K., Löfås, S., Persson, B., Roos, H., Rönnberg, I., Sjölander, S., Stenberg, E., Ståhlberg, R., Urbanicsky, C., Östlin, H., and Malmqvist, M. (1991) Real-time biospecific interaction analysis using surface plasmon resonance and a sensor chip technology. *Biotechniques* 11:620–627.

Jönsson, U. and Malmqvist, M., eds. (1992) Real time biospecific interaction analysis: The integration of surface plasmon resonance detection, general biospecific interface and microfluidics into one analytical system. In *Advances in Biosensors,* vol. 2, ed. A. Turner, pp. 291–336. San Diego: JAI Press Ltd.

Jönsson, U., Malmqvist, M., and Rönnberg, I. (1985a) Adsorption of immunoglobin G, protein A and fibronectin in the submonolayer region evaluated by a combined study of ellipsometry and radiotracer techniques. *J. Colloid Interface Sci.* 103:360–372.

Jönsson, U., Rönnberg, I., and Malmqvist, M. (1985b) Flow injection ellipsometry: an in situ method for the study of biomolecular adsorption and interaction at solid surfaces. *Colloids and Surfaces* 13:333–339.

Karlsson, R., Altschuh, D., and Van Regenmortel, M. H. V. (1992) Measurement of antibody affinity. In *Structure of Antigens,* ed. M. H. V. Van Regenmortal, pp. 127–148. Ann Arbor, Boca Raton, London, Tokyo: CRC Press.

Karlsson, R., Fägerstam, L., Carlsson, H., and Persson, B. (1993) Analysis of active antibody concentration: Separation of affinity and concentration parameters. *J. Immunol. Methods* 166(1):75–84.

Karlsson, R., Michaelsson, A., and Mattsson, L. (1991) Kinetic analysis of monoclonal antibody-antigen interactions with a new biosensor based analytical system. *J. Immunol. Methods* 145:229–240.

Kelley, R. F. and O'Conell, M. P. (1993) Thermodynamic analysis of an antibody functional epitope. *Biochemistry* 32:6828–6835.

Kretschman, E. and Raether, H. (1968) Radiative decay of non-radiative surface plasmons excited by light. *Z. Naturforschung* A23:2135–2136.

Liedberg, B., Nylander, C., and Lundström, I. (1983) Surface plasmon resonance for gas detection and biosensing. *Sensors and Actuators* 4:299–304.

Liedberg, B., Stenberg, E., and Lundström, I. (1993) Principles of biosensing with an extended coupling matrix and surface plasmon resonance. *Sensors and Actuators B* 11, 63–72.

Löfås, S. and Johnsson, B. (1990) A novel hydrogel matrix on gold surfaces in surface plasmon resonance sensors for fast and efficient covalent immobilization of ligands. *J. Chem. Soc., Chem. Commun. Issue* 21:1526–1528.

Löfås, S., Johnsson, B., Tegendal, K., and Rönnberg, I. (1993). Dextran modified gold surfaces for surface plasmon resonance sensors: immunoreactivity of immobilized antibodies and antibody–surface interaction studies. *Colloids and Surfaces B: Biointerfaces* 1:83–89.

Löfås, S., Malmqvist, M., Rönnberg, I., and Stenberg, E. (1991). Bioanalysis with surface plasmon resonance. *Sensors and Actuators B* 5:79–84.

Malmqvist, M. (1994) Kinetic analysis of engineered antibody-antigen interactions. *J. Molecular Recognition* 7:1–7.

Marks, J. D., Griffiths, A. D., Malmqvist, M., Clackson, T., Bye, J. M., and Winter, G. (1992) Bypassing immunisation: building high affinity human antibodies by chain shuffling. *Bio/Technology* 10:779–783.

Matsuda, H. (1967) Theory of the steady-state current potential curves of redox electrode reactions in hydrodynamic voltammetry. II Laminar pip-and channel flow. *J. Electroanal Chem.* 15:325–336.

O'Shannessy, D .J., Brigham-Burke, M., and Peck, K. (1992) Immobilization chemistries suitable for use in the BIAcore surface plasmon resonance detector. *Analytical Biochem.* 205:132–136.

O'Shannessy, D .J., Brigham-Burke, M., Soneson, K. K., Hensley, P., and Brooks, I. (1993) Determination of rate and equilibrium binding constants for macromolecular interactions using surface plasmon resonance: use of non-linear least squares analysis methods. *Analytical Chemistry* 212:457–468.

O'Shannessy, D. J. and Quarles, R. H. (1987) Labeling of the oligosacharide moities of immunoglobulins. *J. Immunol. Methods* 99:153–161.

Panayotou, G., Gish, G., End, P., Troung, O., Gout, I., Dhand, R., Fry, M. J., Hiles, I., Pawson, T., and Waterfield, M. D. (1993) Interactions between SH2 domains and tyrosine-phosphorylated platelet-derived growth factor ß-receptor sequences: analysis of kinetic parameters by a novel biosensor-based approach. *Molecular and Cell Biology* 13(6):3567–3576.

Schuster, S. C., Swanson, R. V., Alex, L. A., Bourret, R., and Simon, M.I. (1993) Assembly and function of a quaternary signal transduction complex monitored by surface plasmon resonance. *Nature* 365:343–346.

Sjölander, S. and Urbaniczky, C. (1991) Integrated fluid handling system for biomolecular interaction analysis. *Analytical Chemistry* 63:2338–2345.

Stenberg, E., Persson, B., Roos, H., and Urbaniczky, C. (1991) Quantitative determination of surface concentration of protein with surface plasmon resonance by using radiolabeled proteins. *J. Colloid and Interface Science* 143:513–526.

van der Merwe, P. A., Brown, M. H., Davis, S. J., and Barclay, A. N. (1993) Affinity and kinetic analysis of the interaction of the cell-adhesion molecules rat CD2 and CD48. *EMBO* J 13(1):4945–4954.

VanCott, T. C., Loomis, L. D., Redfield, R. R., and Birx, D. L. (1992) Real-time biospecific interaction analysis of antibody reactivity to peptides from the envelope glycoprotein, gp160, of HIV-1. *J. Immun. Methods* 146:163–176.

Wood, S .J. (1993) DNA–DNA hybridization in real time using BIAcore. *Microchemical Journal* 47:330–337.

Zeder-Lutz, G., Altschuh, D., Geysen, H. M., Triflieff, E., Sommermeyer, G., and Van Regenmortel, M. H. V. (1993) Monoclonal antipeptide antibodies: affinity and kinetic rate constants measured for the peptide and the cognate protein using a biosensor technology. *Molecular Immunology* 30(2):145–155.

Zeder-Lutz, G., Wenger, R., Van Regenmortel, M. H. V., and Altschuh, D. (1993) Interaction of cyclosporin A with an Fab fragment of cyclophilin. Affinity measurements and time-dependent changes in binding. *FEBS Letters* 326(1,2,3):153–157.

8

Standard test targets for high-resolution light microscopy

RUDOLF OLDENBOURG

Marine Biological Laboratory, Woods Hole, MA 02543 and
Martin Fisher School of Physics, Brandeis University, Waltham, MA 02254

SHINYA INOUÉ

Marine Biological Laboratory, Woods Hole, MA 02543

RICHARD TIBERIO

National Nanofabrication Facility, Cornell University, Ithaca, NY 14853

ANDREAS STEMMER

Marine Biological Laboratory, Woods Hole, MA 02543

GUANG MEI

Marine Biological Laboratory, Woods Hole, MA 02543

MICHAEL SKVARLA

National Nanofabrication Facility, Cornell University, Ithaca, NY 14853

8.1 Introduction

The light microscope, aided by analog and digital image enhancement, is now used to visualize objects, and to measure events, at dimensions (including the third dimension along the microscope axis) near and considerably below the Abbe limit of resolution. Therefore, it is increasingly important that we assess experimentally, and understand quantitatively, images formed by high-resolution microscope optics from simple, well-characterized test objects. This is necessary both to avoid misinterpreting images and to gain further insight into the specimen fine structure. To address this problem, we are developing and fabricating test slides for the following two purposes: (1) to improve our understanding of the optical transfer functions and reliability of the image generated in three dimensions by wide-field and confocal microscope optics in various contrast modes, and (2) to provide a standard for assessing, and to help improve, the quality of microscope optics, electronic imaging equipment, and digital image processors.

Traditional test targets include fluorescent microspheres, pinholes, or jagged edges of lines ruled in metalized slides, silica shells of diatoms, and so forth. Fluorescent microspheres and pinholes act as point sources of light in the specimen plane, whose three-dimensional images – also called the point-spread functions (that are seen as one focuses above and below the in-focus image of the point source) – serve as sensitive indicators of optical aberration. Oblique-illumination images of the jagged edges in thin metal films in the Abbe test plate (which incorporates a slightly wedged coverglass whose graded thickness is calibrated) are also used for detecting aberrations. The silica shells of diatoms are decorated with rows of perforation (the frustule pores) whose spatial periods are, to a rather close approximation, characteristic of the diatom species. These rows of pores, and the pores within a row (which have a somewhat smaller spacing), are used as test targets for determining the resolution and image quality provided by a microscope objective lens.

The point-spread function is also used to calculate a thin optical section, or the "true" intensity distribution in the image attributable to a single specimen plane within a three-dimensional object (Agard, 1984; Carrington et al., 1990). In principle, the contrast transfer characteristics of the optical system, which includes the coverglass, can be calculated from the point-spread function measured on pinholes or the line-spread function measured on sharp edges in metalized slides. However, the calculations are not straightforward, and they involve assumptions that may not be readily quantifiable, especially in the presence of aberrations. We therefore fabricated simple test targets for contrast transfer measurements and image calibrations of object spacings down to, and somewhat beyond, the resolution limit of high NA (numerical aperture), highly corrected microscope objective lenses.

The test targets that we have fabricated contain line gratings with accurate periods down to 100 nm. The gratings allow, for the first time, direct measurement of contrast transfer characteristics of high-resolution microscope objectives with numerical apertures of up to 1.4. We have fabricated test targets for different contrast modes, including transmission and reflection microscopy (metallic targets), phase contrast, polarizing and differential interference microscopy (phase targets made of quartz), and fluorescence microscopy (targets made of fluorescently doped resist). Recently, we made gratings that can be mounted at a tilt angle to the focus plane of the microscope optics. With the tilted gratings we are able to measure three-dimensional imaging characteristics.

All test targets described here were developed jointly at the Marine Biological Laboratory and the National Nanofabrication Facility at Cornell University; therefore, we have named these targets MBL/NNF test targets.

We will first describe the fabrication procedures of the test targets and then report on our contrast transfer measurements imaging these targets with different light microscope optics. At the end, we show an image of a square grid used to document geometric image distortions, for example, in scanning probe (force) microscopy.

8.2 Fabrication of test targets

All test targets were fabricated at the National Nanofabrication Facility at Cornell University, Ithaca, N.Y. We used electron lithography techniques to make test patterns, such as bar gratings with accurate periods from 2.00 μm down to 0.10 μm, which is just beyond the resolution limit of light microscope optics of even the highest numerical aperture. Fabricated test patterns were checked for accuracy and consistency using a scanning electron microscope (Figure 8.1). As seen in Figure 8.1, targets contain bar gratings as well as single object features, such as double lines, single lines, double dots, and single dots. An additional, very useful test object that is incorporated is the Siemens star, which represents bar gratings of continuously varying periodicity and orientation (Figure 8.2). Furthermore, a square grating of 1-μm-spaced horizontal and vertical lines is included to assess geometric image distortions.

Each test target was fabricated on the surface of a microscope coverglass, selected for 0.17 ±0.005-mm thickness (the thickness assumed for lens design calculations to minimize aberrations). The coverglass is a stable substrate for the exceedingly thin

Figure 8.1. Scanning electron micrograph of portion of the MBL/NNF test target. Bar gratings, lines, and dots were etched into the 50-nm-thick aluminum using electron lithography techniques. Numbers above and below the edged features indicate bar grating periods in micrometers. Single and double lines, associated with each bar grating, have the same dimensions as the grating lines. Dot diameters and spacings are equal to the line widths and spacings of the associated line patterns.

test target itself and, at the same time, it is a required optical component for use with many highly corrected microscope objectives, (Inoué and Oldenbourg, 1994).

The targets mentioned so far extend in the focus plane of the microscope, that is, the entire target will be in focus for a given focus position. The targets are only 20- to 100-nm thick, much thinner than the z-resolution of an optical microscope. Therefore, these targets are ideal for measuring two-dimensional transfer characteristics of microscope optics. To determine three-dimensional imaging performance, we have fabricated oblique test targets that are tilted at angles of 45° or 90° to the focus plane. With the tilted targets we obtained experimental contrast transfer values using well-defined test objects that produced out-of-focus information at any given focus level. Next, we will briefly describe the fabrication procedures of the different types of test targets.

8.2.1 *Metallic test target*

Metallic test targets consist of an aluminum film, 50-nm thick, on a microscope coverglass. Etched into the aluminum film were test patterns created by nanolithographic processes: After evaporation of a uniform aluminum film onto the clean coverglass (glass plate, typically 18×18 mm^2, 0.17-mm thick), a thin polymer resist film (polymethyl methacrylate, PMMA, typically 100-nm thick) was spin coated on the aluminum. The resist layer was exposed with the test patterns using electron lithography. The exposed resist was developed, forming a mask for the subsequent reactive ion etching process. During reactive ion etching, the exposed parts of the aluminum were etched away, leaving most of the film intact. Hence, in a standard transmission light microscope the test patterns appear bright against a dark background (Figure 8.2). A method that we reported earlier produced dark test patterns against a white background (Oldenbourg et al., 1993).

8.2.2 *Phase target*

The fabrication of phase targets is very similar to the fabrication of metallic test targets. Instead of the metal film, a 90-nm-thick layer of SiO$_2$ was deposited directly on the clean coverglass surface. A thin PMMA resist layer was spin coated and then topped by a sputtered film of Au/Pd, which served as a conducting layer during the electron lithography exposure. After the lithographic exposure, the sputtered film was dissolved away, the resist was developed, and the SiO$_2$ layer was etched using the reactive ion etching process. It is worth mentioning that pure SiO$_2$ etches very well, whereas regular glass etches very slowly using standard reactive ion etching methods. Thus, the glass provides an accurate edge stop. As an alternative to SiO$_2$, other transparent dielectrics can be deposited onto the glass surface to obtain higher refractive index layers. A particularly simple target to make is the developed polymer resist mask itself which can be used as a

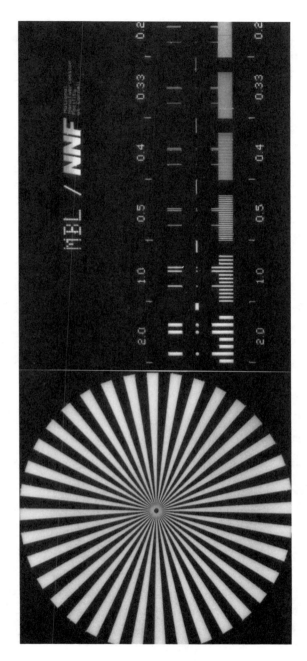

Figure 8.2. Siemens star, and line and dot patterns of the MBL/NNF test target imaged with a wide-field light microscope using transmitted light and a 60x/1.4 NA Plan Apo oil immersion objective lens (Nikon Inc.). The dark background is due to the low transmissivity of the 50-nm-thick aluminum film. Bright features were edged into the film (see text; also see legend to Figure 8.1). The Siemens star consists of 36 wedge pairs, with an outer diameter of 75 μm. The period near the outer edge is 6.5 μm, decreasing continuously toward the center. The smallest period is 0.1 μm and is near the inner black disk (1.2-μm diameter).

dielectric layer containing the test patterns (see Figure 8.3). However, the polymer resist mask is not stable against contact with solvents, including water, which tend to lift the mask off the substrate.

8.2.3 Fluorescent target

Of all the test targets we fabricated, the fluorescent targets were the easiest to make, and the easiest to loose, due to the bleaching of fluorescent dopants. We made fluorescent targets simply by adding a fluorescent, either rhodamine or fluorescein-like dopant (Molecular Probes, Eugene, Ore.), to the polymer resist solution, and then bleaching the test patterns into a 100-nm-thick fluorescent resist layer using electron beam irradiation. The fluorescent resist layer was coated on the coverglass by the standard procedure of dispensing some dissolved resist (PMMA in chlorobenzene) with fluorescent dopant on the coverglass, which was spun at 1,500 rpm for 1 min and then baked at 170°C for 1 h. To make the specimen suitable for electron lithography, a thin conducting layer of gold/palladium was sputtered onto the resist surface. None of these procedures noticeably affected the fluorescence efficiency of the dopant. The exposure of the test patterns by electron lithography bleached the coarse and fine gratings, Siemens star, and other parts of the pattern into the hardened, uniformly fluorescent layer (see Figure 8.4).

Figure 8.3. Siemens star fabricated as phase target (resist mask on coverglass) and imaged with a differential interference contrast microscope using a 60x/1.4 NA Plan Apo oil immersion objective lens.

Figure 8.4. Fluorescent target with rhodamine-like dopant in resist layer imaged with epi-illumination microscope using 60x/1.4 NA Plan Apo objective lens and silicon intensified video camera (SIT). Wedges of the star pattern were bleached by electron lithography. Geometric distortions in the image are due to the intensifier stage of the video camera.

Although the fabrication is relatively simple and straightforward, the solid, fluorescent layer as a whole is progressively bleached when imaged in a regular fluorescence microscope. By keeping the light exposure to a minimum, however, we could preserve our fluorescent targets for several measurement cycles.

8.2.4 Tilted test target

Tilted test targets were fabricated on specially prepared edges of microscope coverglass. Several coverglass panels were glued together to form a glass block, which was firmly secured and glued to an aluminum chuck used for grinding and polishing one edge surface of the block (Figure 8.5a). The grinding surface was oriented either at 45° or 90° to the individual glass panels in the block. After the grinding, lapping, and cleaning of the edge surface, a 50-nm-thick aluminum layer was evaporated onto it. Bar gratings were sputtered from the aluminum film using a focused ion beam. The focused ion beam removes the aluminum directly on the sample surface during the exposure, at a resolution of about 100 nm, eliminating many steps of the electron lithography process described earlier. Furthermore, patterns that are written with high ion-beam intensity can be imaged immediately after writing using a low beam intensity. (Compared to electron lithography, however, ion-beam "milling"

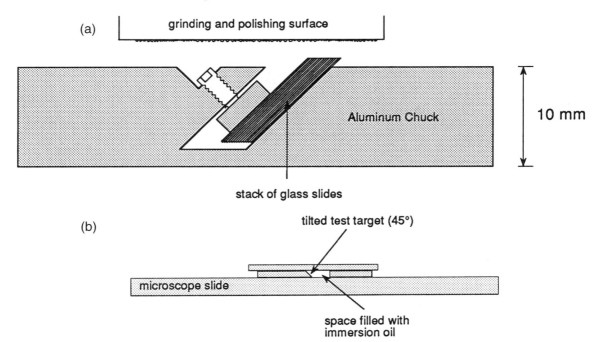

Figure 8.5. (a) Cross section showing block of coverglass panels secured in aluminum chuck for grinding and polishing. The edge surface of the coverglass stack is tilted at 45° to the grinding surface. (b) Schematic of tilted test target mounted on microscope slide and with top coverglass for observation by light microscope.

is still in an experimental stage, has lower resolution by at least a factor of two, and takes longer machine time; therefore, it is more costly in the exposure process.) After patterns were sputtered into the aluminum film, the glass block with aluminum chuck was soaked in acetone overnight to dissolve the glue ("Krazy Glue," cyanoacrylate ester) and recover the individual coverglass pieces with aluminized edges and test patterns. Individual coverglass pieces were mounted as shown in Figure 8.5b for contrast transfer measurements with the light microscope.

8.3 Applications of test targets

8.3.1 Contrast transfer of coherent confocal vs. incoherent nonconfocal optics

Figure 8.6 shows an intensity scan taken through a portion of the MBL/NNF test target as seen in Figure 8.2. The gratings with large periods are well resolved, but the ones with smaller periods are barely resolved. As the intensity scan shows, the signal from the well-resolved pattern is well modulated (the intensity varies between a large maximum, I_{max}, to a small minimum, I_{min}). As the spacing approaches the limit of resolution, the contrast transfer value $(I_{max} - I_{min})/(I_{max} + I_{min} - 2I_g)$

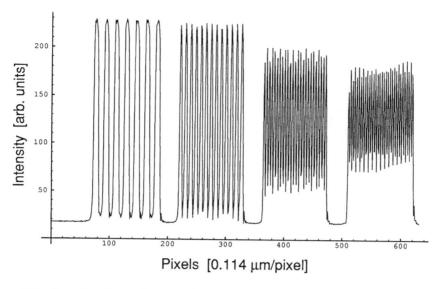

Figure 8.6. Sample of intensity scan through gratings of the MBL/NNF test target with grating periods (from left to right) of 2.0, 1.0, 0.5, and 0.4 μm. The target was imaged with a 40x/1.3 NA Fluor objective lens (Nikon Inc.) using light of 546-nm wavelength. Intensities were averaged over the dimension parallel to the grating lines.

decreases. At the background level, I_g, there is only weak transmission because of the low transmissivity of the aluminum film.

In a recent study (Oldenbourg et al., 1993) we have measured contrast transfer values using both confocal and nonconfocal microscope optics to investigate why images recorded in the confocal reflection mode seemed to be crisper and to resolve finer detail than images of the same specimens taken with nonconfocal optics. This is in contrary to the theoretical prediction that the confocal reflection mode has the same limit of resolution as the incoherent nonconfocal imaging mode (Wilson and Sheppard, 1984). Our results are summarized in Figure 8.7, which compares the ability of confocal versus nonconfocal optics to retain image contrast as a function of specimen spacing. These "contrast transfer characteristics" curves were measured for several plan apochromatic objective lenses on a single confocal microscope (prototype instrument developed by Hamamatsu Photonics K.K. of Hamamatsu City, Japan). The confocal transfer characteristic was measured in the reflection contrast mode with the exit pinhole closed down to a fraction of the Airy disk image diameter. The nonconfocal characteristic was measured in the transmission mode, which lacks an exit pinhole. As Figure 8.7 shows, our measurements confirm the resolution limit to be equal for both imaging modes. However, the contrast transfer for fine specimen detail that is close to the resolution limit is up to twice as efficient in the confocal compared to the nonconfocal imaging mode. This explains the improved contrast and resolution of the image in reflection confocal microscopy.

Contrast Transfer Characteristics

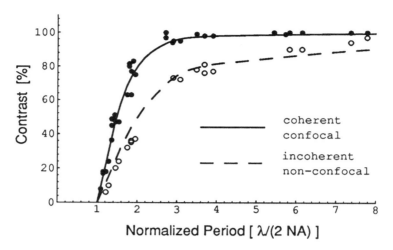

Figure 8.7. Contrast transfer characteristics measured in coherent confocal (●) and incoherent nonconfocal (○) imaging. The gratings were imaged with Plan Apo objective lenses (Nikon Inc., Melville, N.Y.) with numerical apertures (NAs) ranging from 0.45 to 1.4; laser wavelengths (λ) of 514.5 nm or 488 nm were used. Grating periods are expressed in units of the limiting wavelength, λ/2(NA), to normalize the data taken with different laser λ and lenses of different NA. (From Oldenbourg et al., 1993)

8.3.2 Contrast transfer of oil-immersion versus water-immersion objective

Objective lenses with the highest numerical aperture (NA > 1.3) are designed for use with homogeneous immersion; that is, the oil-immersion medium contacting the front element of the objective lens and the coverglass, the coverglass itself, and the medium imbibing the specimen are all expected to possess a refractive index (~1.515) equal to that of the front element of the objective lens. With the refractive indexes of all these layers equaling each other, the rays pass from the specimen to the hemispherical rear surface of the front element in the objective without being refracted or lost by total internal reflection. Thus, the front element of the objective lens is designed to capture the highest NA rays and to satisfy the sine condition that leads to correction for spherical aberration and coma (Inoué and Oldenbourg, 1994)

If one wishes to view living cells or functional cell-free extracts, the specimen must generally be imbibed in an aqueous medium. If an oil-immersion lens is used to view such a specimen, the lower refractive index (1.33–1.35) of the imbibing medium voids the assumption used in designing a homogeneous immersion lens. This gives rise to three undesirable optical effects: (1) The highest NA rays from the specimen are lost by being totally internally reflected at the water–glass interface; (2) the incidence angle of the rays that do enter the objective lens is restricted by refraction at the water–glass interface; and (3) the sine condition no longer holds for the objective lens front element. This third effect introduces spherical aberration and coma unless the region of interest in the specimen is positioned directly against the coverglass. The aberration becomes worse the further one focuses into the aqueous medium. The consequence is loss of resolution and image contrast, as clearly shown in the lower panel of Figure 8.8.

One should be able to avoid these losses by using a well-corrected, high-NA water-immersion lens, that is, an objective lens that is designed to use water

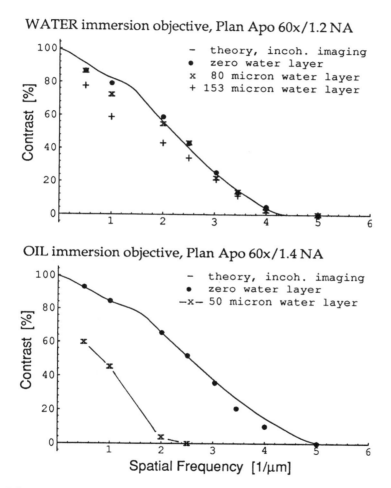

Figure 8.8. Contrast transfer values of (**upper panel**) 60x/1.2 NA water-immersion and (**lower panel**) 60x/1.4 NA oil-immersion objectives (both Plan Apochromats), measured with and without water layer between coverglass and test gratings. The graphs demonstrate that the water-immersion objective performs at the theoretical limit of contrast transfer even with water layers as thick as 153 μm, whereas the contrast transfer of the oil-immersion objective is dramatically reduced and the resolution limit is cut by half when the test grating is imaged through a 50-μm-thick water layer. The continuous lines are theoretically calculated contrast transfer functions of aberration-free objective lenses of corresponding NA at a wavelength of 546 nm (Oldenbourg and Inoué, 1994, in Brenner, 1994).

instead of immersion oil to fill the space between the objective front element and the coverglass. The upper panel in Figure 8.8 shows the contrast transfer property measured on a prototype of a recently developed Nikon Plan Apo 60x/1.2 NA water-immersion objective lens equipped with a correction collar (to accommodate slight variations in medium refractive index or coverglass thickness). Clearly the resolution and contrast transfer capability remain high even when the specimen needs to be focused through a thick layer of water between it and the coverglass. (For a sample DIC [differential interference contrast] image of a diatom frustule located under a 220-μm water layer beneath the coverglass, see Inoué and Stemmer [1994] in Brenner [1994].)

Although correction for the aberration introduced by the aqueous layer is important for observing details in living cells and tissue by conventional modes of microscopy, it is especially significant in confocal microscopy. In the latter mode, confocal efficiency is affected by the imaging quality not only of the imaging rays, but also of the illuminating rays that form the scanning image of the source pinhole. In the presence of aberration, the specimen is not only imaged with lower contrast and resolution, but scanned by a fuzzy volume of light instead of by a tight diffraction-limited image of the source pinhole. Thus with confocal microscopy in the presence of aberration, there is a dual source of loss in resolution and contrast as well as of loss in luminous efficiency.

8.3.3 *Contrast transfer of bright-field microscope using tilted test targets*

Recently, we recorded bright-field images of test targets tilted at 45° to the focus plane (Figure 8.9). We used a Nikon Microphot-SA microscope and a 40x Fluor oil-immersion objective lens with aperture iris (Nikon). The bar gratings with periods of 5.0, 2.0, 1.0, and 0.5 μm were illuminated with 546-nm light and images were recorded with a charge coupled device video camera (Dage-MTI CCD C72). Computer-assisted image analysis (NIH-Image, a public domain image analysis program for Macintosh computers, available by anonymous ftp from zippy.nimh.nih.gov/pub/nih-image) provided measurements of contrast transfer values at various grating periods (Figure 8.10).

The left panel in Figure 8.9 shows the image of a grating of the tilted test target recorded at a single focus level. Only one row of pixels near the center shows a small portion of the test grating in focus; the other rows represent out-of-focus images of other portions. The right panel of Figure 8.9 shows an extended-focus image of the same grating. The extended-focus image was assembled using the in-focus rows of images taken at different focus levels. Extended-focus images were used to measure contrast transfer values by the same procedure as discussed earlier. As can be seen in Figure 8.10, contrast transfer values for gratings tilted at 45° to the focus plane are generally lower than for horizontal gratings of the same peri-

Figure 8.9. Tilted test target (45°) imaged at a single focus level (left panel) and with extended focus (right panel). Grating lines run parallel to the tilt axis. Grating period, measured in the plane of the grating, is 2.0 μm. The image was recorded with a 40× oil-immersion objective with both objective and condenser NA adjusted to 1.0 (wavelength 546 nm). In the left panel the focus level is near the center part of the grating. Due to the shallow depth of field of the high-NA objective, grating parts with increasing distance from the central, in-focus part are progressively more out of focus. The extended-focus image is a composite of image parts taken at single focus levels.

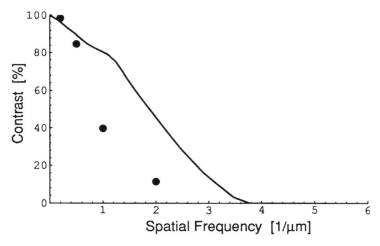

Figure 8.10. Contrast transfer values measured with tilted (45°) test gratings using extended-focus images such as the one shown in the right panel of Figure 8.9. The horizontal axis indicates the spatial frequency of the bar gratings as measured in the plane of the gratings. The continuous curve represents the theoretical contrast transfer function for *horizontal* bar gratings, which are located entirely in the focus plane, assuming imaging conditions that are identical to the ones used for the tilted targets (see legend to Figure 8.9).

od. We are conducting further experimental and theoretical studies to develop a clear understanding of the nature of the three-dimensional contrast transfer function of bright-field and other imaging modes in light microscopy.

8.3.4 *Imaging square grids with the scanning force microscope*

Square grids are very useful for assessing and correcting geometric distortions in images. Our phase gratings with grid lines etched into SiO_2 also provide suitable test targets, for example, for scanning force microscopy. In addition to calibrating lateral scan range, square gratings allow us to check orthogonality and linearity of the scanning process. Figure 8.11 shows such a test scan of a 1-μm-square grid. Because the grid lines are etched into the substrate, this kind of test target is not sensitive to tip shape and produces sharp grid lines even with fairly blunt probes.

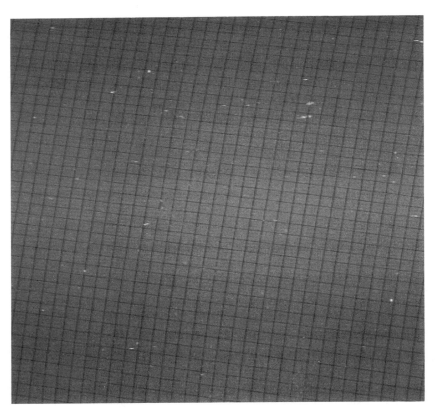

Figure 8.11. Scanning force microscope image of square grid of 1-μm-spaced lines edged into SiO_2 layer (phase target). Gray values represent sample topography, with bright indicating high topographic feature and dark indicating low topographic feature. The curvature and the variable spacing between lines in the image manifest uneven scanning motions.

8.4 Conclusion

We have fabricated well-defined, accurate test targets for light microscopes of different contrast modes, such as bright-field, fluorescent, and phase contrast microscopes. By analyzing images of these targets we were able, for the first time, to determine quantitatively the optical performance of real microscope optics of the highest numerical aperture. We have demonstrated that modern, highly corrected microscope objective lenses, when properly used, can perform at their theoretical diffraction limit in a confocal as well as wide-field microscope set-up. However, for achieving diffraction-limited performance, the optical set-up needs to match the stringent conditions assumed by lens design calculations to minimize aberrations, especially of high-NA objective lenses. As demonstrated, a water layer of 20 µm or more between specimen and oil-immersion objective can seriously degrade the image. However, we found that a high-NA water-immersion lens can provide diffraction-limited images even when focusing through a water layer of 200 µm underneath the coverglass.

We have started to explore the three-dimensional imaging performance of microscope optics using tilted test targets. These measurements are the first of their kind, and they promise to provide important information toward a more complete understanding of the three-dimensional imaging process. This process is widely used but still poorly understood, in particular with regard to phase-sensitive imaging modes such as phase contrast, differential interference contrast, and polarized light microscopy.

Acknowledgments

We wish to thank Robert A. Knudson of the Marine Biological Laboratory (MBL) for assisting in the design and fabrication of mechanical parts, in particular for the tilted test target project. We are grateful to Louie Kerr of the Central Microscopy Facility at the MBL for the electron microscopy of test target samples. We acknowledge support by NIH grant R01 GM49210 to R. O., NIH grant R-37 GM31617 and NSF grant DCB-8908169 to S. I., NSF grant ECS-8619049 to the National Nanofabrication Facility at Cornell University, and a Swiss National Science Foundation Fellowship to A. S.

References

Agard, D. A. (1984) Optical sectioning microscopy: cellular architecture in three dimensions. *Ann. Rev. Biophys. Bioeng.* 13:191–219.

Brenner, M. (1994) Imaging dynamic events in living tissue using water immersion objectives. *Am. Lab.* 26(April):14–19.

Carrington, W. A., Fogarty, K. E., Lifschitz, L., and Fay, F. S. (1990) Three dimensional imaging on confocal and wide-field microscopes. In *Handbook of Biological Confocal Microscopy*, ed. J. B. Pawley, pp 151–161. New York: Plenum Press.

Inoué, S. and Oldenbourg, R. (1994) Microscopes. In *Handbook of Optics*, 2nd Edition, ed. M. Bass, pp 17.1–17.45. New York: McGraw-Hill, Inc.

Oldenbourg, R., Terada, H., Tiberio, R., and Inoué, S. (1993) Image sharpness and contrast transfer in coherent confocal microscopy. *J. Microsc.* 172:31–39.

Wilson, T. and Sheppard, C. (1984) *Theory and Practice of Scanning Optical Microscopy*. London: Academic Press.

9

Development of voltammetric methods, capillary electrophoresis and TOF SIMS imaging for constituent analysis of single cells

PAULA BEYER HIETPAS, S. DOUGLAS GILMAN, ROXANE A. LEE,
MARK R. WOOD, NICK WINOGRAD, and ANDREW G. EWING

152 Davey Laboratory, Department of Chemistry
Pennsylvania State University,
University Park, PA 16802

9.1 Introduction

The adaptation of methodologies from one scientific field to another is not uncommon. Technological advances in one field can often lead to great insight in solving the problems of another. One very successful overlap can be observed between biochemistry and analytical chemistry. In particular, methods developed in analytical chemistry have been quite beneficial to the field of neurochemistry. This interaction between the two disciplines has resulted from the miniaturization of existing analytical techniques and the development of new methods for analyzing minute environments. These miniaturized techniques can be applied to the study of cellular environments (Adams, 1976; Ewing, 1993; Ewing et al., 1992; Wightman, 1981).

Several of these cellular environments are heterogeneous, where each cell has its own function. Thus, the role of each cell must be determined individually. Once the specific function of each cell is understood, then its relation to other cells and to the entire organism can be determined. Only in this way can a true understanding of the chemistry of an entire organism be realized.

The considerable interest in studying single-cell chemistry has resulted in the development of a number of analytical techniques. These include enzyme activity measurements (Giacobini, 1987), immunoassay (Giacobini, 1987), microgel electrophoresis (Matioli and Niewisch, 1965), fluorescence imaging techniques (Tank et al., 1988), microscale ion-selective electrodes (Nicholson and Rice, 1988), voltammetric microelectrodes (Ewing et al., 1992), microcolumn separation techniques (Ewing et al., 1992; Hogan and Yeung, 1992; Kennedy et al., 1989), optical and electron microscope techniques (Betzig et al., 1991; Sossin and Scheller, 1989), and secondary ion mass spectrometry (Mantus et al., 1991). Although these methods have provided valuable information, they have important limitations. Most suffer from inadequate sensitivity, poor quantitative capabilities, or an inability to monitor chemical dynamics on a time scale similar to the neurotransmission process.

This chapter describes two techniques that have successfully been used to analyze single cells with either great selectivity or rapid response time. The first technique, electrochemical monitoring, is useful for detecting chemical dynamics in or at single cells (Ewing et al., 1992). Recent advances in miniaturization of electrodes and several significant applications from biological environments are discussed. The second technique is narrow bore capillary electrophoresis (CE). CE can be used to determine a great deal about the actual chemical composition of a system, because it provides rapid and highly efficient separations of ionic species in extremely small volume samples (Ewing et al., 1989). CE has been used to analyze single invertebrate and mammalian cells (Olefirowicz and Ewing, 1990a).

9.2 Electrochemical analysis of single cells

Electrochemical experiments measure the current generated by a charge transfer reaction, such as an oxidation or reduction, when a potential is applied across two electrodes. The electrochemical experiments presented here are carried out in one of two modes, voltammetry or amperometry. In voltammetry, the potential across the electrodes is scanned from a voltage where no reaction occurs past the reaction potential while the current is monitored. Because the reaction potential is different for different analytes, qualitative information can be obtained. In amperometry, the electrode is held at a constant potential, and any species that react at this potential will be detected as an oxidation or reduction current. Electrochemical analysis is limited to compounds that are easily oxidized or reduced. Fortunately, several cell components in the neurons of the brain are easily oxidized. These include dopamine (DA), norepinephrine (NE), epinephrine (E), and serotonin (5-HT) (Wightman, 1981).

9.2.1 Electrodes

Recent advances in the miniaturization of electrode design has enabled analyses to be performed in small environments. There are several advantages to these small electrodes. First, voltammetry at ultrasmall electrodes is rapid and sensitive. In addition, smaller electrodes pass less current and are virtually nondestructive to the species (Wightman, 1981). Finally, their size allows them to be used in microenvironments including single cells.

Electrodes with ultrasmall electroactive areas have been constructed from many different materials, including noble metals and carbon (Kim et al., 1986; Meulemans et al., 1986; Pendley and Abruna, 1990; Penner et al., 1989, 1990; Saraceno and Ewing, 1988; Wightman et al., 1988). For use in biological microenvironments, electrodes must have not only small electroactive areas, but also total tip structures of micrometer or submicrometer dimensions (Ewing et al., 1992). In addition, electrodes must also be biologically compatible, because the proteins, carbohydrates, and lipids in biological media will adsorb to electrodes and render them inactive. Carbon electrodes are the most resistant to this fouling and therefore have been most widely used (Ewing et al., 1992).

Carbon electrodes have been fabricated in the forms of disks (Wightman et al., 1988), rings (Kim et al., 1986), and cylinders (Saraceno and Ewing, 1988). The carbon disk electrodes are fabricated from thin carbon fibers that are housed in glass capillaries. These electrodes generally have a total structural tip diameter of 15–20 μm with electroactive diameters of 5–11 μm. This size allows them to be used outside of single cells (Wightman et al., 1988). Smaller, carbon ring electrodes have been fabricated by pyrolyzing methane inside drawn silica tubes yielding total structural tip diameters as small as 1–5 μm. The inside of the tip is insulated with epoxy, leaving a ring of carbon between the silica and the epoxy. A schematic diagram of one of these electrodes is shown at the top of Figure 9.1.

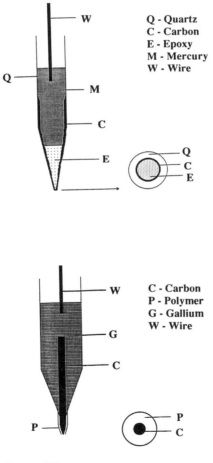

Figure 9.1. Schematics of two different carbon voltammetric electrodes. (Top) Carbon ring electrode. (Bottom) Etched and insulated carbon fiber electrode.

These electrodes are useful for both intracellular and extracellular voltammetry (Kim et al., 1986).

Cylindrical carbon fiber electrodes have also been fabricated from carbon fibers with diameters 5–11 μm. In this case, however, individual fibers protrude from the glass housing and can be etched to tips as small as 200 nm. The sides of the fiber must still be insulated in order to constrain the electroactive area to the tip of the electrode. This insulation tends to increase the overall structural diameter. Insulating films of 100-nm thickness have been obtained by electropolymerization of phenol and 2-allylphenol (Strein and Ewing, 1992). Electrodes with total structural diameters as small as 400 nm have been constructed. A schematic of this type of electrode is shown at the bottom of Figure 9.1.

9.2.2 Monitoring exocytosis

The release of neurotransmitters from single cells has also been examined using electrochemical detection. In this type of analysis, a microelectrode is placed inside of, on top of, or directly next to, a cell. The cell is stimulated and the release of the easily oxidizable neurochemicals, the catecholamines, is monitored. The rapid response times of these electrodes are well suited for the detection of these discrete neurochemical events (Wightman et al., 1991).

Mammalian neurons are difficult to work with in the laboratory, because they stop proliferating once they have matured. Green and Tischler (1976) established a nerve-growth-factor–responding clonal line of rat Pheochromocytoma (PC12) cells. This cell line shares many physiological properties with primary cultures of sympathetic ganglion neurons, and has been studied as a model for the developing sympathetic nerve (Wagner, 1985). PC12 cells can synthesize, store, and release catecholamines in a manner similar to that used by sympathetic ganglion neurons (Green and Tischler, 1976; Wagner, 1985). PC12 cells are more similar to sympathetic ganglion neurons than to adrenal chromaffin cells, because they contain more dopamine than norepinephrine with no detectable level of epinephrine (Clift-O'Grady et al., 1990). Most important, PC12 cell vesicles are valid analogs of brain synaptic vesicles for four criteria: size, density, protein composition, and endocytotic origin (Clift-O'Grady et al., 1990; Wagner, 1985).

Carbon fiber electrodes have been used to monitor single exocytotic events from individual PC12 cells. The electrode is placed against the top of a cell in culture and the cell is stimulated in a manner as previously described for adrenal cells with either KCl or nicotine as stimulant. Figure 9.2 shows the amperometric response following three successive stimulations with both 1 mM nicotine and 105 mM KCl. Stimulated release is not observed from these cells if calcium is omitted from the medium. In addition, preliminary data suggest that exposure to nicotine alone results in minimal exocytosis, and only after a substantial delay.

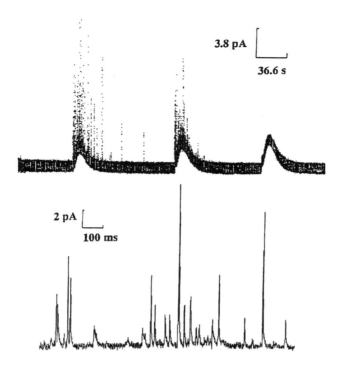

Figure 9.2. Amperograms of single vesicular exocytosis from PC12 cells. (Top) Amperogram of vesicular exocytosis induced by 1 mM nicotine in 105 mM K^+ balanced salt solution. (Bottom) Enlargement of a 1-sec period of the amperogram from the first stimulation. Data displayed correspond to the time period near the middle of the first baseline disturbance in top panel.

Exposure to elevated KCl causes most of the exocytosis observed. Individual current spikes have an average half-width of 9.3 ± 0.1 ms (n = 1912 from 13 cells) and the average catecholamine level calculated with Faraday's law is 190 zmol (114,000 molecules) per vesicle. Monitoring zeptomole levels of catecholamine from single exocytotic events provides the means to monitor physiological and pharmacological alterations of quantal size. Each vesicle has attoliter volume, so these measurements represent an extreme example of measurements concerning small environments.

9.3 Microcolumn separations of single cells

9.3.1 Open tubular liquid chromatography and microbore LC

Open tubular liquid chromatography (OTLC) and microbore LC separations are carried out in 1–5-μm inner-diameter (i.d.) capillaries 1–3 m in length with a stationary phase bound to the inner wall (OTLC) or to micrometer-sized particles (microbore LC) (Ewing et al., 1992). Separation of analytes is then based on the differential retention of analytes by the stationary phase. This type of separation exhibits high resolution with low sample volumes and is also useful for both quantitative and qualitative analysis. Electrochemical and fluorescence detection methods are the most sensitive for OTLC (Cooper et al., 1992; Jorgenson, 1989; Kennedy and Jorgenson, 1989; Kennedy et al., 1987, 1989; Oates et al., 1990).

9.3.2 *Capillary electrophoresis*

Capillary electrophoresis is a relatively new separation technique that is well suited for the analysis of single cells. In CE, a solution-filled capillary is suspended across two buffer reservoirs with a potential applied across them. A sample CE apparatus is shown in Figure 9.3. When a potential is applied, ionic solutes begin to migrate in the capillary; positive ions will migrate toward the negative end of the capillary, and vice versa. This movement is known as electromigration. The rate at which the species migrate is proportional to their mass-to-charge ratio. If this were the only force acting in CE, anions and cations would elute at opposite ends of the capillary; the presence of another force, electroosmotic flow (EOF), however, results in all ions eluting at the anodic end of the capillary. Electroosmotic flow is the bulk movement of liquid through the capillary when a potential field is applied. This is due to the build-up of cations next to the negatively charged wall (Rice and Whitehead, 1965). As a potential is applied, the cations migrate toward the anode and drag bulk solvent along with them. This results in a flow profile that has a flat velocity distribution across the center and causes highly efficient separations (Sloss and Ewing, 1993a). One of the main advantages of EOF is that it is strong enough to cause elution of cations, neutrals, and anions at the same end of the capillary.

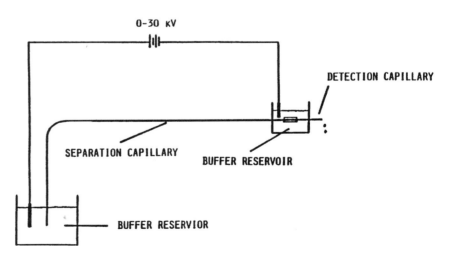

Figure 9.3. Schematic of CE system. From A. G. Ewing, R. A. Wallingford, and T. M. Olefirowicz. (1989) *Anal. Chem. 61*:294A. With permission.

The capillaries used for CE range from 10 to 100 cm in length and 2 to 100 μm i.d. Such small capillary bores have very high surface area–volume ratios, which allow for rapid heat dissipation through the capillary wall; therefore, high separation potentials (30 kV/m) can be used to expedite the analysis (Sloss and Ewing, 1993a).

9.3.3 Injection techniques

Because the total volume of a 5-μm-i.d., 100-cm-long capillary is only on the order of 20 nL, the introduction of very narrow sample plugs is necessary. Two of the injection techniques used with CE that are effective for introduction of such low volumes are hydrodynamic injection and electrokinetic injection. Hydrodynamic injection is implemented through the application of a pressure differential across the length of the capillary. The pressure differential can be generated by raising the sample end of the capillary relative to the detection end, by pressurizing the sample container, or by creating a vacuum at the detection end of the capillary. The other method, electrokinetic injection, utilizes the same electromigration process that is used to separate the ions. In this process, the cathodic end of the capillary is placed directly in the sample as a voltage is applied for a few seconds. Electrokinetic injection is particularly useful for biological samples, due to the minimization of sample handling and dilution. The simplicity of this technique has caused it to be used frequently; it does have some drawbacks, however. One problem is that a charge bias is induced. During the injection time, cations will move farther ahead than anions. This causes a difference in the apparent injection volume of the two species; the actual injection volume, however, can be determined if the electrophoretic mobility is known.

9.3.4 Detection techniques

Many different detection methods have been used with CE including absorption, fluorescence, mass spectrometry, and electrochemistry (Ewing et al., 1989). Of these, the two most sensitive methods are electrochemical and fluorescence detection. Electrochemical detection (EC) has been demonstrated in both the amperometric and voltammetric modes (Curry et al., 1991). Amperometric methods are more sensitive; however, voltammetric detection can result in more qualitative information. In CE–EC experiments, carbon microelectrodes are placed either directly outside of the capillary or in the anodic end of the separation capillary (Curry et al., 1991; Huang et al., 1991). Capillaries as small as 2-μm i.d. have been used with electrochemical detection after a conical section has been etched in the detection end of the capillary with hydrofluoric acid (Sloss and Ewing, 1993b). Detection limits of 19 amol of catechol have been obtained under these conditions. The best detection limits for catechol using a 5-μm-i.d. capillary for off-column and end-column detection are 0.6 and 56 amol, respectively (1 amol = 10^{-18} mol) (Olefirowicz and Ewing, 1990a; Sloss and Ewing, 1993b).

Laser-induced fluorescence detection (LIF) is also compatible with CE of biological samples (Albin et al., 1991; Hernandez et al., 1991; Lee and Yeung, 1992b; Yeung et al., 1992). In CE–LIF, excitation is induced by focusing a laser to a small spot on a window in the separation capillary. The lasers used most frequently are helium cadmium lasers and argon ion lasers at wavelengths of 325 and 488 nm, respectively. Fluorescence emission is collected perpendicular to the incident light to minimize the influence of stray light. Because most species in a cell are not inherently fluorescent, derivatization is necessary. Pre- and post-column derivatization schemes have been demonstrated using the fluorophores naphthalenedicarboxyaldehyde (NDA), *o*-phthaldialdehyde (OPA), fluorescein isothiocyanate (FITC), and 3-(4-carboxybenzoyl)-2-quinoline carboxaldehyde (CBQCA) (Amankwa et al., 1992).

Pre-column derivatization is the simplest method; however, multiple derivatization sites of a single analyte may lead to band broadening or even to several peaks for the same analyte. In addition, migration times for the analyte and the derivatized product are different and may cause difficulty in identification. Post-column derivatization avoids these problems by allowing the separation of analytes to occur before derivatization. In this case, however, band broadening and derivatization reaction time become issues (Amankwa et al., 1992; Lee and Yeung, 1992a; Yeung et al., 1992). Detection limits of $10^{-15} - 10^{-21}$ mol have been reported depending upon the fluorescent tag used.

Indirect fluorescence is a possibility for detection of species that cannot be derivatized. In this case, the sample buffer is fluorescent and the analyte is recognized by a decrease in signal as the fluorophores are displaced. Indirect detection is less sensitive than direct LIF, but it does offer the opportunity to detect inorganic species such as sodium or potassium (Hogan and Yeung, 1992).

9.4 Whole cell analysis

9.4.1 OTLC and microbore column analysis

Jorgenson and co-workers have demonstrated single cell analysis using OTLC with voltammetric detection (Kennedy et al., 1987; Kennedy and Jorgenson, 1989; Oates et al., 1990). They separated the components of homogenized giant neurons of the land snail *Helix aspersa* and identified tyrosine, tryptophan, dopamine, and serotonin in these cells (Kennedy and Jorgenson, 1989). These experiments have been used to substantiate coexistence of more than one neurotransmitter in a single neuron. OTLC with amperometric detection and NDA derivatization has also been used for single-cell analysis (Oates et al., 1990). This analysis proved useful for amino acids; however, the neurotransmitters DA and 5-HT were lost during sample preparation.

Single mammalian cells from adrenomedullary tissue have also been homogenized and separated with liquid chromatography combined with electrochemical detection. However, in this separation, packed microcolumns were used rather than open columns, because the open columns do not provide enough retention for polar compounds such as NE (Cooper et al., 1992). The existence of two types of cate-cholamine-secreting cells, NE-rich and E-rich, in mammalian adrenomedullary cells is generally accepted (Cooper et al., 1992). The contents of 22 cells from five cell preparations were quantified with OTLC–EC. It is interesting to note that all of these E-rich cells contained NE (detection limit 46 amol) although several of the NE-rich cells did not contain E (detection limit 75 amol). Several cells contained significant amounts of both neurotransmitters, which could indicate a third type of cell in this tissue or that cells can convert back and forth between the two cell types (Cooper et al., 1992).

9.4.2 CE of invertebrate cells

Jorgenson and co-workers used CE with LIF detection and derivatization with NDA to analyze homogenized neurons from the land snail, *Helix aspersa*. In these experiments, only about 20% of the cell was analyzed at a time. This allowed an assessment of the reproducibility of the results, because multiple runs of the same cell could be performed (Kennedy et al., 1989). However, this requires a significant amount of sample preparation.

Whole cells can also be injected directly onto the capillary by electrokinetic injection then broken down, or lysed, inside the capillary. The lysing procedure involves injection of a nonphysiological buffer immediately after cell injection, followed by a 60-sec or longer lysing time. In this way, the entire contents of a cell can be analyzed at once.

Kristensen et al. (1994) used CE–EC with electrokinetic injection to analyze the giant dopamine cell from the pond snail, *Planorbis corneus*. Separations of cell components resulted in peaks for uric acid (UA) and dehydroxyphenylacetic acid (DOPAC), and what appears to be two peaks for dopamine. The second dopamine peak is dependent upon the lyse time and is diminished by the vesicle-depleting drug, reserpine. These results indicate a two-compartment model for dopamine in these cells. The first peak represents the functional, or releasable, dopamine seen on stimulation of the cell. The second peak represents the dopamine found in storage vesicles that are nonfunctional and are not released until the cell is completely lysed (Kristensen et al., 1994).

9.4.3 CE of red blood cells

One of the smallest cells analyzed to date by CE is the single human erythrocyte, or red blood cell (Hogan and Yeung, 1992, 1993; Lee and Yeung, 1992b). This cell is about 8–9 μm in diameter and has a total cell volume of about 90 fL. Yeung and co-workers isolated these cells from whole blood and examined them using CE with both direct (Lee and Yeung, 1992b) and indirect fluorescence detection (Hogan and Yeung, 1992). The direct fluorescence scheme was used to determine the level of the peptide glutathione (GSH). GSH is suspected to play a role in the cellular response to various types of drugs and radiation. Because GSH is not inherently fluorescent, the cells were derivatized with monobromobimane (mBBr) before sampling. Derivatization was initiated by allowing the mBBr to diffuse through the semipermeable cell membrane (Lee and Yeung, 1992b). (The derivatized product is unable to migrate back through the membrane, due either to size or change in hydrophobicity). A derivatized cell was then hydrodynamically injected into a 10-μm-i.d. capillary, and lysed and separated. Levels of 68 ±48 amol of GSH have been reported. The large error shows the great variability between cells and reiterates the importance of looking at such phenomena on the single-cell level.

Fluorescence detection without derivatization has been demonstrated for the natively fluorescent proteins, hemoglobin A_0 (HemA) and carbonic anhydrase (CAH), in red blood cells. Values of 5 amol (CAH) and 0.47 fmol (HemA) have been determined with this method, and these values are very similar to the standard literature values (Lee and Yeung, 1992b).

Indirect fluorescence detection has been used to determine the amount of sodium and potassium in single erythrocyte cells (Figure 9.4), because these inorganic ions have no good derivatization schemes. A fluorescent buffer containing 6 aminoquinoline and cetyltrimethylammonium bromide is used and the Na^+ and K^+ ions are detected by the charge-displacement effects of the buffer. This method

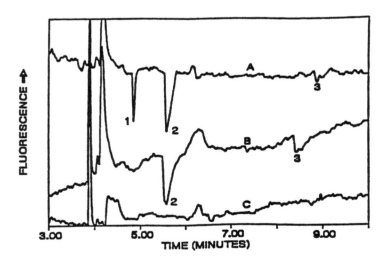

Figure 9.4. Electropherogram using indirect fluorescence detection. (A) Injection of 45 μM standards. Peak 1 is Li (11.7 fmol injected), peak 2 is Na (10.7 fmol), and peak 3 is K (6.5 fmol). (B) Injection of one human erythrocyte. (C) Blank injection of extracellular matrix. From B. L. Hogan and E. S. Yeung. (1992) *Anal. Chem. 64*:3045. With permission.

of detection is not as sensitive as direct detection, and therefore it cannot currently be used for analysis of less-concentrated species (Hogan and Yeung, 1992).

9.4.4 CE of mammalian lymphocytes

Using CE–EC, the presence of catecholamines and their metabolites has been detected in single lymphocytes and extracts of T-cell clones (Bergquist et al., 1994). A CE separation of a single lymphocyte contains a peak with a migration time of 10.75 min that has tentatively been identified as dopamine. Similar electropherograms are obtained for cloned CD4$^+$ T-cells. Cerebrospinal fluid (CSF) lymphocytes have been observed to contain 2.3 ±1.7 amol of catecholamine and cloned CD4$^+$ T-cells have been observed to contain 31 ±29 amol. In one lymphocyte, 310 zmol of catecholamine were detected, demonstrating the extreme sensitivity of this methodology.

To study the uptake and metabolism of catecholamines in clones of CD4$^+$ T lymphocytes, the effects of incubating T-cell clones with either dopamine or the dopamine-synthesis inhibitor, α-methyl-*p*-tyrosine (α–MPT), were examined. Following incubation, the cultured cells were extracted and a sample was examined by capillary electrophoresis. For the cells incubated with dopamine, a large increase in the level of catecholamine per cell is observed. Conversely, incubation with α-MPT decreases the level of catecholamine in cloned CD4$^+$ T lymphocytes. The data for dihydroxyphenylacetic acid follow this general trend. Experimentally, these data strongly suggest that the catecholamine peak is indeed dopamine, and that this catecholamine is both synthesized and accumulated by lymphocytes (Bergquist et al., 1994). This is the first evidence of the presence of catecholamines in lymphocytes.

9.5 Cytoplasmic Analysis

Neurotransmitters can be found in two different compartments in neurons – free (cytoplasmic) and bound (vesicularized). Previous analyses have focused on the bound neurotransmitters, so investigations of the neurotransmitter content in the cytoplasm should lead to a better understanding of the role of the cell in neurotransmitter uptake, storage, and metabolism. Detection of cytoplasmic levels of the neurotransmitter, dopamine, became possible with implementation of direct electrokinetic injection with 2- and 5-μm-i.d. etched capillaries (Chien et al., 1990; Olefirowicz and Ewing, 1990a, 1990b). Capillaries were etched in hydrofluoric acid to outer diameters of 6–10 μm, which made them small enough to place through the cell membrane directly into a neuron. A small amount (50 pL) of the cytoplasm was sampled and separated. The systems examined were the giant dopamine and serotonin nerve cells of the pond snail, *Planorbis corneus.* The DA cell is about 200 μm in diameter with a total cell volume of about 5 nL. The injection volume of 50 pL is only about 1% of the total cell volume.

Cytoplasmic levels of 5-HT were found to be 3.1 ± 0.57 µm and those of DA, 2.2 ± 0.52 µm (Chien et al., 1990; Olefirowicz and Ewing, 1990a, 1990b).

9.6 Future directions

9.6.1 Picoliter reaction vials

Future directions for single-cell analysis include further improvements in sample handling and derivatization schemes. One possibility is the use of picoliter vials as reaction vessels and sampling wells. Square pyramidal vials ranging in size from 8.5 to 95 pL have been fabricated with the assistance of the National Nanofabrication Facility at Cornell University. These vials are inverse square pyramids etched in silicon using standard photolithographic techniques. A scanning electron micrograph of a 50-pL vial is shown in Figure 9.5. Arrays of 2,500 wells of four different sizes have been etched on a 3-inch silicon wafer. These are spaced 1 mm apart to allow for eventual automation of injection procedures. A thin layer (1,000 Å) of gold with a 70-Å chrome adhesion layer is deposited over the entire structure to serve as an electrophoresis anode for injection. Larger vials (118 nL) have been used previously for similar purposes (Jansson et al., 1992).

One purpose of these vials is to limit the dilution of the sample either upon derivatization or upon stimulated release of the cell. By confining such release to a known volume, more accurate quantitation of the release should be possible.

Figure 9.5. Reproduction of a scanning electron micrograph of a 50-pL microvial. Scale bar is 10 µm.

Using CE and the smaller vials, the entire volume can be injected and separated.

Initial characterization of the vials to determine the effects of evaporation, capillary action, contamination, and adsorption has been accomplished. Evaporation from these vials can be a significant problem. Complete evaporation from the 95-pL vials occurs in minutes in the open laboratory. Because the smaller vials have the same surface area–volume ratio, the evaporation rate should be constant, and the smaller vials will be evaporated to dryness even more quickly. We have found that adding 15% glycerol by volume to the solutions successfully slows evaporation to a usable rate, and all successful experiments carried out to date have used this protocol. The successful implementation of picoliter microvials affords the possibility of localizing extremely small quantities of material with great spatial resolution.

Picoliter microvials have been used in conjunction with time-of-flight secondary ion mass spectrometry imaging to localize and quantitate test substances with micrometer coordinates and submicrometer resolution. In these experiments, a 10^{-4} M solution of crystal violet was injected into a 50-pL vial and the solvent allowed to evaporate completely. A time-of-flight secondary ion mass spectrometry image (TOF SIMS) of the crystal violet molecular ion (m/e = 372) is shown in Figure 9.6. The square shape of the vial can be clearly observed with very little crystal violet located outside the vial. The total imaged area is 100 x 84 µm with 0.33 x 0.33 µm pixel resolution. This level of resolution is obtained by use of a

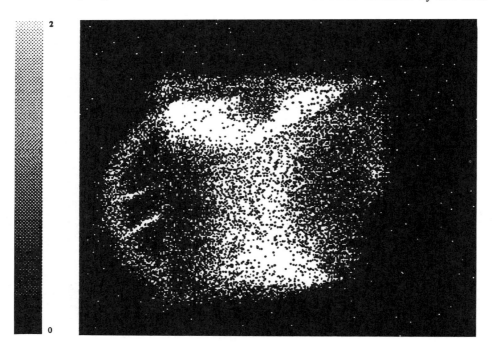

Figure 9.6. SIMS image of 10^{-4} M crystal violet in a 50-pL microvial. Imaged area is 100 x 84 µm taken with pixel resolution of 0.33 µm x 0.33 µm. The ion dose to the surface was 1.7×10^{12} ions/cm.

gallium liquid-metal ion gun with a tightly focused ion beam. TOF SIMS images have also been obtained inside and outside microvials monitoring total ion count, sodium ion (contamination), and the gold coating that overlays the vials and the surrounding silicon wafer (data not shown).

When gold is used as a substrate, its contamination must be considered, because gold is contaminated fairly readily in the atmosphere. Adsorption of organic species from the air occurs on the order of a few minutes (Gaines, 1981). This causes the gold surface to become more hydrophobic and suggests the possibility that the sides of the vials might repel the solution. Observations by optical microscopy indicate that, particularly when glycerol is added, this is not the case. Hence, the microvials have been used for liquid-phase sampling at the picoliter level for subsequent capillary electrophoresis separation.

A preliminary separation of five easily oxidized catecholamine standards from a 95-pL vial is shown in Figure 9.7. A solution consisting of 10^{-5} M dopamine, epinephrine, and catechol with 10^{-4} M uric acid and DOPAC (15% glycerol by volume) was injected into a vial using pressure injection and then electrokinetically injected into the etched end of a 5-μm-i.d. separation capillary. Electrochemical detection was used to identify the species by comparisons with the electromigration times of the standards from a larger vial under identical

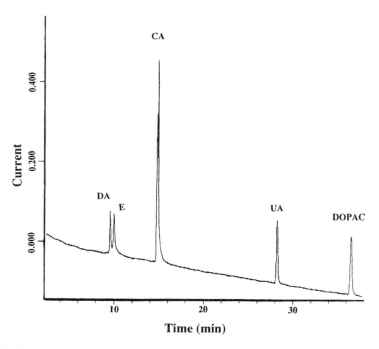

Figure 9.7. Trial separation of standards sampled from a 95-pL microvial with electrochemical detection. Injection: 10 s, 10 kV; 84.1-cm, 5-μm-i.d. capillary; separation voltage: 30 kV.

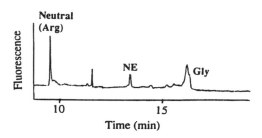

Figure 9.8. Trial separation of standards sampled from a 95-pL microvial with fluorescence detection. Injection: 8 s, 2 kV (60 pL); 100-cm, 10-μm-i.d. capillary etched to 25-μm tip; separation voltage: 30 kV; derivatization: 15-min reaction of 45 pL NDA (1 mM); KCN (1 mM) with 45 pL amine sample.

conditions. The peaks from the picoliter vials in all cases were smaller than those from the pure standards. This might result from analyte adsorption onto the gold surface of the vial. The sample injection volumes were estimated to be about 75% of the total vial volume. The pyramidal shape of the vials prevents the injector from reaching the bottom, because the opening narrows with depth. Therefore, the solution in the point at the bottom of the pyramid is not necessarily sampled.

Derivatization in a picoliter vial has also been attempted. In these experiments, 45 pL of an amino acid solution was injected into a 95-pL vial immediately followed by a 45-pL injection of NDA. The amount of NDA was twice the amount needed for complete reaction, and a reaction time of approximately 10 min was used before separation. The results of this separation are shown in Figure 9.8.

Experiments with picoliter reaction vials are still in the characterization stage. The eventual goal is to use these vials to facilitate single-cell measurements. Electrochemical monitoring of the neurotransmitter release from single vesicles upon stimulation have been previously described. We plan to carry out these experiments on cells immobilized in picoliter sampling wells. This should permit the total quantitation of easily oxidized substances released via exocytosis from a single nerve cell in culture. Fast-scan voltammetry (300 V/sec) in ferrocene-containing solutions has shown that these voltammetric analyses will be plausible in the future.

9.6.2 Post-column reactors for CE–LIF

Another future development is the optimization of post-column reactors for CE–LIF. As previously mentioned, post-column derivatization allows the analytes to migrate at their own electrophoretic mobilities without interference by the derivatization agent. Post-column techniques have generally used o-phthaldialdehyde

(OPA) or fluorescamine to derivatize primary amines in amino acids, proteins, and biological polyamines (Kuhr et al., 1993; Nickerson and Jorgenson, 1989). Designs for post-column reactors using pressure-driven mixing or electroosmotic-flow mixing have been reported. Most of these designs have used capillaries in the range of 25–50-μm i.d. Because smaller capillary is generally used for single-cell analysis, a reactor using a 10-μm-i.d. capillary was constructed (Gilman et al., 1994).

This reactor used the same 10-μm-i.d. capillary for the separation and detection segments of the capillary, with a narrow gap in between. The gap distance between the two capillary segments ranged from 4 to 160 μm. The best results were obtained with the smaller gap distance (<10 μm). A reagent reservoir was formed over the gap to allow more control in the introduction of the derivatizing reagent. A schematic of this design is shown in Figure 9.9, and a separation of separated and derivatized amino acids and proteins is shown in Figure 9.10. Introduction of derivatizing reagent is controlled mainly by diffusion in the gap. The mass sensitivity for amino acids and proteins is better in this system than in other post-column derivatization systems (Gilman et al., 1994).

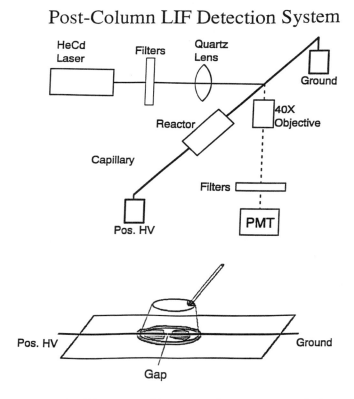

Figure 9.9. Drawing of a CE system with a post-column reactor and a laser fluorescence detector.

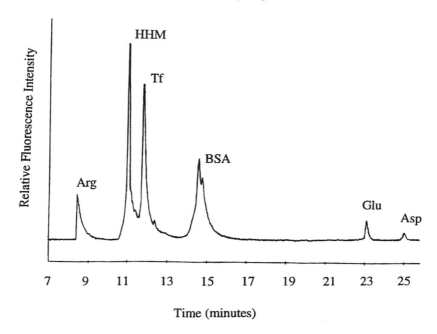

Figure 9.10. Separation of three proteins and three amino acids using post-column derivatization and LIF detection. Peak identities and injected amounts are Arg: DL-arginine hydrochloride (5×10^{-5} M, 9.8 fmol); HHM: horse heart myoglobin (0.5 mg/mL, 4.1 fmol); Tf: iron-free human transferrin (0.25 mg/mL, 450 amol); BSA: bovine serum albumin (0.5 mg/mL, 880 amol); Glu: (L+)-glutamic acid (5×10^{-5} M, 3.5 fmol); Asp: DL-aspartic acid, (5×10^{-5} M, 3.2 fmol). Conditions: capillary, 10-μm i.d., 100-cm length, 80 cm to gap and 82.5 cm to detection window; 4-μm gap; injection, 5 s, 10 kV (182) pL; separation potential, 30 kV (300 V/cm); buffer, 100 mM borate, pH 9.5; current 1.0 μA.

In summary, many advances have been made in developing nanoscale tools to understand what occurs in biological microenvironments. Advances still need to be made in detection sensitivity, selectivity, and response time. In addition, electrodes with considerably smaller tip diameters (approximately 50 nm) will be required. These constitute the major challenges in nanofabrication for analytical chemistry in biological microenvironments.

9.7 Acknowledgments

We would like to thank our co-workers whose work is cited in this manuscript. This work was supported in part by grants from the Office of Naval Research, the National Science Foundation, and the National Institutes of Health. A. G .E. is a Camille and Henry Dreyfus Teacher-Scholar. We are particularly grateful for the assistance of Dr. Beth Hall and Leanne Keim and the use of the National Nanofabrication Facility at Cornell University.

References

Adams, R. N. (1976) Probing brain chemistry with electroanalytical techniques. *Anal. Chem.* 48:1128A–1138A.

Albin, M., Weinberger, R., Sapp, E., and Moring, S. (1991) Fluorescence detection in capillary electrophoresis: evaluation of derivatizing reagents and techniques. *Anal. Chem.* 63:417–422.

Amankwa, L. N., Albin, M., and Kuhr, W. G. (1992) Fluorescence detection in capillary electrophoresis. *Trends in Anal. Chem.* 11:114–120.

Bergquist, J., Tarkowski, A., Ekman, R., and Ewing, A. (1994) Discovery of endogenous catecholamines in lymphocytes and evidence for catecholamine regulation of lymphocyte function via an autocrine loop. *Proc. Natl. Acad. Sci. U.S.* (submitted).

Betzig, E., Trautman, J. K., Harris, T. D., Weiner, J. S., and Kostelak, R. L. (1991) Breaking the diffraction barrier: optical microscopy on a nanometric scale. *Science* 251:1468–1470.

Chien, J. B., Wallingford, R. A., and Ewing, A. G. (1990) Estimation of free dopamine cell of planorbis corenus voltammetry and capillary electrophoresis. *J. Neurochem.* 54:633–638.

Clift-O'Grady, L., Linstedt, A. D., Lowe, A. W., Grote, E., and Kelly, R. B. (1990) Biogenesis of synaptic vesicle-like structures in a pheochromocytoma cell line. *J. Cell. Biol.* 110:1693–1703.

Cooper, B. R., Jankowski, J. A., Leszczyszyn, D. J. Wightman, R. M., and Jorgenson, J. W. (1992) Quantitative determination of catecholamines in individual bovine adrenomedullary cells by reversed-phase microcolumn liquid chromatography with electrochemical detection. *Anal. Chem.* 64:691–694.

Curry, P. D., Engstron-Silverman, C. E., and Ewing, A. G. (1991) Electrochemical Detection for capillary electrophoresis. *Electroanalysis* 3:587–596.

Ewing, A. G. (1993) Microcolumn separations of single cell components. *J. Neurosci. Meth.* 48:215–224.

Ewing, A. G., Strein, T. G., and Lau, Y. Y. (1992) Analytical chemistry in microenvironments: single nerve cells. *Accts. Chem. Res.* 25:440–447.

Ewing, A. G., Wallingford, R. A., and Olefirowicz, T. M. (1989) Capillary electrophoresis. *Anal. Chem.* 61:292A–303A.

Gaines, G. L. (1981) On the water wettability of gold. *J. Colloid and Interface Sci.* 79:295.

Giacobini, E. (1987) Neurochemical analysis of single neurons. *J. Neurosci. Res.* 18:632–637.

Gilman, S. D., Pietron, J. J., and Ewing, A. G. (1994) Post-column derivatization in narrow-bore capillaries for the analysis of peptides and proteins by capillary electrophoresis with fluorescence detection. *J. Microcol. Sepn.* (in press).

Green, L. A. and Tischler, A. S. (1976) Establishment of a noradrenergic clonal line of rat adrenal pheochromocytoma cells which respond to nerve growth factor. *Proc. Natl. Acad. Sci. U.S.A.* 73:2424–2428.

Hernandez, L., Escalona, J., Joshi, N., and Guzman, N. A. (1991) Laser-induced fluorescence and fluorescence microscopy for capillary electrophoresis zone detection. *J. Chromatogr.* 559:183–196.

Hogan, B. L. and Yeung, E. S. (1992) Determination of intracellular species at the level of a single erythrocyte via capillary electrophoresis. *Anal. Chem.* 64:2841–°2845.

Hogan, B. L. and Yeung, E. S. (1993) Single-cell analysis at the level of a single human erythrocyte. *Trends in Anal. Chem.* 12:4–9.

Huang, X., Zare, R. N., Sloss, S., and Ewing, A. G. (1991) End-column detection for capillary zone electrophoresis. *Anal. Chem.* 63:189–192.

Jansson, M., Emmer, A., Roeraade, J., Lindberg, U., and Hok, B. (1992) Microvials on a silicon wafer for sample introduction in capillary electrophoresis. *J. Chromatogr.* 626:310–314.

Kennedy, R. T. and Jorgenson, J. W. (1989) Quantitative analysis of individual neurons by open tubular liquid chromatography with voltammetric detection. *Anal. Chem.* 661:436–441.

Kennedy, R. T. Oates, M. D., Cooper, B. R., Nickerson, B., and Jorgenson, J. W. (1989) Microcolumn separations and the analysis of single cells. *Science.* 246:57–63.

Kennedy, R. T., St.Claire III, R. L., White, J. G., and Jorgenson, J. W. (1987) Chemical analysis of single neurons by open tubular liquid chromatography. *Mikrochimica Acta* II:37–45.

Kim, Y.T., Scarnulis, D. M., and Ewing, A. G. (1986) Carbon ring electrodes with 1 micrometer tip diameter. *Anal. Chem.* 58:1782–1786.

Kristensen, H.K., Lau, Y. Y., and Ewing, A. G. (1994) Capillary electrophoresis of single cells: observation of two compartments of neurotransmitter vesicles. *J. Neurosci. Meth.* 51:183–188.

Kuhr, W. G., Licklider, L., and Amankwa, L. (1993) Imaging of electrophoretic flow across a capillary junction. *Anal. Chem.* 65:277–282.

Lee, T. T. and Yeung, E. S. (1992a) High-sensitivity laser-induced fluorescence detection of native proteins in capillary electrophoresis. *J. Chromatogr.* 595:3199–325.

Lee, T. T. and Yeung, E. S. (1992b) Quantitative determination of native proteins in individual human erythrocytes by capillary zone electrophoresis with laser-induced fluorescence detection. *Anal. Chem.* 64:3045–3051.

Mantus, D. S., Valaskovic, G. A., and Morrison, G. H. (1991) High mass resolution secondary ion mass spectrometry via simultaneous detection with a charge coupled device. *Anal. Chem.* 63:788–792.

Matioli, G. T. and Niewisch, H. B. (1965) Electrophoresis of hemoglobin in single erythrocytes. *Science* 150:1824–1826.

Meulemans, A., Poulain, B., Baux, G., Tauc, L., and Henzel, D. (1986) Microcarbon electrode for intracellular voltammetry. *Anal. Chem.* 58:2088–2091.

Nickerson, B. and Jorgenson, J. W. (1989) Characterization of a post-column reaction-laser-induced fluorescence detector for capillary zone electrophoresis. *J. Chromatogr.* 480:157–168.

Nicholson, C. and Rice, M. E. (1988) Use of ion-selective microelectrodes and voltammetric microsensors to study brain cell microenvironment. In *Neuromethods: Neuronal Microenvironment*, eds. Boulton, A. A., Baker, G. B., Walz, W., pp. 247–361. New Jersey: Humana Press.

Oates, M. D., Cooper, B. R., and Jorgenson, J. W. (1990) Quantitative amino acid analysis of individual snail neurons by open tubular liquid chromatography. *Anal. Chem.* 62:1573–1577.

Olefirowicz, T. M. and Ewing, A. G. (1990a) Capillary electrophoresis in 2 and 5 diameter capillaries: application to cytoplasmic analysis. *Anal. Chem.* 62:1872–1876.

Olefirowicz, T. M. and Ewing, A. G. (1990b) Dopamine, concentration in the cytoplasmic compartment of single neurons determined by capillary electrophoresis. *J. Neurosci. Meth.* 34:11–15.

Pendley, B. D, and Abruna, H. D. (1990) Construction of submicrometer voltammetric electrodes. *Anal. Chem.* 62:782–784.

Penner, R. M., Heben, M. J., and Lewis, N. S. (1989) Preparation and electrochemical characterization of conical and hemispherical ultramicroelectrodes. *Anal. Chem.* 61:1630–1636.

Penner, R. M., Heben, M. J., Longin, T. L., and Lewis, N. S. (1990) Fabrication and use of nanometer sized electrodes in electrochemistry. *Science* 250:1118–1121.

Rice, C. L. and Whitehead, R. (1965) Electrokinetic flow in a narrow cylindrical capillary. *J. Phys. Chem.* 69:4017–4024.

Saraceno, R. A. and Ewing, A. G. (1988) Electron transfer reactions of catecholes at ultrasmall carbon ring electrodes. *Anal. Chem.* 60:2016–2020.

Sloss, S. E. and Ewing, A. G. (1993) Improved method for end-column detection for capillary electrophoresis. *Anal. Chem.* 65:577–581.

Sloss, S. and Ewing , A. G. (1993) Capillary electrophoresis for the analysis of single cells. In *Handbook of Capillary Electrophoresis: Principles, Methods, and Practice*, ed., J. Landers, pp. 391–417. CRC Press

Sossin, W. S. and Scheller, R. H. (1989) A bag cell neuron-specific antigen localizes to a subset of dense core vesicles in Aplysia californica. *Brain Res.* 494:205–214.

Strein, T. G. and Ewing, A. G. (1992) Characterization of submicron-sized carbon electrodes insulated with a phenol-allylphenol copolymer. *Anal. Chem.* 64:1368-1373.

Tank, D. W., Sugimori, M., Conner, J. A., and Llinas, R. R. (1988) Spatially resolved calcium dynamics of mammalian purkinje cells in cerebellar slice. *Science* 242:773–776.

Wagner, J. A. (1985) Structure of catecholamine secretory vesciles from PC12 cells. *J. Neurochem.* 45:1244–1253.

Wightman, R. M. (1981) Microvoltammetric electrodes. *Anal. Chem.* 53:1125A-1134A.

Wightman, R. M., Jankowski, J. A., Kennedy, R. T., Kawagoe, K. T., Schroeder, T. J., Leszczyszyn, D. J., Near, J. A., Diliberto, Jr., E. J., and Viveros, O. H. (1991) Temporally resolved catecholamine spiles correspond to single vesicles release from individual chromaffin cells. *Proc. Natl. Acad. Sci. U.S.A.* 88:10754–10758.

Wightman, R. M., May, L. J., and Michael, A. C. (1988) Detection of dopamine dynamics in the brain. *Anal. Chem.* 60:769A–779A.

Yeung, E. S., Wang, P., Li, W., and Giese, R. W. (1992) Laser fluorescence detector for capillary electrophoresis. *J. Chromatogr.* 608:73–77.

10

Scanning force microscopy of biological macromolecules: present and future

CARLOS BUSTAMANTE, DOROTHY A. ERIE,
and GUOLIANG YANG

Institute of Molecular Biology and
Howard Hughes Medical Institute
University of Oregon
Eugene, Oregon
97403

10.1 Introduction

In recent years, a new generation of powerful microscopes, based on a series of physical principles, has been invented. These *scanning probe microscopes* operate by scanning a tip or probe over the sample to detect a property, such as height, conductivity, or paramagnetism, at every location on the surface. The image is constructed by mapping this property at every point on the surface. The first of these instruments, the scanning tunneling microscope (STM) (Binnig et al., 1982), generated a great deal of interest in the possibility of high-resolution imaging of biological macromolecules in environments more closely resembling the milieu of the cell. Many publications have appeared on the early attempts by several groups to use the STM to image nucleic acids, proteins, and so forth (Engel, 1991). In the last two years, however, it has become increasingly clear that biological macromolecules do not possess sui~ficient conductivity to permit reliable and reproducible imaging under the STM (Dunlap et al., 1993). Instead, an instrument related to the STM, the scanning force microscope (SFM) – (also known as the atomic force microscope (AFM) – has demonstrated great potential for imaging biological systems.

Biological applications of the SFM rely on several inherent properties of the instrument. (1) It has a resolution equivalent to that of the electron microscope, and this resolution is likely to improve in the near future. (2) It yields topographic images directly. (3) It can operate in vacuum, in air, and in aqueous buffers (Drake et al., 1989), thus raising the possibility of imaging structures in physiological conditions and of following dynamic molecular processes in almost real time. These capabilities should be particularly valuable in structural and functional studies of complex macromolecular assemblies such as protein–DNA complexes, eukaryotic chromosomes, ribosomes, and integral membrane protein assemblies.

In the past three years, several groups have demonstrated the general applicability of the SFM to a wide range of biological systems, from single molecules to whole, living cells (Hoh and Hansma, 1992). The studies carried out during this period have revealed many of the general issues that are important in the applications of the SFM to biological systems. In this chapter, we present a review of some of the demonstrated capabilities of the SFM and discuss likely future developments as well as some of the present limitations in this technique.

10.2 The scanning force microscope

A block diagram of a typical SFM is shown in Figure 10.1. The heart of any SFM is a sharp probe tip mounted on the end of a flexible cantilever. As the tip approaches the sample, forces begin to act that either attract the tip toward the sample and cause a downward deflection of the cantilever or, at short distances, repel the tip and cause an upward deflection of the cantilever. Although the dimensions of the apex of the tip are usually very small, ~100 Å or less, the forces acting on it deflect the entire, macroscopic cantilever (typically about 0.1 mm in length). The cantilever, therefore, acts as a high-gain tip–sample force transducer, converting small, highly localized forces into the mechanical motion of a large object. As the sample scans under the tip, the cantilever is deflected in correspondence with the topography of the surface. The cantilever and tip together, therefore, form an instrument for probing samples on very small scales.

Very small cantilever deflections (a few Å) can be detected by a variety of methods (Sarid, 1991). The most common approach, called an optical lever, is sim-

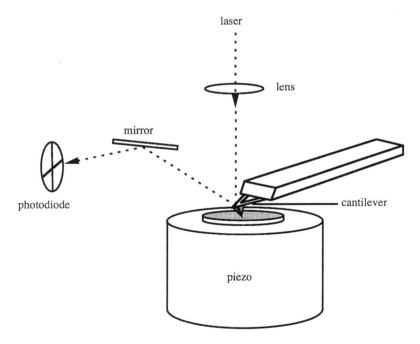

Figure 10.1. Schematic diagram of a contact scanning force rnicroscope (SFM). The sample is scanned relative to the tip by a piezoelectric positioner.

ply to reflect a laser beam off the back side of the cantilever into a two-segment photodetector (Sarid, 1991). The difference in output between the two detectors is proportional to the deflection amplitude. The optical lever is essentially a motion amplifier. For a cantilever of length L and a distance of L between the cantilever and the photodetector, the deflection of the laser spot at the photodetector, Δz, is proportional to the deflection of the cantilever, Δx, according to Sarid (1991):

$$\Delta z = (2kL/L)\,\Delta x, \tag{10.1}$$

where k is a constant that depends on the design of the cantilever and has a value typically between 312 and 5. The ratio L/L is usually ~100, so the overall gain factor is about 500 to 1,000. Accordingly, a 1-Å cantilever deflection results in a laser spot deflection of 500 to 1,000 Å, which is easily detected by the differential photodiode. In fact, the limiting factor in probe microscopy is the spontaneous vibration of the cantilever caused by thermal energy (that is, the *thermal noise*) and not the ability to monitor small cantilever deflections (motion detectors much more sensitive than the optical lever are possible).

In all types of probe microscopes the motion of the sample (and sometimes the tip) is controlled by piezoelectric ceramic elements. Piezoelectric materials expand or contract by a small percentage of their total length when subjected to an electric field. Piezoelectrics in probe microscopes typically have a sensitivity of about 30 Å/V, which allows movement of the sample in increments of 0.1 Å or less. This capability is crucial to the operation of probe microscopes at the nanometer scale.

SFMs are generally classified as contact, noncontact, and tapping microscopes. The contact microscopes, which require the tip to touch the sample during scanning, are the most commonly employed. These microscopes can operate in one of the three imaging modes: constant height mode, constant force mode, and error signal mode (Figure 10.2). In most SFM applications, especially in biological ones, it is important to keep the force between the tip and the sample at a minimum. Presently, the minimum attainable force is a few nano-Newtons when imaging in air, and approximately a tenth of a nano-Newton when imaging in liquids. This

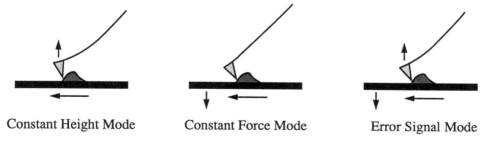

| Constant Height Mode | Constant Force Mode | Error Signal Mode |

Figure 10.2. Schematic diagram showiog the three modes of operation of a contact scanning force microscope.

force is one of the main causes of sample distortion and damage. For example, if the tip–sample force is 10^{-8} Newton and the area of contact is 100 $Å^2$, the pressure beneath the tip is 10^{10} Pascals, or about 10^5 atm. Efforts are under way to reduce this force, since it is one of the main causes of lost resolution.

The most common mode of operation is the constant force mode, in which the force of contact between the tip and the sample is held constant by an electronic feedback loop. That is, whenever the deflection of the cantilever is upward, indicating that the tip is beginning to encounter an obstacle on the surface, the sample retracts so as to cancel the deflection and keep the force constant. Similarly, when the deflection is downward, the sample is raised. The tip–sample force (and the pressure beneath the tip), therefore, can be approximately controlled and is usually set to the minimum value possible without loss of contact with the surface. In constant force mode, the topography of the surface is obtained by displaying the advance and withdrawal of the sample required to keep the force (or the cantilever deflection) constant. If the sample is so hard that there is no sample deformation, this image is a topographical map of the sample surface.

In constant height mode, the sample is held at an average constant height, and the cantilever is allowed to deflect upward and downward as it passes over sample features. In practice, this mode is achieved by turning down the feedback gain so that the sample height changes very slowly when the cantilever deflects in response to a surface feature. In this case, the feedback system responds only to changes in the surface topography that vary slowly, such as changes caused by thermal expansions and contractions in the SFM or overall sample tilt. Instead of recording the upward and downward motion of the sample, as in the constant force mode, the SFM records tbe cantilever deflection directly. Because the force exerted by the cantilever on the sample is proportional to the cantilever deflection, the SFM image in the constant height mode is a map of the tip-sample force at each position.

Finally, in the error signal mode, the feedback is set for fast response, to follow the sample closely as in the constant force mode, but cantilever deflection is recorded instead of sample height. This method produces an image in which all broad features are suppressed and sharp features are enhanced. The resulting images are equivalent to applying a high-pass filter to the data. Unlike the high pass filter, which is an after-the-fact image-processing operation, the error signal mode works in real time. For certain types of samples, this method allows otherwise invisible features to be seen. For example, the features of interest on cell surfaces are often shallow details (that is, lines caused by submembrane cytoskeletal structure) on a large, steep, rolling surface. Such shallow features often are not visible in real time in either the constant force or the constant height modes but are visible in real time in the error signal mode.

During the scanning operation with contact microscopes, the sarnple experiences a tangential force in addition to the force normal to the sample. This shearing force can deform the sample and remove it from its attachment to the substrate. The magnitude of this force is again a function of the tip dimensions and depends on the adhesive interactions between the tip and the sample. Generally, resolution is reduced by the elastic deformation of the sample and the lateral motion induced on the sample by these shearing forces. One possible solution to these problems is offered by noncontact microscopes (Ohnesorge and Binnig, 1993). These microscopes require the tip or probe to hover over the sample at a few angstroms from the surface as the tip and sample are scanned relative to each other. Better sensitivity is obtained if the cantilever is oscillated above the sample with a small amplitude (~ 1 nm) near its resonance frequency (typically a few hundred kHz). The attractive forces between the tip and the sample manifest themselves as a shift in the resonance frequency of the cantilever. A topographic map of the surface can be obtained with a feedback circuit tbat raises or lowers the sample relative to the tip so as to keep the set-point frequency shift constant. Noncontact microscopes are difficult to operate because of the dominant, long-range attractive forces that tend to draw the cantilever toward the surface. These attractive forces are due mainly to the water meniscus developed between the tip and the sample during imaging in air at finite humidities (Bustamante et al., 1992; Thundat et al., 1992) and are also due to attractive electrostatic and van der Waals forces (Butt, 1991a, 1991b; Hartmann, 1991) during imaging in liquids. As the sample approaches the tip, the presence of this large, attractive regime in the potential energy of the cantilever leads to mechanical instability and the cantilever snaps onto the surface. This is one of the main difficulties that prevent stable, reproducible noncontact imaging.

10.3 Recent developments

During the last three years, the applications of scanning force microscopy in biology have benefited from progress in five areas: (1) development of reliable deposition methods, (2) development of consistently sharper tips, (3) implementation of new modes of operation that make it possible to greatly reduce tangential and shear forces on the sample, thus minimizing sample deformation and dragging by the scanning tip, (4) demonstration of biomolecular imaging in aqueous buffers and of the potential to observe molecular processes in buffer, and (5) implementation of methods of image reconstruction and analysis.

10.3.1 Methods of deposition

Because the SFM is a surface technique, the samples must be placed on relatively flat, smooth surfaces. Mica (Bailey, 1984) has proven to be both a reliable and an inexpensive substrate fulfilling this requirement. Mica can easily be peeled to obtain an atomically flat surface with a well-defined chemical composition. Furthermore, it is a relatively inert, negatively charged, hydrophilic substrate; thus it favors the attachment of biological macromolecules. Most of the methods of deposition developed on mica rely on electrostatic binding of the molecules to the surface. Since mica is negatively charged, proteins bind to it more efficiently than do the highly negatively charged nucleic acids. Reliable deposition for nucleic acid can be accomplished by pretreating the mica surface with Mg^{++} (Bustamante et al., 1992; Vesenka et al., 1992) or by simply including Mg^{++} in the deposition buffer. Although mica is relatively inert, Lyubchenko et al. (1992, 1993) have been able to modify it with 3-aminopropyltriethoxy silane (APTES), which permits covalent attachment of nucleic acids to the mica surface. Another method, demonstrated recently by Schaper et al. (1993), involves the adaptation of a transmission electron microscopy (TEM) technique to spread nucleic acids by means of a cationic detergent, the quaternary ammonium salt benzyldimethylalkylammonium chloride (BAC). These latter methods relieve the constraint on the use of Mg^{++} in those cases where the presence of this cation is undesirable.

Another substrate commonly used in biological applications is glass. Although it is not as smooth or consistently flat as freshly cleaved mica, ultraclean glass coverslips used in optical microscopy often display a few square-micron areas with root mean-square roughness of only a few angstroms. The advantage of glass is its chemical reactivity and the many chemical modification reactions known for it (Bhatia et al., 1989). Specific modification of glass surfaces for the attachment of macromolecules, therefore, may become useful in the near future.

10.3.2 Development of sharper tips; applications to imaging protein-DNA complexes

In contact SFM, the spatial resolution depends on the dimensions of the tip. Hard, atomically flat surfaces can be imaged with atomic resolution by relatively blunt tips (Grutter et al., 1992), because the interaction is mediated only by the most apical atoms on the tip and therefore decays very rapidly. By contrast, if the surface has topographic details or is soft and can deform (as in the case of biological samples), the tip–surface interactions occur over a larger area. Both the spatial resolution and the strength of these interactions depend on the size of the tip. There is no generally agreed-upon definition of resolution in the SFM, but intuitively the smallest discernible features in an SFM image cannot be much smaller than the dimensions of the tip. Therefore, the radius of curvature of the tip, R_c, largely determines the spatial resolution in the SFM. In addition, the taper of the tip affects how well the SFM renders steep features, such as the edges of molecules.

Finite tip sizes often produce a distortion of the lateral dimensions of the molecules (Bustamante et al., 1992; Keller and Chou, 1992; Vesenka et al., 1992). This distortion is caused by *tip-induced deformation* and by the geometric exclusion between the tip and the molecule, the *envelope effect* (Figure 10.3). A blunt

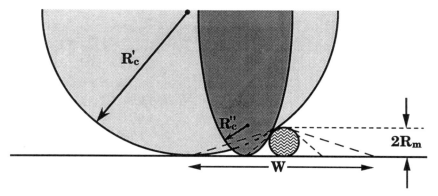

Figure 10.3. Schematic diagram illustrating the tip-broadening effect. The lateral dimen
sions of a molecule are overestimated in the SPM image due to the finite size of the tip.
A molecule of radius R_m is imaged to an apparent width W by a tip with a radius of curvature R'_c. A sharper tip reduces this effect.

tip with a radius of curvature R'_c can distort the lateral dimensions of a molecule
having a circular cross section of radius R_m. A simple geometric argument
(Bustamante et al., 1992; Vesenka et al., 1992) shows that, for a parabolic tip, the
molecule will be imaged with an apparent width, W, given by:

$$W = \frac{4(R_c + R_m)\, R_m(Rc - R_m)}{R_c}, \; R_c > R_m \tag{10.2}$$

Thus, with a blunt tip having $R_c = 50$ nm (a typical value for commercially available tips), a molecule that has $R_m = 1$ nm will be imaged with a width of 30 nm!
Figure 10.3 also shows the improvement obtained by using a sharper tip of radius
of curvature R''_c. The envelope effect is reduced but not eliminated.

Using an earlier observation, Akama et al. (1990) showed that the electron beam
from a scanning electron microscope (SEM) could be used to deposit a small
amount of carbonaceous material on a surface. By varying the electron flux, the
electron energy, and the time of deposition, it is possible to grow structures with
radii of curvature between 80 and 120 Å onto regular pyramidal tips (Figure 10.4).
Recently, electron-beam tips and tips made by other processes with equivalent
dimensions have become commercially available. One of the main advantages of
these carbon tips appears to be their hydrophobic composition. In general,
hydrophilic tips should have strong adhesive interactions with most biological samples. In the contact mode, these adhesive interactions can easily destroy the sample.
Moreover, the tip may be easily fouled by loose molecules. On the other hand,
hydrophobic tips should develop smaller adhesion forces and be less susceptible to
fouling (Hansma et al., 1993; Lyubchenko et al., 1993; and Yang and Shao, 1993).

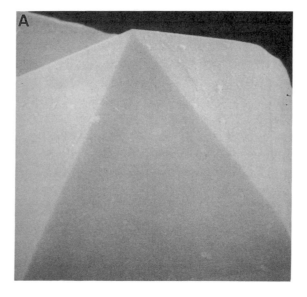

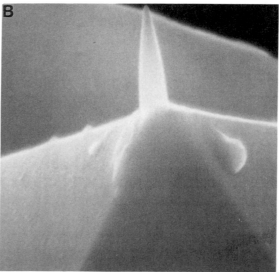

Figure 10.4. SEM photos of a commercially available silicon nitride tip (A) before and (B) after a microtip was deposited by focused electron beam on its apex.

A variety of aps of different shapes, compositions, and dimensions are now available commercially. Table 10.1 lists some of the most commonly used tips, their geometry, composition, characteristics, and manufacturers.

Cro–DNA Complexes. Cro, a transcription regulatory protein from bacteriophage λ, binds as a dimer (MW = 14.7 kDa) to three operator sites, O_R1, O_R2, and O_R3, in

Table 10.1. *Characteristics of some of the most commonly used tips for SFM and their manufacturers.*

Chemical Composition	Tip Geometry	Radius of Curvature at Tip Apex	Manufacturers
Silicon Nitride	Pyramidal	~ 30 mn	Digital Instruments;[1] TopoMetrix[2]
Silicon Nitride[a]	Pyramidal	< 20 nm	Digital Instruments; Park Scientific[3]
Silicon[b]	Conical	< 5 mn	Park Scientific
Silicon	Conical	< 10 nm	IBM German Manufacturing Center[4]
Amorphous Carbon[c]	Paraboloidal	~ 10 nm	TopoMetrix; Materials Analytical Services[5]

[a] Sharpened by oxidation
[b] For noncontact SFM
[c] Electron-beam deposited tip on the apex of pyramidal tip

[1] 520 E. Montecito St., Santa Barbara, CA 93103
[2] 5403 Betsy Ross Dr., Santa Clara, CA 95054
[3] 1171 Borregas Ave., Sunnyvale, CA 94089
[4] P.O. Box 266, 7032 Sindelfingen, Gennany
[5] 616 Hutton St., Raleigh, NC 27606

the O_R region to repress transcription from the divergent promoters λP_R and λP_R (Ptashne, 1992). Cro binds to all three O_R binding sites, whose centers are separated by 23 bp, with very high affinities (10^{-10} to 10^{-12} M^{-1}) and bends DNA upon binding each of these sites (Brennan et al., 1990; Kim et al., 1989; Lyubchenko, 1991). Since Cro is very small and has closely spaced binding sites, Cro-DNA complexes provide an ideal system for demonstrating the high resolu tion of the SFM.

DNA deposited in the absence of Cro is essentially free of sharp bends and high features. In contrast, DNA deposited in the presence of Cro consistently reveals sharp bends with a discrete high feature at the locus of each bend. In addi tion, at stoichiometric Cro concentrations, the locations of the high features on the DNA are consistent with the position of the OR binding sites (two-fifths from one end) and are identified as Cro-DNA complexes. The image of a Cro–DNA com plex shown in Figure 10.5(a) reveals the expected Cro-induced DNA bending.

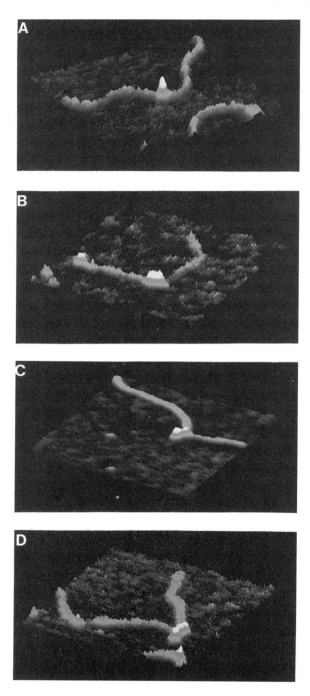

Figure 10.5. Surface plots of Cro–DNA complexes in which one, two, and three Cro molecules are bound specifically. The plane of the mica is inclined 30° to display the topography of the surface. The scan sizes are 250 nm. Cro molecules bound to (A) a single operator site, (B) adjacent sites, (C) O_R3 and O_R1, and (D) O_R3, O_R2, and O_R1. These conformations are consistently observed for the multiply occupied species.

Two Cro dimers specifically bound in adjacent sites (FIgure 10.5B) are partially resolved from one another and appear as an elongated feature of a size consistent with two Cro dimers. Cro dimers at adjacent sites are not expected to be completely resolved because each dimer should cover approximately 20 base pairs. Two Cro dimers occupying sites O_R1 and O_R3 (Figure 10.5C) are well resolved, indicating a spatial resolution of ~50 Å. The conformational effect of adjacently bound Cro dimers appears to be simply the sum of the bend angles for the individual Cro dimers (compare Figures 10.5A and 10.5B). By contrast, Cro dimers bound at sites O_R1 and O_R3 simultaneously form more complicated structures on the DNA (Figure 10.5C) and cannot be explained by a simple addition of bend angles. Interestingly, Cro bound in all three operator sites (Figure 10.5D) results in an overall conformation that appears to be a composite of those observed for the two classes of doubly occupied sites (Figures 10.5B and 10.5C). These results demonstrate the potential of the SFM for imaging a range of biological molecules that have not been easily accessible to direct visualization. In addition, the high spatial resolution indicates that it is possible to resolve individual proteins in a multiprotein complex.

10.3.3 New modes of operation: tapping in air and its applications

An alternative method that possesses most of the advantages of the noncontact mode but none of its disadvantages recently has been developed by Digital Instruments (1993) and is called the *tapping mode,* In this operation, a stiff cantilever is oscillated with a high amplitude (typically 10–20 nm) near its resonance frequency as the sample is scanned laterally. The amplitude of the oscillation is clipped and reduced by the tip's contact with the sample. A feedback loop is used to keep this amplitude reduction constant during scanning and tracking of the topography of the sample. The image is generated by mapping, at every point on the surface, the advance and withdrawal of the sample necessary to keep the amplitude dampening constant. This mode of operation has many advantages. First, contact forces are very light and there are no shear forces. Second, this operation is comparable to the noncontact mode in terms of sample damage, but its operation is more stable. Third, the tapping mode eliminates the detnmental effect of the capillary forces when imaging in air because the large energies associated with the higher amplitude oscillation are enough to overcome these attractive forces. Perhaps the most important advantage of the tapping mode is its high spatial resolution and its minimal sample deformation. Resolution in the tapping mode is limited *only* by the tip dimensions.

Chromatin Fibers. Tapping mode SFM was used to obtain high-resolution images of chicken erythrocyte chromatin fibers deposited in low salt concentrations. Images of fixed and unfixed chromatin deposited in very low salt were obtained both on mica and on glass. Fixed and unfixed chromatin fibers exhibit very simi-

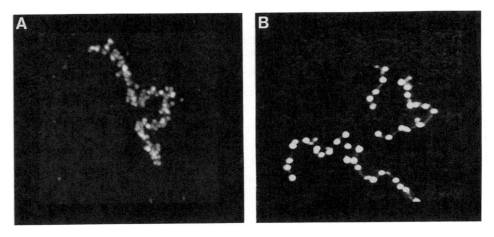

Figure 10.6. Tapping mode SFM images of (A) fixed and (B) H1- and H5- depleted chromatin fibers deposited in 5 mM TEA-HCl, pH7.0, on mica and imaged in air. Scan size is 400 nm x 400 nm.

lar structures when deposited on mica and glass. The image of a gluteraldehyde-fixed chromatin fiber deposited on mica is shown in Figure 10.6A. Images like this reveal higher-order structures than the "zigzag" and "beads-on-a-string" conformations that have been frequently observed using electron microscopy (Olins and Olins, 1974; Thoma et al.,1979; Thoma and Koller, 1977,1981; Woodcock, 1973). In the SFM images, these lower-order conformations are seen only with H1- and H5-depleted chromatin fibers (Figure 10.6B). These results indicate that H1 and H5 are necessary for maintenance of higher-order chromatin structures at low salt concentrations. In addition, these results support the proposal that H1 and H5 fix the angles at which the DNA enters and exits the nucleosome (van Holde, 1988).

Woodcock et al. (1993) have proposed that fixed entry and exit angles of the DNA would produce irregular helices if the linker DNA were stiff and had a variable length. At low salt concentrations, the linker DNA is expected to be stiff; in higher salt concentrations, the linker DNA should become more flexible. If chromatin coils into a helical structure, these salt-dependent properties would produce an extended irregular helix at low salt, and the helix would become more compact as the salt concentration is increased. The results shown here are consistent with such a model of chromatin folding.

10.3.4 Imaging in liquids

Since the 1950s, direct visualization of biological structures has involved a compromise between imaging environment and spatial resolution. Optical microscopy offers the possibility of imaging structures in physiologically relevant environments but at relatively low spatial resolution. Conversely, the electron microscope offers the advantages of a high resolution, but in artificial, dehydrated conditions. In recent years, this last limitation has been partially relieved by the development of cryo-electron microscopy, in which the structures are trapped in their physiologically relevant conformations by quick freezing (Adrian et al., 1984). Although cryo-electron microscopy can image structures at high resolution, all images are static because imaging cannot be done in aqueous buffers. One of the most attractive features of the SFM is its capability to operate not only in air but also in liquids. This capability is particularly useful in biology, where it raises the possibility of imaging biological molecules in ionic buffer solutions, that is, in conditions more closely resembling the intracellular environment. In particular, it may be possible to follow biologically relevant processes as they occur but at resolutions comparable to that of electron microscopy.

During the past few years, several groups have demonstrated the capability of the contact SFM to image DNA molecules in propanol (Hansma et al., 1992), in water (Lyubchenko et al., 1993), and in aqueous buffers (Bezanukka et al., 1994). Similarly, both membrane and nonmembrane proteins have been imaged in aqueous environments (Yang et al., 1993). These studies have already provided a wealth of information regarding imaging in liquid environments. Imaging under liquids eliminates the capillary force associated with the water meniscus formed between the tip and the surface when imaging in air (Bustamante et al., 1993). When samples are imaged in air, a water layer of varying thickness is present on the sample. This layer forms a meniscus of liquid between the tip and the sample and generates an *attractive capillary force* that pins the tip to the surface. This force can dominate the tip–sample interaction and lead to deformation and/or detachment of the macromolecules from the surface. Bustamante et al. (1992), Thundat et al. (1992), and Vesenka et al. (1992) have conducted independent studies on the effect of humidity on the quality of images obtained in air. The capillary force depends on the radius of curvature and the taper of the tip. Imaging in liquids usually involves a substantial decrease in the forces acting on the sample. Typical operational forces in air are in the range of 1–10 nN, whereas imaging in liquids is typically done in the range of 0.1–1.0 nN. In addition, imaging of proteins appears to be more reliable than imaging of nucleic acids. The reason is that proteins generally bind more strongly to mica and glass than do nucleic acids. Stably bound proteins are not so easily removed by the tip during scanning, thus providing the mechanical stability required to image under liquids. On the contrary, the highly negatively charged nucleic acids are more difficult to attach to these substrates and can be easily removed by the tip.

Assembly of protein–nucleic complexes. One of the most exciting advances that has been made in the SFM is the ability to follow molecular processes in real time. Because it is possible to image in aqueous buffers, the assembly of multiprotein or protein–DNA complexes can be monitored if the molecules are sufficiently well attached to the surface. The first of such processes to be observed was the binding of *E. coli* RNA polymerase to DNA.

To image the assembly of RNA polymerase-DNA complexes in physiological buffers, the DNA molecules must first be bound stably to the mica surface. A 1.3-kbp DNA fragment containing the λP_R promoter at 1/5 from one end is deposited on freshly cleaved mica, then thoroughly rinsed and dried. The dried DNA sample is placed in the liquid cell chamber of tbe SFM and submerged in doubly distilled water. After stable images of the DNA in water are obtained, the liquid cell is filled with a HEPES/MgCl$_2$ buffer. As can be seen in Figure 10.7, clear and highly resolved images of DNA fragments can be also obtained in ionic solutions.

Once stable imaging of the DNA fragment is obtained in buffer, 300μL of 1 nM RNA polymerase holoenzyme in HEPES/MgCl$_2$ buffer is injected manually into the liquid cell. Two consecutive 1,500 nm x 1,500 nm frames obtained within 2 min after polymerase injection are shown in Figures 10.8A and 10.8B. In the earliest image, (Figure 10.8A) streaking is observed across the field along the direction of scanning. This streaking is presumably a result of the polymerase molecules' being dragged over the surface by the scanning tip. In the next frame, obtained approximately 1 min later (Figure 10.8B), high features about 200 Å in diameter, identified as RNA polymerase molecules (Darst et al., 1989), can be seen attached loosely to the mica surface or being dragged by the tip. However, most polymerase molecules are seen interacting with DNA fragments. Significantly, several DNA molecules can be recognized both before (Figure 10.8A) and after (Figure 10.8B).

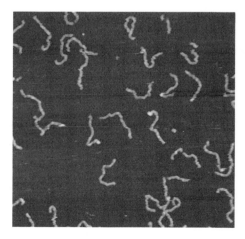

Figure 10.7. SFM image of 1,258-bp DNA fragments in 10 mM HEPES/1 mM MgCl$_2$, pH = 7.4 buffer. The average length of the fragments is 419 ± 11 nm in water and 416 ± 11 nm in buffer. The image size is 2,000 nm x 2,000 nm and the height color scale ranges from 0 nm (dark brown) to 10 nm (white). Both images are flattened and the scan rate is 7.6 scan lines/sec.

polymerase binding has occurred. Close-up views of some of the DNA fragments before and after complex assembly (Figure 10.9) reveal that the DNA becomes distorted upon binding the polymerase molecule. These results demonstrate the great potential of SFM for following molecular processes in real time.

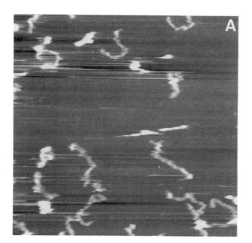 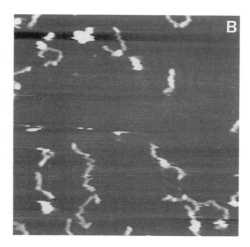

Figure 10.8. (A) SFM image of the first scan after flushing the liquid cell containing the DNA sample wilh 300 µl of 1 nM *E. coli* RNA polymerase (Eipicentre Technologies) solution and a 2-min waiting period for the tip to settle. The protein was diluted 2,000-fold into 10 mM HEPES/l mM MgCl₂, pH = 7.4, just prior to injection. (B) Second scan after protein injection (about 3 min after injection). Several DNA molecules can be recognized in both frames; (A) is prior to and (B) is after binding of the RNA polymerase. RNA polymerase molecules are also binding to the surface. Both images are 1,500 nm x 1,500 nm in size and the height scale is the same as in Figure 10.7.

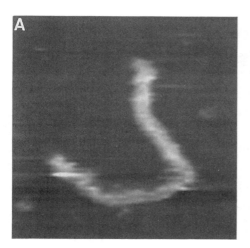 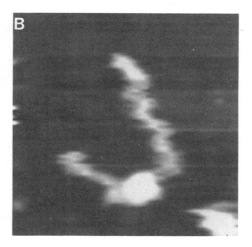

Figure 10.9. Close-up view of two DNA molecules: (A) before binding and (B) after binding of RNA polymerase. Image sizes are 300 nm x 300 nm.

10.3.5 *Image reconstruction and simulation*

As previously discussed, SFM images would depict the true topography of the sample surface only if the tip were infinitely thin and the sample were perfectly rigid. In practice, these conditions are not fulfilled and the SFM images of macromolecules often display distorted dimensions. To interpret SFM images correctly, it is essential to understand the cause and the extent of the distortions. Two methods, image reconstruction and image simulation, can be used to elucidate the tip effects on the images. In the first method, the image is reconstructed using a tip of known geometry. This procedure partially eliminates the distortions due to the finite tip size. In the second method, an SFM image is generated by simulating the effect of scanning a tip of known geometry over a model of the object. Comparison of the images before and after the scanning reveals the effects of the tip and helps to interpret the experimental images.

In the constant force mode, an SFM image of a surface is generated by recording the height of the surface at each position of the tip during scanning. This mode of operation is equivalent to recording the height of the tip apex while keeping the sample stationary; that is, the SFM image is composed of the vertical positions of the tip apex at each point on the surface. This observation forms the basis of the image reconstruction method of Keller and Franke (1993). In this method, the apex of a simulated tip is placed at each point on the image surface to be reconstructed. The envelope of all the tip surfaces constitutes the reconstructed image. This process partially removes the apparent broadening of the sample due to the finite dimensions of the tip.

The reconstruction is useful for improving the fidelity of the SFM images, but it cannot eliminate the image distortion entirely for a number of reasons. First, the scanning tips used in experiments have neither uniform dimensions nor well-defined shapes so, they cannot be accurately represented by simple mathematical functions. Moreover, due to tbe small dimensions of the tips, electron microscopy examination usually does not provide an exact profile of individual tips, and sometimes the tips get damaged during the process. Second, there are regions in SFM images that cannot be reconstructed. These regions occur when the tip contacts the sample at more than one point simultaneously. Since the tip never touches the region between the contact points, the image contains no information about this region, and the features in this region cannot be reconstructed. The third difficulty in this image reconstruction method is the noise in the experimental images. Some noise can be eliminated in the reconstruction process but other noise is treated as true features, thus causing artifacts in the reconstructed image.

Computer-generated artificial SFM images of model objects are very useful for interpreting experimental images. Comparison between the simulated and experimental images determines whether the model of the object is appropriate or must be improved further. Figure 10.10A shows a simulated SFM image of a model chromatin fiber of 50 nucleosomes. The chromatin fiber was generated using a model

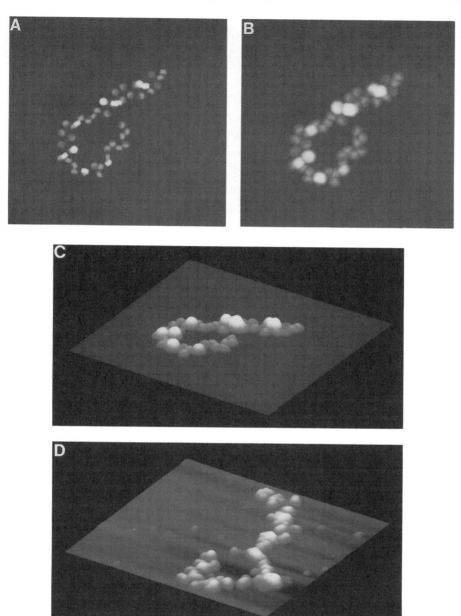

Figure 10.10. (A) Calculated SFM image of a model chromatin fiber containing 50 nucleosomes. (B) The chromatin fiber in (A) after being scanned by a parabolic tip with a 10-nm radius of curvature at the end. (C) The same image as in (B), viewed at a 30° angle. (D) An experimental SFM image of a fixed chicken erythrocyte chromatin fiber. All images are 400 nm x 400 nm in size.

similar to that of Woodcock et al. (1993), in which small random variations in nucleosome spacing are allowed. The model also assumes that the presence of histones
H1 and H5 fixes the angles at which the DNA enters and exits the nucleosome. The
artificial SFM image of the model chromatin fiber was generated by projecting the
disc-shaped nucleosomes and the linker DNA onto a surface without changing their
orientations. The projection plane was chosen by minimizing the root mean square
(rms) of the distances from the plane to all the nucleosomes. The nucleosomes were
placed on the surface in the order of tbeir distance to the surface so that each nucleosome touches either the surface or other nucleosomes already on the surface.

Figures 10.10B and C show the same chromatin fiber as in Figure 10A, after
being scanned by a geometric tip. Figure 10.10D shows an experimental SFM
image of a fixed chicken erythrocyte chromatin fiber. To make the mean height of
the nucleosomes in the calculated image agree with that in the experimental image,
the height of all the features in the model image was compressed by a factor of 2
as the tip passed. For the calculated image, the tip was assumed to be a parabo-
loid with a 10-nm radius of curvature at the end. The size and shape of the tips
used in the experiments varied, but electron microscope examination revealed that
most of the tips had end radii of curvature around 10 nm and were approximately parabolic in shape near the end.

10.4 Prospects for the future

A great deal of technical development has occurred in scanning force microscopy
during the past few years, driven in part by concurrent developments in the biological applications of the technique. As a result of these growing applications, it
is increasingly apparent that SFM will likely become a versatile tool of structural
characterization of biological systems. It also appears certain that future advance
ments in this method will be shaped by the unique requirements of investigating
biological systems. These requirements are (1) reduced tip–sample forces and
sample deformation, (2) improved spatial resolution by use of sharper tips, and (3)
reliable methods to image samples and molecular processes in physiologically
relevant environments.

As it may be apparent from this list, these capabilities are not independent but
are closely related to one another. For example, reduced tip–sample forces can be
achieved only in liquids, in which the dominant attractive forces between the tip
and the sample (that is, capillary and electrostatic forces) can be reduced and
screened Similarly, improved spatial resolution by use of sharper tips will be possible only if the tip–sample forces are simultaneously decreased. The reason is
twofold. First, as the tip dimensions are reduced, the local pressure experienced
by the sample increases roughly proportionally to the inverse of the tip's radius of
curvature. This dependence arises because the area of contact between the tip and
the sample is proportional to the square of the radius of curvature of the tip,

whereas tip–sample forces (capillary and electrostatic) scale roughly with the tip end dimensions. Second, the integrity of these sharper tips during scanning also requires reduction of tip–sample forces.

To image biological samples in a stable manner, tip–sample forces must be smaller than those holding the sample to the substrate. There are two reasons, however, why it is important to keep attachment forces relatively small. First, strong adhesion forces could potentially affect the native configuration of the macromolecules and, second, many molecular processes, such as catalysis, ligand binding, or complex assembly, may require the molecules to be only transiently or locally attached to the surface to facilitate the accessibility and proper positioning of the interacting molecular surfaces.

All these requirements will likely be met in the near future. New methods of tip fabrication will probably generate tips with end radii of curvature below 5 nm. On the other hand, the recent development of tapping in liquids (Hansma et al., 1994; Putnam, 1994) will greatly reduce both normal and lateral forces on the samples. In turn, lower forces will minimize the requirement of strong sample attachment to the substrate, while it will facilitate the use of more fragile, but sharper tips. These developments may lead eventually to the development of true noncontact force microscopy that could fulfill the requirements for studying biological systems with minimal or no sample damage.

References

Adrian, M., Dubochet, J., Lepault, J., and McDowall, A. W. (1984) Cryo-electron microscopy of viruses. *Nature* 308:32–36.

Akama, Y., Nishimura, E., Sakai, A., and Murakami, H. (1990) New scanning tunneling microscopy tip for measuring surface topography. *J. Vac. Sci. Technol. A* 8:429–33.

Bailey, S. W. (ed.). (1984) Micas. Chelsea BookCrafters, Inc. (Reviews in Mineralogy, 13).

Bezanilla, M., Bustamante, C., and Hansma, H. G. (1993) Improved visualization of DNA in aqueous buffer with the atomic force microscope. *Scanning Microscopy 7*: 1145–1148.

Bhatia, S. K., Shriver-Lake, L. C., Prior, K. J., Georger, J. H., Calvert, J. M., Bredehorst, R., and Ligler, F. S. (1989) Use of thiol-terminal and heterobifunctional crosslinkers for immobilization of antibodies on silica surfaces. *Anal. Biochem.* 178:408–413.

Binnig, G., Rohrer, H., Gerber, Ch., and Weibel, E. (1982) Surface studies by scannin-tunneling microscopy. *Phys. Rev. Lett.* 49:57–61.

Brennan, R. G., Roderick, S. L., Takeda, Y., and Matthews, B. W. (1990) Protein-DNA conformational changes in the crystal structure of a lamda Cro-operator complex. *Proc. Natl. Acad. Sci., USA* 87:8165–8169.

Bustamante, C., Keller, D., and Yang, G. (1993) Scanning force microscopy of nucleic acids and nucleoprotein assemblies. *Current Opinion Struct. Bio.* 3:363–372.

Bustamante, C., Vesenka, J., Tang, C. L., Rees, W., Guthold, M., & Keller, R. (1992) Circular DNA molecules imaged in air by scanning force microscopy. *Biochemistry* 13:22–26.

Butt, H. J. (1991a) Measuring electrostatic, van der Waals, and hydration forces in electrolyte solutions with an atomic force microscope. *Biophys. J.* 60:1438–1444.

Butt, H. J. (1991b) Electrostatic interaction in atomic force microscopy. *Biophys. J.* 60:777–785.

Darst, S. A., Kubaleck, E. W., and Kornberg, R. D. (1989) Three-dimensional structure of Escherichia coli RNA polymerase holoenzyme determined by electron crystallography. *Nature* 340:730–732.

Digital Instruments, Nanotips 5, No. 1 (1993).

Drake, B., Prater C. B., Weisenhorn, A. L., Gould, S. A. C., Albrecht, T. R., Quate, C. F., Cannel D. S., Hansma, H. G., and Hansma, P. K. (1989) Imaging crystal polymers, and processes in water with the atomic force microscope. *Science* 243:1586–1589.

Dunlap, D. D., Garcia, R., Schabtach, E., and Bustamante, C. (1993) Masking generates contiguous segments of metal-coated and bare DNA for scanning tunneling microscope image. *Proc. Natl, Acad. Sci. USA* 90:7652–7655.

Engel, A. (1991) Biological applications of scanning probe microscopes. *Annu. Rev. Biophys. Biophys. Chem.* 20:79–108.

Grutter, P., Zimmermann-Edling, W., and Brodbeck, D. (1992) Tip artifacts of microfabricated force sensors for atomic force microscopy. *Appl. Phys. Lett.* 60:2741–2743.

Hansma, H. G., Bezanilla, M., Zenhausern, F., Adrian, M., and Sinsheimer, R L. (1993) Atomic force microscopy of DNA in aqueous solutions. *Nucleic Acids Res.* 21:505–512.

Hansma, H. G., Vesenka, J., Siegerist, C., Kelderman, G., Morret, H., Sinsheimer, R. L., Elings, V., Bustamante, C., and Hansma, P. K. (1992) Reproducible imaging and dissection of plasmid DNA under liquid with atomic force microscope. *Science* 256:1180–1184.

Hansma, P. K., Cleveland, J. P., Radmacher, M., Walters, D. A., Hillner, P., Bezanilla, M., Fritz, M., Vie, D., and Hansma, H. G., Prater C. B., Massie, J., Fukunaga, L., Gurley., & Elings, V. (1994) Tapping mode atomic force m croscow in liquids. *Appl. Phys. Lett.* 64:1738–1740.

Hartmann, U. (1991) van der Waals interactions between sharp probes and flat sample surfaces. *Phys. Rev. B* 43:2404–2407.

Hoh, J. H. and Hansma, P. K. (1992) Atomic force microscopy for high-resolution imaging in cell biology. *Trends Cell Bio.* 2:208–213.

Keller, D. J. and Franke, F. S. (1993) Envelope reconstruction of probe microscope images. *Surface Science* 294:409–419.

Keller, D. and Chou, C. C. (1992). Imaging ste~, high structures by scanning force microscopy with electron beam deposited tips. *Surface Sci.* 268:333–339.

Kim, J., Zwieb, C., Wu, C., and Adhya, S. (1989) Bending of DNA bu gene regulatory proteins: construction and use of a DNA bending vector. *Gene* 85:1523.

Lyubchenko, Y. L., Gall, A. A., Shlyakhtenko, L., Harrington, R., L. Jacobs, B. L., Oden, P. I., and Lindsay, S. M. (1992) Atomic force microscopy imaging of double stranded DNA and RNA. *J. Biomol. Struct. Dyn.* 10:589–606.

Lyubchenko, Y. L., Oden, P. I., Lampner, D., Lindsay, S. M., & Dunker, K. A. (1993) Atomic force microscopy of DNA and bacteriophage in air, water and propanol: the role of adhesion forces. *Nucleic Acids Res.* 21:1117–1123.

Lyubchenko, Y., Shlyakhtenko, L., Chernov, B., & Harrington, R. E., (1991) DNA bending induced by Cro protein binding as demonstrated by gel electrophoresis. *Proc. Natl. Acad. Sci. USA* 88:5331–5334.

Lyubchenko, Y., Shlyakhtenko, L., Harrington, R., Oden, P., & Lindsay, S. M. (1993) Atomic force microscopy of long DNA: imaging in air and under water. *Proc. Natl. Acad. Sci. USA* 90:2137–2140.

Ohnesorge, G. and Binnig, G. (1993) True atomic resolution by atomic force microscopy through repulsive and attractive forces. *Science* 260:1451–1456.

Olins, A. L. and Olins, D. E. (1974) Spheroid chromatin units (v-bodies). *Science* 183:330–332.

Ptashne, M. (1992) *A Genetic Switch, Phage Lambda and Higher Organisms*, 2nd ed. Cell Press.

Putman, C. A. J. (1994) *Development of an atomic force microscope for biological applications.* Ph.D. dissertation, University of Twente, the Netherlands.

Sarid, D. (1991) *Scanning force microscopy with applications to electric, magnetic, and atomic forces.* New York: Oxford University Press.

Schaper, A., Pietrasanta, L. I., and Jovin, T. M. (1993) Scanning force microscopy of circular and linear plasmid DNA spread on mica with a quaternary ammonium salt. *Nucleic Acids Res.* 21:6004–6009.

Thoma, F., and Koller, Th. (1977) Influence of histone Hl on chromatin structure. *Cell* 12:101–107.

Thoma, F. and Koller, Th. (1981) Unravelled nucleosomes, nucleosome beads and higher order structures of chromatin: influence of non-histone components and histone H1. *J. Molec. Biol.* 149:709–733.

Thoma, F. Koller, Th., and Klug, A. (1979) Involvement of histone Hl in the organization of the nucleosome and of the salt-dependent superstructures of chromatin. *J. Cell Biol.* 83:403–427.

Thundat, T., Warmack, R. J., Alison, D. P., Bottomley, L. A., Lourenco, A. J., and Ferrel T. L. (1992) Atomic force microscopy of deoxiribonucleic acid strands adsorbed on mica the effect of humidity on apparent width and image contrast. *J. Vac. Sci. Technol. A* 10:630–635.

Vesenka, J., Guthold, M., Tang, C. L., Keller, D., Delaine, E., and Bustamante, C. (1992) Substrate preparation for reliable imaging of DNA molecules with the scanning force microscope. *Ultramicroscopy* 4244:1243–1248.

von Holde, K. E. (1988) *Chromatin.* New York:Springer Verlag.

Woodcock, C. L. F. (1973) Ultrastructure of inactive chromatin. *J. Cell Biol.* 59:368a.

Woodcock, C. L., Grigoryev, S. A., Horowitz, R. A., and Whitaker, N. A. (1993). Chromatin folding model that incorporates linker variability generates fibers resembling the native structures. *Proc. Natl. Acad. Sci. USA* 90:9021–9025.

Yang, J. and Shao, Z. (1993) The effect of probe force on the resolution of atomic force microscopy of DNA. *Ultramicroscopy* 50: 157–170.

Yang, J., Tamm, L. K., Somlyo, A. P., and Shao, Z. (1993) Promises and problems of biological atomic force microscopy. *J. Microscopy* 171:183–198.

11

Nanoscale structures engineered by molecular self-assembly of functionalized monolayers

DAVID L. ALLARA

Department of Materials Science and Department of Chemistry
Pennsylvania State University
University Park, Pennsylvania
16802

11.1 Introduction

Because of extensive preparative and structural studies carried out for many years, it is well known that molecules can be assembled into densely packed, highly organized monolayers at both solid and liquid surfaces (Ulman, 1991). Over the past decade the interest in these organized monolayer films has increased tremendously because of their potential for practical applications. For example, many of the useful properties of a solid object, such as wetting and acceptance by biological systems, are strongly dependent on the surface condition; accordingly, modification of the surface by a monolayer coating can dramatically alter these properties, even though the changes induced in the corresponding bulk properties will appear insignificant. Most often these monolayers are comprised of alkyl-chain compounds because of the inherent ability of alkyl chains to form very densely packed structures. This phenomenon is well appreciated by biologists who are familiar with the critical role that lipids, which typically contain mid-length alkyl chains in the range of 16 carbons, play in the formation of cell membranes and other functional structures. Figure 11.1 shows a schematic diagram of an ideal, ordered monolayer assembly of molecules with all-*trans* alkyl chains. The figure shows a few common terms used to describe these structures. The term "headgroup" refers to the functional group that attaches the molecules to the substrate surface, "tail" or "terminal group" refers to the outermost functional group on the alkyl chain that contacts the ambient phase, and the angle ϕ refers to the tilt angle of the chains with respect to the surface normal. Generally, ϕ refers to an average or a collective value for the entire chain assembly. Because the thickness d of the layer is less than or equal to the molecular length, the structure can be considered as a nanometer-size material (nanoscale) in the vertical direction but a macrosopic material in the horizontal (substrate surface) plane. If the monolayer becomes inhomogeneously distributed across the surface, in composition and/or structure, then three-dimensional,

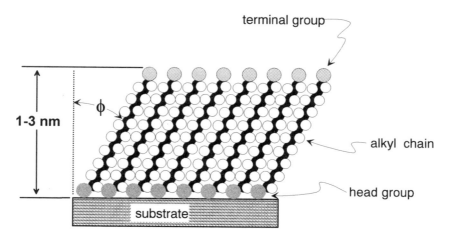

Figure 11.1. Schematic diagram of an assembly of terminally functionalized, all-*trans* alkyl-chain molecules attached to a substrate. The typical thickness of such a monolayer is shown as 1–3 nm and the tilt angle of the chains is indicated as ϕ.

nanometer-scale structures can arise. As will be seen later, such structures can arise by lithographic techniques as well as through natural compositional and density fluctuations.

Figure 11.2 depicts three of the typical ways to assemble these monolayers: Langmuir–Blodgett (LB) film transfer at the air–water interface, gas-phase self-assembly, and solution-phase self-assembly. The LB technique applies surface pressure to the exterior of the molecular ensemble in order to "squeeze" the molecules into an organized state prior to and during withdrawal of the substrate from the water bath. Although it may appear that the application of such mechanical pressure should always result in far higher organization and packing density in LB films as compared to self-assembled monolayers (SAMs), experimental observation shows otherwise. Figure 11.3 shows an experimental plot of surface pressure vs. area/molecule for dipalmitoyl-phosphatidylcholine (DPPC), a common lecithin, at the air–water interface (Hietpas, 1994). The inset in Figure 11.3 shows the infrared (IR) reflection spectra in the C–H stretching mode region of LB monolayer films on oxidized silicon substrates withdrawn from the water surface at pressures of 50 and 5 mN/m and the spectrum of a monolayer of DPPC self-assembled onto a silicon substrate from chloroform solution. The antisymmetric mode (d^-), located in the vicinity of ~2,920 cm^{-1}, is a diagnostic for the conformational ordering of the alkyl chains. The low value of the peak frequency for the LB film taken at high surface pressure indicates high conformational ordering of the alkyl chains, and the higher value of the low–surface pressure film indicates decreased ordering. However, the self-assembled DPPC spectrum shows a d^- peak at lower frequency than either of the

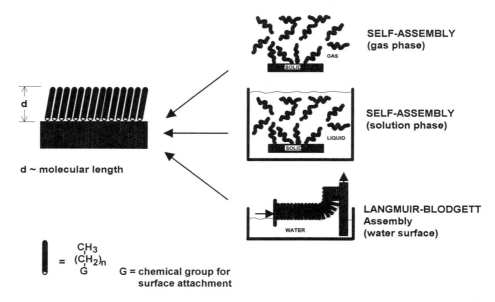

Figure 11.2. Schematic diagram of three methods for forming a monolayer films. The diagram shows that self-assembly can be carried out in either solution or gas phase; Langmuir–Blodgett deposition utilized a monolayer of molecules compressed by a surrounding barrier at an air–water interface.

other two LB films. This result indicates that the conformational ordering that exists in the self-assembled film is equivalent to the ordering in an LB film with a very high applied surface pressure. Thus the self-assembly process can create high internal film pressures just by natural chain–chain interactions arising from van der Waals attractive forces.

This example of self-assembly involves attachment to the substrate by weak physisorption forces that lead to film instability, for example, with respect to removal by washing with water. In terms of practical applications, it is necessary to prepare films that exhibit much higher degrees of stability. This can be accomplished in two ways: (1) chemical bonding between the substrate and the monolayer and (2) chemical bonding between the adsorbate molecules (crosslinking). Figure 11.4 depicts two limiting modes of SAM formation based on the monolayer stability. In spite of the generality of these bonding interactions, very few distinctly different examples of SAM systems have been developed. Figure 11.5 summarizes the major systems that have been well characterized and gives some indication of their stability. Details about these monolayers will be discussed later.

In order to understand the stability and properties of SAMs, the structures can be analyzed in terms of the component parts, shown in Figure 11.6. For SAMs that are chemically bonded to the substrate, the strength of these bonds will be a dominant contribution to the film stability. Furthermore, the spatial

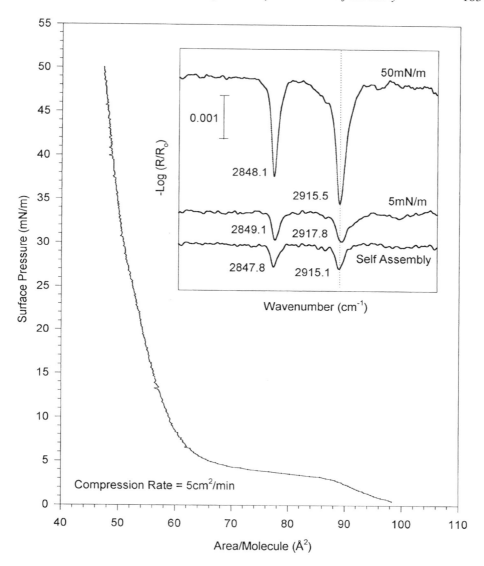

Figure 11.3. The main plot shows a surface pressure–area per molecule isotherm for a monolayer of DPPC at the air–water interface. The compression rate is 5 cm²/min and the data were taken at 20 °C. The inset shows the C–H stretching mode infrared spectra of each of two monolayer Langmuir–Blodgett films of DPPC deposited at surface pressures of 5 and 50 mn/m (top two spectra). The observation of the peak frequency of the antisymmetric stretching mode at a lower value for the higher surface pressure monolayer (2915.5 vs. 2917.8 cm⁻¹) is an indication of higher conformational ordering of the alkyl chains in the high–surface pressure film. The self-assembled monolayer (bottom spectrum) shows the lowest value of the antisymmetric stretching mode peak frequency (2915.1 cm⁻¹), which indicates that the alkyl chains in the self-assembled film are conformationally ordered at least as well as those in the high–surface pressure Langmuir–Blodgett film.

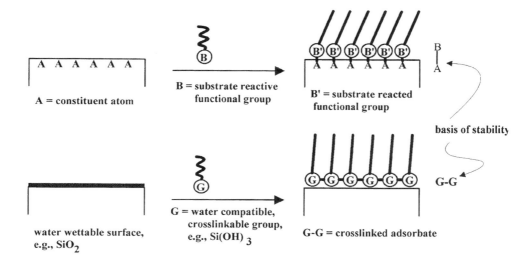

Figure 11.4. Schematic diagram of two limiting modes of self-assembly of monolayer films. The top mode involves direct chemical bonding of the adsorbed molecules to specific substrate sites; the bottom mode involves non-site-specific adsorbate–substrate interactions. An example of the latter film type is given approximately by self-assembled alkylsiloxane monolayers on highly hydrated substrates in which the water layer decouples the monolayer structure from that of the substrate.

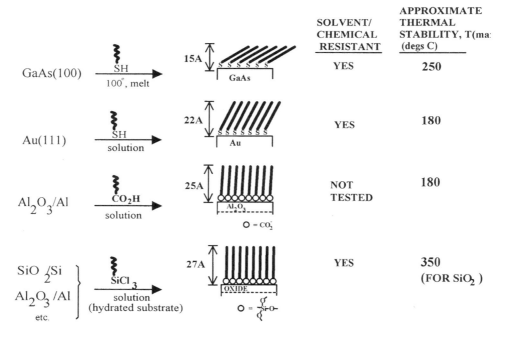

Figure 11.5. Known types of self-assembled monolayers. The table schematically shows the approximate chain tilt angles an film thicknesses. Also listed are simple characteristics related to monolayer stability.

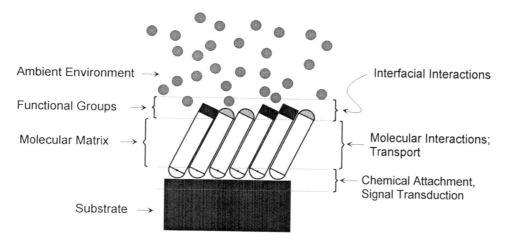

Ambient Environment →

Functional Groups →

Molecular Matrix →

Substrate →

Interfacial Interactions

Molecular Interactions; Transport

Chemical Attachment, Signal Transduction

Figure 11.6. Schematic drawing of a self-assembled monolayer exposed to an ambient environment such as liquid solution. The diagram shows the important structural parts of the monolayer and corresponding properties dependent upon these structural units.

arrangement of the chains, in terms of collective tilt angle and translational symmetry, will be determined mainly by the corresponding arrangement of the substrate sites upon which the chains are pinned. For SAMs with extremely weak substrate interactions and strong intermolecular bonding, the strength and symmetry of these bonding patterns will provide the main basis for the film stability and organization. The interior matrix of the alkyl chains can provide a sizable fraction of the internal energy of the SAM through the van der Waals interaction of the $(CH_2)_n$ segments, particularly for long chains such as $n > 16$. The ambient surface of the SAM interacts with the surrounding environment. For a gas-phase environment, the interaction energy is small, but for a condensed medium the interfacial energy can be expected to influence the orientation and energetics of the terminal groups. Furthermore, this outer layer provides the major basis of the chemical reactivity of the SAM with respect to reactants or biological materials in the ambient phase, and susceptibility to degradation in a harsh environment can be expected to start at this ambient interface. The structure shown in Figure 11.6 also illustrates the possibility of having different types of terminal groups in the monolayer. Preparation of such mixed monolayers with different proportions of the component adsorbates can provide surfaces with a variety of chemical, wetting, and biological responses. Thus, in principle, one could prepare surfaces of mixed hydrophobic and hydrophilic character by mixing, for example, CH_3 and OH groups. Such possibilities have been realized, particularly in the alkanethiolate/Au SAM system, where mixed SAMs are quite easy to prepare; the specific case of the

CH$_3$/OH SAM has been studied in some detail (Bain and Whitesides, 1988, Bertilsson and Leidberg, 1993; Folkers et al., 1992a, 1992b; Sanassy and Evans, 1993).

In addition to preparation of mixed monolayers of equal chain length, it is also possible to prepare monolayers with different length molecules. Such possibilities are depicted in Figure 11.7. These structures add a vertical dimension to the distribution of surface functional groups. A limited number of studies on such mixed chain length structures have revealed interesting properties with respect to both wetting of liquids (Atre et al., 1994; Bain and Whitesides, 1988b, 1988c; Troughton et al., 1988) and adsorption of biomolecules (Pale-Grosdemange et al., 1991; Prime and Whitesides, 1991, 1993).

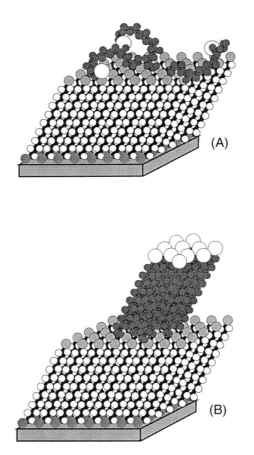

The discussion

Figure 11.7. Schematic drawing of two very different structural possibilities for self-assembled monolayers comprised of two components having different terminal functional groups and chain lengths. In (A) the longer chains are highly disordered and partially cover the lower surface; in (B) the longer chains are bunched together and maintain a highly all-*trans* conformational structure.

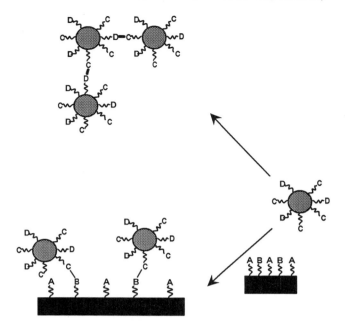

Figure 11.8. Schematic drawing of microscopic objects, such as biomolecules or inorganic clusters, attached to surfaces or to one another by bifunctionally terminated alkylchain molecules. The drawing implies strategies for making self-assembled structures of these objects using the "linker" molecules.

so far has implied that the SAM substrates are planar. However, if one considers other material shapes, such as particles or rods, it is clear that a number of additional possibilities exist for complex structures as illustrated in Figure 11.8, where clusters or spherical particles are shown linked to planar surfaces as well as to each other. In these cases, the organization of the SAMs may be of secondary importance to the type of chemical functionality at the head and tail sites on the alkyl chain, and the molecules can be considered more as simple linker layers or coupling agents than organized assemblies. The spheres also could represent large molecules such as proteins or other biomolecules.

Most of the work focusing on the fundamental structures and properties of SAMs has occurred during the past five years or so. From this work a qualitative understanding of the correlation between component materials and the final structures has evolved, but a number of important issues remain unresolved; some of the most signifcant ones are listed in Table 11.1. Those issues relating to the static aspects of SAM structures will be discussed in the following sections of the paper. The issues of dynamics have received little attention up to this time, primarily because of the experimental difficulties in such studies; consequently, they will not be discussed here.

Table 11.1. *Summary of critical issues in self-assembled monolayers.*

Prominent Issues in Self-Assembled Monolayers	Applications in Which the Prominent Issues Are Important
Coupling of substrate–film structures (pinning lattice effects)	* all

Spatial distribution in mixed monolayers
  * lateral: phase segregation and lithography
  * vertical: functional group screening
	* bioselectivity
	* chemical and biosensors
	* molecular recognition

Defects	* electrochemistry * electronic devices * nanolithography

Dynamics
  * monolayer formation
  * defect mobility
  * pattern stability
  * adsorbate–solution interchange
	* all

11.2. Examples of two self-assembled monolayer systems

11.2.1 Alkanethiolates on gold

In the past few years the most intensively studied SAM system has been alkanethi-olates on Au(111) surfaces. The latter surfaces are easily provided by the vacuum deposition of thin gold films, and the monolayers are deposited by simple immer-sion of the gold surfaces in a dilute solution of the thiol compound in a suitable sol-vent. An excellent review of these SAMs has been published (Dubois and Nuzzo, 1992). One of the major questions in the formation of SAMs is the effect of the sub-strate surface structure on the structure of the adsorbed monolayer. Figure 11.9A shows the top view of an Au(111) surface with all-trans conformation $(CH_2)_n$ chains located on top and oriented with the chain axes perpendicular to the gold surface plane. The figure shows that the diameter of the alkyl chain is larger than the near-est neighbor spacings of the gold atoms. As a consequence, if the S atoms of the alkanethiols are to be associated with discrete gold atoms on the surface, then the closest spacing possible is associated with the next-nearest neighbor gold atoms. This pattern will produce an overlayer on a $(\sqrt{3} \times \sqrt{3})$, R30° superlattice. As shown in Figure 11.9, this arrangement leaves space between the alkyl chains, and it might be expected that the chains would collectively tilt in order to achieve maximum van der Waals interactions. A simple geometric calculation predicts an ~30° chain tilt (ϕ in Figure 11.1); this value is about within experimental error of the values determined from both infrared spectroscopy (Laibinis et al., 1991) and x-ray diffraction (Fenter et al., 1993) studies. However, recent scanning tunneling microscopy (Poirer and

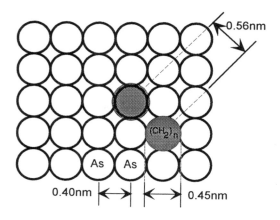

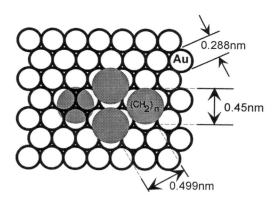

Figure 11.9. A top view, made to approximate scale, of crystalline substrates covered with adsorbed alkyl chains in which the chain axes are vertical to the substrate surface plane. The dimensions of the substrate lattice atoms and of the alkyl chain diameters are shown. (A) and (B) represent the cases of Au(111) and GaAs(100) surfaces, respectively. The diagrams show that in both cases the alkyl chains are too large to occupy adjacent substrate lattice sites but can occupy next-nearest neighbor sites.

Tarlov, 1994) studies, together with earlier x-ray (Fenter et al., 1993) and helium diffraction (Camillone et al., 1993) analyses, indicate that the overlayer structure is more complicated than the simple ($\sqrt{3} \times \sqrt{3}$), R30° overlayer and actually is better represented as a rectangular superlattice, in which the ($\sqrt{3} \times \sqrt{3}$), R30° structure is distorted to give two distinct S–S distances, as opposed to the uniform-distance hexagonal packing of the $\sqrt{3}$ structure. In this case, chemical effects of substrate–adsorbate bonding add to the simple geometric effects of chain packing.

This example of the alkanethiolate SAM on Au illustrates what appears to be a general rule of thumb: For SAMs of simple alkyl compounds that involve direct chemical bonding to substrate atoms, the overlayer structure can be predicted approximately from a pattern that allows closest packing of the alkyl chains with direct bonding to the surface substrate atoms. It appears that the terminally substituted alkanethiols also follow these rules. A study (Dubois et al., 1990) of SAMs of $X-(CH_2)_{15}S$ on gold has shown that for $X = CO_2CH_3$, CO_2H, CH_2OH, and $CONH_2$, the average tilt angles and chain conformational structures are all similar to those of the *n*-alkanethiolate SAMs.

11.2.2 Alkanethiolates on GaAs(100)

Although the formation of alkanethiolates on Au represents an example of SAMs on a metallic surface, it is also of technological interest to form SAMs directly on semiconductor surfaces. Treatment of bare GaAs(100) surfaces with molten alkanethiols at ~100 °C under an inert atmosphere results in formation of an organized monolayer in which the alkyl chains are bonded directly to the oxide-free GaAs surface and, for the case of C_{18} chains, exhibit highly *trans* conformational sequences (Sheen et al., 1992). Figure 11.9B shows the top view of a GaAs(100) surface (arsenic terminated) with adsorbed, vertically oriented alkyl chains. As with alkanethiolate/Au(111) (Figure 11.9A), the chain diameters dictate that the closest spacing allowed for bonding of the S atoms to the substrate atoms is at next-nearest neighbor sites. Maximization of the chain–chain attractive energies at these spacings would give a collective chain tilt angle of 55–60°, exactly the range of values observed experimentally (Sheen et al., 1992). With reference to Figure 11.1, it can be seen that this large cant of the chains provides a distribution of surface groups with more vertical modulation than for the case of alkanethiolates on Au(111). Such a property could be of potential use in "tuning" surface properties, such as wetting, by keeping terminal functionality constant while changing substrates.

Far less work has been done on exploring the types of substituents that are possible with the GaAs SAMs than with the Au SAMs, but it has been shown that CO_2H-terminated SAMs can be prepared in the GaAs system and that, as expected, these SAMs exhibit an acidic surface (Sheen et al., 1994). This indicates that possibilities exist for modification of the chemistry of GaAs surfaces, and it suggests that opportunities exist for developing tailored biological responses at semiconductor surfaces such as might be required for chemical sensors and biosensors.

11.3 Selected applications

Although the basic scientific issues of the formation and structures of SAM, as described in the previous section, are of great interest, it is the potential applications that propel much of the research efforts. From the point of view of biologically related applications (Swalen et al., 1987), the major phenomena of interest are the abilities of these films to control the surface chemical and biological responses in highly specific ways and to control transport of charged and/or neutral species through the films. Much of the recent interest involves the development of strategies for forming chemically patterned films on the micrometer scale in order to form arrays of biological moieties by highly selective, chemically controlled adsorption.

The most most widely studied aspect of SAMs in general has been the wetting properties, primarily determined by contact angle measurements of wetting liquids. In some sense, these studies have been viewed by the biologically oriented community as precursor studies to bio-adsorption phenomena. Because of the flexibility of controlling the terminal functional group in alkanethiolate/gold monolayers, most of the wetting studies have concentrated on these SAMs. For example, terminal groups such as CH_3, OH, CO_2H, CO_2CH_3, $CONH_2$, CN, and OC_nH_{2n+1} can be readily incorporated into SAMs by self-assembling thiols containing these groups (Prime and Whitesides, 1991, 1993).

The ability to vary such functionality in these SAMs in an extremely well controlled way provides a means to study the effects of chemical surface properties on complex biological responses such as cell growth and adhesion (Mastandrea et al., 1989). An example of this is given in Figure 11.10, which shows the results of growth of bovine aortic cells on alkanethiolate/gold SAM surfaces (Ertel et al., 1994). The figure shows that the ability of the monocomposition surfaces to sustain uniform, dense growth of cells increases in the order OH $< CO_2CH_3 < CH_3 < CO_2H$.

Studies of protein adsorption have been carried out on CH_3-, CH_2OH-, and $(OCH_2CH_2)_nOH$-terminated alkanethiolate/gold SAMs (Pale-Grosdemange et al., 1991; Prime and Whitesides, 1991, 1993) as well as on carboxylic- and amino-acid–terminated SAMs (Tengvall et al., 1992). The first two studies utilized SAMs prepared from mixtures of thiol molecules containing different functional groups, for example, the hydrophobic CH_3 group plus a hydrophilic group, and with different chain lengths. Such surfaces contain both mixed functionality and varied chain flexibility, factors that will be of great importance in the eventual control of biological response at surfaces. SAMs incorporating phospholipid chains with the biologically important phosphocholine group exposed at the surface have been synthesized and utilized in cell adhesion studies (Coyle et al., 1989). It is also possible to directly modify the surface of an existing SAM by direct chemical reaction; such a strategy has been utilized as a means to prepare biologically active surfaces from precursor SAMs (Bertilsson and Leidberg, 1993; Neogi et al., 1993).

It is clear from this discussion that a large variety of polyfunctional, flexible chain molecules can be incorporated into surface monolayer films by the self-assembly method. In addition to adsorption and wetting studies, a very simple preparative use of these monolayers is as linker layers or coupling agents for

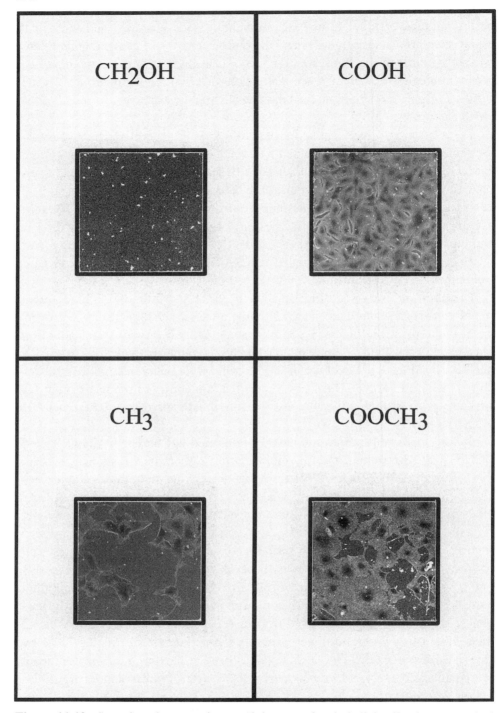

Figure 11.10. Scanning electron micrograph images of endothelial cell cultures attached to four types of terminally functionalized alkanethiolate self-assembled monolayers on gold substrates. The images show that the most robust cell confluence occurs on the CO_2H surface and the poorest surface is the OH-terminated one.

grafting of another layer or microscopic object (such as a cluster or biomolecule) onto the surface, for example, as suggested in Figure 11.7. The variability of the chemical groups in the SAM allows a wide variability of choice in promoting bonding to different kinds of overlayers or objects.

Another application relevant to control of nanometer-scale features is the use of SAMs as resists for electron beam lithography with ultrahigh spatial resolution. As feature resolutions approach the size scale of tens of nanometers, electron beams become the tool of choice because of their ability to be focused readily at this scale. One of the major problems in extending the well-known resist technology from the standard micrometer scale to the difficult nanometer scale has been the problem of reducing electron-beam resist coating thicknesses from the many-tens-of-nanometer scale into the nanometer scale. The central problem is that the electron-beam damage area increases with increasing depth of penetration into the resist film because of an attendant increase in secondary electron scattering with penetration. Typical polymeric resists do not form coherent coatings at nanometer thicknesses, so they do not provide good prospects for solving the resolution problem. However, SAMs can be extremely uniform and dense even at 1–2 nm thicknesses. The major requirements for these applications are that the unirradiated SAM shows a high resistance to the etching treatment designed to remove the substrate material, and that electron-beam irradiation removes the SAM or significantly lowers its resistance to the etching treatment, thus allowing fast etching of the substrate. Such a strategy can be readily realized for aqueous etches by using SAMs with extremely high hydrophobicity, such as those prepared from long-chain alkyl groups to provide a CH_3-terminated surface. Recently it has been demonstrated that, for Si and GaAs substrates coated with only 1.5–2.5-nm-thick resist films, pattern formation at linewidths approaching the 10-nm scale is possible with typical lithographic electron-beam doses (Lercel et al., 1993; Tiberio et al., 1993). Solution etching of these irradiated areas has allowed pattern transfer into the semiconductor substrate with formation of well-defined trenches of depths approaching the 200-nm scale.

Another example of the ability of SAMs to provide dense, protective films is given by their use as blocking layers on electrodes for electrochemistry experiments. It was shown in initial studies that alkanethiolate/gold SAMs could reduce Faradaic current by approximately two orders of magnitude (Porter et al., 1987). One important motivation for continuing interest in this area is the development of chemically and biologically selective electrodes (Swalen et al., 1987).

11.4 Issues

Based on the demands placed on film structures and properties for a number of applications involving SAMs, as partially outlined, several critical issues have arisen. Regarding applications to biological systems, the issues primarily focus on film defects and the ability to control chemical structure and patterns at the nanometer scale. The ability to control defects will have significant impact on the eventual design of SAM-based electrochemical biosensors and on the ability to fabricate low-defect, nanometer-scale patterns by SAM-based electron-beam lithography. The fabrication of chemically patterned surfaces is possible using mixed-composition SAMs, but workable strategies for controlling such patterns at the nanometer scale are not yet available.

11.4.1 Monocomposition films

Alkanethiolates/Au. One current issue in monocomposition SAMs centers on the nature of structural defects. With the advent of scanning tunneling microscopes (STM) it has become possible to image these features at the molecular scale. In the case of n-alkanethiolate/Au(111) SAMs, possible defects include those that could be associated with defects in the gold substrate atoms, as well as adsorbate defects such as missing chains and tilt-phase boundaries. At partial coverages it is obvious that there will be regions of missing chains, but as coverages approach the maximum values possible with typical self-assembly conditions, recent STM studies (Schoenberger et al., 1994) show that a residual number of small defects appear to exist on the size scale of ~2–5 nm in diameter and ~0.2 nm deep. A lateral STM profile (Strannick et al., 1994a) through one such defect observed in a $C18H37S/Au(111)$ SAM is shown in Figure 11.11. These defects appear intrinsic to fully formed monolayers and do not seem to function as transport channels, for example, in current blocking at electrodes. The most reasonable assignment of the structure in Figure 11.11 involves a missing gold atom in the (111) terrace. A recent study indicates that such defects are prevalent but can be removed by thermal annealing (Bucher et al., 1994). As coverages diminish from the highest values obtainable, there obviously will be areas of the surface with missing chains; the major issue is whether the chains are uniformly distributed across the surface or are bunched in some fashion. At partial coverages above ~60%, recent infrared and wetting data (Dunbar et al., unpublished) indicate that the adsorbate molecules are bunched together in islands with void regions between. Thus formation of partial coverage SAMs may be one way to produce chemically heterogenous surfaces at the nanometer level.

Alkylsiloxanes/SiO2. In contrast to the alkanethiolate/Au SAMs that exhibit point defects at high coverages, the highest coverage films of alkylsiloxane/SiO_2 SAMs appear to form line defects (Rondelez et al., 1994). Line defects appear to be a consequence of film formation via island growth. In this mechanism (Rondelez et al., 1994), mobile precursor tri-hydroxyalkylsilane species self-organize into islands on the surface of an adsorbed water film on the hydrated SiO_2 substrate. As high coverages are approached via continued adsorption of solution molecules, the islands form into condensed phases with aligned chains and the siloxy headgroups begin to crosslink, forming rigid structures. At this stage, perfect alignment of adja-

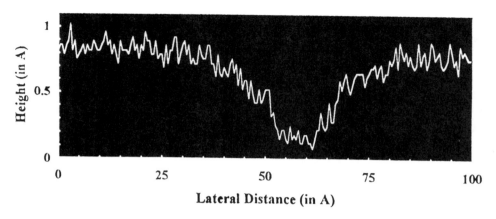

Figure 11.11. Plot of a lateral line scan across a defect in the surface of a C_{18} alkanethiolate monolayer self-assembled on the surface of an evaporated film of gold on mica. The depth of the defect is ~0.1 nm and the span is ~3.5 nm.

cent island edges that have grown together is not possible, because the crosslinking makes motion of individual molecules impossible. At present, the sizes of the islands are not known, but they are estimated to be as small as 5 nm in fully formed films (Rondelez et al., 1994). In general, such mechanisms are expected for the general class of silane coupling agents when reacted with hydrated substrate surfaces.

11.4.2 Multicomponent films and patterns

It is of great interest to prepare surfaces with varied, multiple functionality, particularly for biologically oriented applications. In addition to the presence of multiple surface groups, it is also desirable to be able to control the state of mixing and the lateral distribution of such groups. Conceptually, this could be accomplished through natural segregation processes that occur under conditions of self-assembly of multiple components or through deliberate patterning using lithographic strategies.

In self-assembled multicomponent SAMs, one would expect to observe the same general types of defects as are observed in the single-component films (point and line defects); in addition, defects could also be expected related to the relative distributions of the two components. Ideally, for biological studies one would like to form a variety of nanometer-scale chemical patterns, in order to observe effects on such phenomena as cell growth and protein adsorption. Recent STM studies indicate that mixtures of very similar alkanethiol molecules can form SAMs on gold that exhibit phase segregation of the component molecules (Strannick et al., 1994b). In particular, mixtures of X–$(CH_2)_{15}$S, where X = CH_3 and CO_2CH_3, over the fractional composition range of 1/3 and 3/1 show STM images with islands of pure components on the size scale of a few nanometers. The occurrence of phase segregation is somewhat surprising, because the molecules are within ~1 Å of the same length and neither of the two functional groups is capable of the strong self-interaction that would arise with hydrogen-bonding groups such as OH. These results, which suggest that a variety of different functional groups can lead to phase segregation, thus point to a general method for preparing nanometer-scale chemical heterogeneity involving diverse combinations of hydrophilic and hydrophobic groups, which would be of interest in studying and controlling biological responses of surfaces.

Another type of heterogeneity that can arise, even in ideally mixed multicomponent SAMs, is a lateral surface composition gradient. Such samples are of interest because of their ability to provide a spectrum of surface-composition-dependent results on one sample. Such gradient samples have been prepared recently on gold surfaces from component adsorbates such as OH- and CH_3-terminated alkanethiols (Tengvall and Liedberg, 1994). For these films, both imaging x-ray photoelectron spectroscopic and secondary ion mass spectrometric analyses show that the entire range of binary compositions can be dispersed across a millimeter-scale length.

This discussion has centered on mixed alkanethiolate/gold SAMs, but one can easily envision similar possibilities of phase segregation and gradients in other types of SAMs. In addition, it is also possible to create patterned films by mixing different types of film preparation chemistry. For example, recent results (Seshadri et al., 1994) indicate that it is possible to prepare films of isolated polymethylene (PM) clusters on the tens-of-nanometers scale by exposing evaporated gold film surfaces to diazomethane/ether solutions and that subsequent exposure of the bare gold surface regions to organothiol solutions leads to formation of SAMs between the PM regions. Since PM is so extremely hydrophobic, the filling of the bare regions with hydrophilic SAMs, such as CO_2H-terminated alkanethiolates, gives rise to potentially useful nanometer-scale chemical heterogeneity. An atomic force microscopy image showing the presence of PM clusters on an otherwise bare gold surface is shown in Figure 11.12.

Recently there has been intense interest in lithographic patterning of SAMs for applications to preparing surface arrays of biological molecules on the micrometer scale. The general strategies involve removing molecular groups from targeted areas of the SAM by irradiating with focused beams of light or electrons. After irradiation, the altered areas can be modified further by adsorption of another molecular group or by reaction with some chemical reagent in order to provide a surface with one type of chemical character patterned in a background of a different chemical character. An additional strategy involves mechanical micromachining of the SAM–substrate surface with a hard tool tip in order to expose bare substrate areas for further processing. Finally, it is also possible to write chemical lines directly on a substrate using controlled delivery of solution reagents.

The application of photolithography using light of visible wavelengths can provide pattern resolution down to the micrometer scale as limited by the resolv-

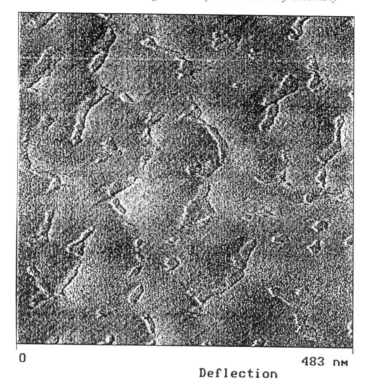

0   483 nm
Deflection

Figure 11.12. Tapping mode atomic force microscope image of a thin film of polymethylene chemically formed by diazomethane decomposition on the surface of a gold film evaporated onto mica. The image shows chainlike structure of length ~100 nm and widths of ~10–20 nm. The structures appear to grow in grain boundaries between the nominally {111} terraces and are absent in unreacted samples. The equivalent planar film thickness of the polymethylene is about 4 nm.

ing power of typical optical equipment. One recent example of this strategy has been reported for alkanethiolate SAMs on gold (Huang et al., 1994). Another approach has been reported in which photolithographic techniques have been applied to alkylsiloxane films on SiO_2 surfaces with the fabrication of patterns of fluoroalkyl groups in an amino background (Stenger et al., 1992). This surface has been utilized for selective adsorption of neuronal and endothelial cells (Stenger et al., 1992). Patterned alkanethiolate SAMs on gold have been produced on the tens-of-micrometers scale by a photolithographic-based, chemical-solution stamping technique (Singhvi et al., 1994), and these surfaces have been used to selectively adsorb primary rat hepatocyte cells in the patterned areas. Furthermore, the cells were observed to conform to the geometrical shape of the patterned areas, and the cell functioning was observed to depend upon the specific pattern geometry constraining the cell shape. Micromachining-based strategies have also been reported for gold-based SAMs (Kumar et al., 1992).

In order to obtain pattern resolutions in the submicrometer to nanometer scales using lithographic techniques, it is necessary to use higher resolution methods; electron beams appear the most attractive because of their ability to provide beam widths down to the nanometer scale using available instrumentation. As described in an earlier section, ultrahigh resolution lithography may turn out to be most effectively performed using appropriate SAMs as resists.

References

Atre, S. V., Liedberg, B., and Allara, D. L. *Langmuir* 11:3882–3893.

Bain, C. D. and Whitesides, G. M. (1988a) Formation of two-component surfaces by the spontaneous assembly of monolayers on gold from solutions containing mixtures of organic thiols. *J. Am. Chem. Soc.* 110:560–6561.

Bain, C. D. and Whitesides, G. M. (1988b) Correlations between wettability and structure in monolayers of alkanethiols adsorbed on gold. *J. Am. Chem. Soc.* 110:3665–3666.

Bain, C. D. and Whitesides, G. M. (1988c) Molecular-level control over surface order in self-assembled monolayer films of thiols on gold. *Science* 240:62–63.

Bertilsson, L. and Liedberg, B. (1993) Infrared study of thiol monolayer assemblies on gold: preparation, characterization and functionalization of mixed monolayers. *Langmuir* 9:141–149.

Bucher, J. P., Santesson, L., and Kern, K. (1994) Thermal healing of self-assembled organic monolayers -hexadecanethiol and octadecanethiol on Au(111) and Ag(111). *Langmuir* 10:979–983.

Camillone, N., Chidsey, C. E. D., Liu, G. Y. and Scoles, G. (1993) Superlattice structure at the surface of a monolayer of octadecanethiol self-assembled on Au(111). *J. Chem. Phys.* 98:3503–3511.

Coyle, L. C., Danilov, Y. N., Juliano, R. L., and Regen, S. L. (1989) Chemisorbed phospholipid monolayers on gold: well-defined and stable phospholipid surfaces for cell adhesion studies. *Chem. Mater* 1:606–611.

Dubois, L. H., Nuzzo, R. G., and Allara, D. L. (1990) Fundamental studies of microscopic wetting on organic surfaces. 1. formation and structural characterization of a self-consistent series of polyfunctional organic monolayers. *J. Amer. Chem. Soc.* 112:558–569.

Dubois, L. H. and Nuzzo, R. G. (1992) Synthesis, structure, and properties of model organic surfaces. *Annu. Rev. Phys. Chem.* 43:437–463.

Dunbar, T. D., Sheen, C. W., and Allara, D. L. (1994) Unpublished results. Studies of partial coverage self-assembled monolayers.

Ertel, S., Atre, S. V., Ratner, B. D., and Allara, D. L. (1994) Studies of protein adsorption and cell growth on self-assembled monolayer surfaces.

Fenter, P., Eisenberger, P., and Liang, K. S. (1993) Chain length dependence of the structures and phases of $CH_3(CH_2)_{n-1}SH$ self-assembled on gold. *Phys. Rev. Lett.* 70:2447–2450.

Folkers, J. P., Laibinis, P. E., and Whitesides, G. M. (1992a) Self-assembled monolayers of alkanethiols on gold: comparisons of monolayers containing mixtures of short-chain and long-chain constituents with CH_3 and CH_2OH terminal groups. *Langmuir* 8:1330–1341.

Folkers, J. P., Laibinis, P. E., and Whitesides, G. M. (1992b) Self-assembled monolayers of alkanethiols on gold: the adsorption and wetting properties of monolayers derived from two components with alkane chains of different length. *J. Adhesion Sci. Technol.* 6:1397–1410.

Hietpas, G. and Allara, D. L. (1994) Unpublished results. Organized lipid monolayers studied by infrared spectroscopy.

Huang, J., Dahlgren, D. A., and Hemminger, J. C. (1994) Photopatterning of self-assembled alkanethiolate monolayers on gold. A simple monolayer photoresist utilizing aqueous chemistry. *Langmuir* 10:626–628.

Kumar, A., Biebuyck, H. A., Abbot, N. L., and Whitesides, G. M. (1992) The use of self-assembled monolayers and a selective etch to generate patterned gold features. *J. Am. Chem. Soc.* 114:9188–9189.

Laibinis, P., Whitesides, G. M., Parikh, A. N., Tao, Y. T., Allara, D. L., and Nuzzo, R. G. (1991) A comparison of the structures and wetting properties of self-assembled monolayers of *n*-alkanethiols on the coinage metal surfaces Cu, Ag and Au. *J. Am. Chem. Soc.* 113:7152–7167.

Lercel, M. J., Tiberio, R. C., Chapman, P. F., Craighead, H. G., Sheen, C. W., Parikh, A. N., and Allara, D. L., (1993) Self-assembled monolayer electron-beam resists on GaAs and SiO_2. *J. Vac. Sci. Technol.* B11:2823–2828.

Mastandrea, M., Wilson, T., and Bednarski, M. D. (1989) The use of self-assembled organic films for controlling biological adhesion. *J. Mater. Educ.* 11:529–564.

Neogi, P., Neogi, S., and Stirling, J. M. (1993) Reactivity of carboxy esters in gold-thiol monolayers. *J. Chem. Soc. Commun.* 1134–1136.

Pale-Grosdemange, C., Simon, E. S., Prime, K. L., and Whitesides, G. M. (1991) Formation of self-assembled monolayers by chemisorption of derivatives of oligo (ethylene glycol) of structure $HS(CH_2)11(OCH_2CH_2)mOH$ on gold. *J. Am. Chem. Soc.* 113:12-20.

Poirer, G. E. and Tarlov, M. J. (1994) The c(4x2) Superlattice of *n*-alkanethiol monolayers self-assembled on Au(111). *Langmuir* 10:2853–28.

Porter, M. D., Bright, T. B., Chidsey, C. E. D., and Allara, D. L. (1987) Spontaneously organized molecular assemblies, 4. Structural characterization of *n*-alkyl thiol monolayers on gold by optical ellipsometry, infrared spectroscopy and electro-chemistry. *J. Amer. Chem. Soc.* 109:3559–3568.

Prime, K. L. and Whitesides, G. M. (1991) Self-assembled organic monolayers: model systems for studying adsorption of proteins at surfaces. *Science* 252:1164–1167.

Prime, K. L. and Whitesides, G. M. (1993) Adsorption of proteins onto surfaces containing end- attached oligo (ethylene oxide): a model system using self-assembled monolayers. *J. Am. Chem. Soc.* 115:10714–10721.

Rondelez, F .R., Azouz, I., Parikh, A. N., and Allara, D. L. (1994) An intrinsic relationship between molecular structure in self-assembled alkylsiloxane monolayers and deposition temperature. *J. Phys. Chem.* 98:7577–7590.

Sanassy, P. and Evans, S. D. (1993) Mixed alkanethiol monolayers on gold surfaces - substrates for Langmuir-Blodgett film deposition. *Langmuir* 9:1024–1027.

Schoenberger, C., Sondag-Huethorst, J. A. M., Jorritsma, J., and Fokkink, L. G. J. (1994) What are the "holes" in self-assembled monolayers of alkanethiols on gold. *Langmuir* 10:611–614.

Seshadri, K., Atre, S. V., Tao, Y. T., and Allara, D. L. (1995) Manuscript in preparation. The nucleation and growth of nanometer-scale clusters of polymethylene on gold surfaces via decomposition of diazomethane: structural characterization and thermal properties.

Sheen, C. W., Chen, L., and Allara, D. L. (1994) Unpublished results. Studies of self-assembled monolayers on GaAs(100).

Sheen, C. W. J., Parikh, A. N., Martensson, J., and Allara, D. L. (1992) A new class of self-assembled monolayers: alkane thiols on GaAs (100). *J. Amer. Chem. Soc.* 114:1514–1515.

Singhvi, R., Kumar, A., Lopez, G. P., Stephanopoulos, G. N., Wang, D. I., Whitesides, G. M., and Ingber, D. E. (1994) Engineering cell shape and function. *Science* 264:696–698.

Stenger, D. A., Georger, J. H., Dulcey, C. S., Hickman, J. J., Rudolph, A. S., Nielsen, T. B., McCort, S. M., and Calvert, J. M. (1992) Coplanar molecular assemblies of amino- and perluorinated alkylsilanes, characterization and geometric definition of mamalian cell adhesion and growth. *J. Am. Chem. Soc.* 114:8435–8442.

Strannick, S. J., Cygan, M., Dunbar, T. E., Allara, D. L., and Weiss, P. S. (1994) Unpublished results. STM studies of alkanethiolates on Au(111).

Strannick, S. J., Parikh, A. N., Tao, Y. T., Allara, D. L., and Weiss, P. S. (1994) Phase separation of mixed-composition self-assembled monolayers into nanometer scale molecular domains. *J. Phys. Chem.* 98:7636–7645.

Swalen, J. D., Allara, D. L., Andrade, J. D., Chandross, E. A., Garoff, S., Israelachvilli, J., McCarthy, T. J., Murray, R., Pease, R. F., Rabolt, J. F., Wynne, K. J., and Yu, H. (1987) Molecular monolayers and films. *Langmuir* 3:932–950.

Tengvall, P., Lestelius, M., Liedberg, B., and Lundström, I. (1992) Plasma protein and antisera interactions with L-cysteine and 3-mercaptoproprionic acid monolayers on gold surfaces. *Langmuir* 8:1236–1238.

Tiberio, R. C., Craighead, H. G., Lercel, M. T., Lau, C. W., Sheen, C. W., and Allara, D. L. (1993) Self-assembled monolayer electron beam resist on GaAs. *Appl. Phys. Lett.* 62:476–478.

Troughton, E. B., Bain, C. D., Whitesides, G. M., Nuzzo, R. G., Allara, D. L., and Porter, M. D. (1988) Monolayer films prepared by the spontaneous self-assembly of symmetrical and unsymmetrical dialkyl sulfides from solution onto gold substrates: structure, properties, and reactivity of constituent functional groups. *Langmuir* 4:365–385.

Ulman, A. (1991) *An Introduction to Ultrathin Organic Films, from Langmuir-Blodgett to Self-Assembly.* Boston, MA: Academic Press.

12

Biofunctionalized membranes on solid surfaces

ROBERT TAMPÉ, CHRISTIAN DIETRICH, STEFAN GRITSCH,
GUNTHER ELENDER, and LUTZ SCHMITT

Lehrstuhl für Biophysik
Technische Universität München
D-85748 Garching, Germany
and
Max-Planck-Institut für Biochemie
Am Klopferspitz 18a
D-82152 Martinsried, Germany

12.1 Introduction

The development of the plasma membrane was a crucial step in the generation of the earliest forms of life. By virtue of the inner compartment formed within the membrane vesicle, cells were able to establish and maintain differences between their content and the environment. The resulting gradients became the motor of life and they still drive all biological energy generation via the electron transfer chains of bacterial, mitochondrial, and choloroplast membranes. At the same time, membranes isolated cells from their surroundings – a deadly threat that led to the development of a wide range of membrane-associated channels, transporters, receptors, signaling molecules, and anchors that restored communication with the outside. As a result, energy coupling, cell recognition and stimulation, immune reactions, biocompatibility, and many other processes have become indissolubly associated with membranes.

Although biological membranes assume a wide range of functions and structures (both in their components and in their overall organization), they are all organized according to a common underlying principle described in the fluid mosaic model of Singer and Nicholson (1972). In this model, lipids create a two-dimensional fluid matrix, in which the embedded proteins possess a high grade of mobility. It is important to realize that in all cellular structures, even ones as complex as the membranes, molecules are held together by a combination of weak, noncovalent forces: hydrophobic interaction, ionic bonds, van der Waals forces, and hydrogen bonds. These forces also mediate binding and recognition processes that must combine a requirement for specificity with a certain level of reversibility. These forces must be considered when developing artificial biofunctional surfaces. In contrast to chemical reactive surfaces, these not only need to fix the reactive groups, but also need to maintain critical dynamic properties for an optimal performance of their functional units. Although these difficulties can be by-passed by studying biochemical mechanisms in their natural environment, this requires the difficult transformation of biochemical signals to measurable or observable physical effects. Spectacular examples are the use of cells as indicators (McConnell et al., 1992), the transduction of cell membrane potentials to electric circuits (Fromherz et al., 1991), and other examples given in this book.

In this chapter, we will focus on artificial biofunctionalized surfaces as interfaces between nature and physical detection apparatus. For example, cell–cell recognition in the immune system has been studied with the help of solid supported lipid layers (McConnell et al., 1986). There are many physical methods, such as reflection interference contrast microscopy (RICM), ellipsometry, surface plasmon spectroscopy (SPS), or attenuated total reflection–fourier transform infrared spectroscopy (ATR–FTIR), that are highly surface-sensitive and well adapted for planar systems. Furthermore, a solid matrix offers the possibility of exchanging the bulk solution and varying experimental parameters in a fast and defined way. Although artificial biofunctionalized interfaces may lack compatibility to nature, they open a wide field of possibilities for investigating biological systems. Thus they can help to gain insight into biochemical processes and serve as a tool to perform measurements. In addition, they are of interest for technical applications (for example, biosensors, biocompatibility, and molecular analysis).

As already mentioned, the optimal performance of the functional units depends critically on the physical properties of the membrane. Lipids play an essential role in the formation of microscopic regions of well-defined composition. Together with proteins, they build a cooperative system. The lipid matrices with biofunctionalized surfaces have the following properties:

1) They spontaneously organize into bilayers. This self-assembly guarantees a certain level of stability.
2) Their components are amphipathic molecules formed by a hydrophobic and a hydrophilic part. In the case of synthetic molecules, an adapted design of the hydrophobic and hydrophilic parts allows a well-defined micromolecular orientation in the lipid film.
3) The overall structures created are strongly dominated by dynamic qualities. For example, they posses a high molecular mobility that is also a specific property of biomembranes. Dynamic systems can react to signals with high cooperativity; therefore, very effective amplification mechanisms, such as phase separation processes or structural changes induced by segregation, are possible. Furthermore, the application of external fields allows a spatial enrichment and separation of different molecules (two-dimensional microelectrophoresis, see Section 12.4).

4) The large number of possible membrane components allows for great variability in the bilayer composition. The choice of micromolecular constituents determines the macromolecular qualities of a bilayer, making it possible to influence properties such as fluidity, elasticity, homogeneity, resistance, capacity, or structure.
5) The choice of natural membrane components suppresses phenomena of denaturation or unspecific adsorption at interfaces that would otherwise mask specific binding effects.

To create a sensitive surface it is necessary to support the thin lipid film. But, two contradictory requirements have to be taken into account, which need to be well-balanced by the support. Since a macroscopic planar membrane is an unstable system, the film must be fixed and stabilized on a surface; this requires sufficient adhesion force. Furthermore, a strong contact with the solid can increase the sensitivity of the physical measuring method employed. On the other hand, a strong interaction leads to epitactic coupling, which induces defects or restricts the essential dynamic properties of the membrane. As already emphasized, an undisturbed condition is of high importance for the functionality of the film. A preparation that separates the support from the film avoids any epitactic coupling and enables the reconstitution of transmembrane proteins (see Figure 12.5).

Finding an acceptable compromise between the two requirements demands a knowledge of modification techniques for solid surfaces, a well-adapted architecture of the model membrane, and knowledge of the relevant biochemical and physical features. We illustrate the complexity of the interaction between substrate and membrane by a short overview of the relevant forces.

The electrostatic force has its origin in the surface charge of chemical groups. The electric field interacts with solvent ions, building an electric double layer (Gouy–Chapman–Stern [GCS] theory). In our case, we need to consider the interaction between the double layers and their corresponding potentials of the substrate and the membrane. By changing the charge density of the surfaces, or the molarity or pH of the electrolyte, the strength of the electrostatic force can be modified. A few multivalent ions (calcium for example) exhibit a strong electrostatic behavior that cannot be explained simply by the GCS–theory (Israelachvili, 1992).

Another important, always attractive, long-range component is the van der Waals force. The van der Waals force between a solid surface and a membrane in a third medium can be calculated by the Lifschitz theory (Mahanty and Ninham, 1976) and has an effective range of ≈ 15 nm for biological conditions. The interaction energy between two surfaces is proportional to

$$\approx \frac{A\,(r)}{r^2} \tag{12.1}$$

where $A(r)$ is the Hamaker constant.

A combined treatment of the van der Waals and electrostatic forces is described in the Derjaguin–Landau–Verwey–Overbeck (DLVO) theory (Derjaguin and Landau, 1941; Verwey and Overbeck, 1948).

A further component is the repulsive hydration force, which has a very short-range exponential effect of a few water layers. As two surfaces approach each other, the dehydration of the surface groups requires free energy. This is influenced by the degree of hydration of the surfaces and the presence of adsorbed hydrated ions. The interaction of a fluctuating membrane with a surface also gives rise to a steric repulsion (Helfrich, 1978). If the distance between substrate and membrane is in the range of thermal fluctuations, the membrane will be repelled from the substrate surface. The interaction energy is proportional to

$$\approx \frac{kT}{r^2 k_c} , \tag{12.2}$$

where the bending constant k_c is a property of the membrane.

In the following sections we will describe several commonly used preparation techniques and membrane functions on solid support investigated in our group.

12.2 Membranes on solid support

The first example of a lipid bilayer supported on a glass surface was reported by Tamm and McConnell (1985). More recent work has shown that the properties of the substrate, such as transparency, conductivity, and reflectivity, are constrained according to the measurement method employed. Microscopic techniques often require a transparent medium, whereas electric measurements and plasmon spectroscopy depend on a conductive material. Another important property of the substrate is the roughness. Some surfaces have intrinsic properties that stabilize the membrane directly in a suitable way, but others need some modification to improve or reduce the stabilizing forces. The modification of the surface consists of making a defined molecular or polymeric film to cover the surface in order to change its chemical and physical properties.

All experiments presented in this section use vesicle fusion for deposition of bilayers onto substrates. The Langmuir–Blodgett (LB) technique is more widely used to deposit monolayers on substrates. In the LB technique, a drop of a solution of an amphiphilic molecule is spread at the air–water interface. After evaporation of the solvent, a monolayer of the molecules remains on the water surface; this monolayer is then transferred to the substrate by dipping the substrate through the water surface. More information about this technique can be found in Ulman (1991). Bilayers made from vesicles are often nonuniform compared with those made by LB, but the membranes obtained are still sufficiently tight to be characterized.

Some substrates of interest, such as indium-tin oxide (ITO) and magnesium fluoride, have sufficient surface charge to create strong electrostatic forces that stabilize the membrane–substrate system (Gritsch et al., 1995). As an example, the formation of a stable supported lipid bilayer on a ITO surface is demonstrated in Figure 12.1.

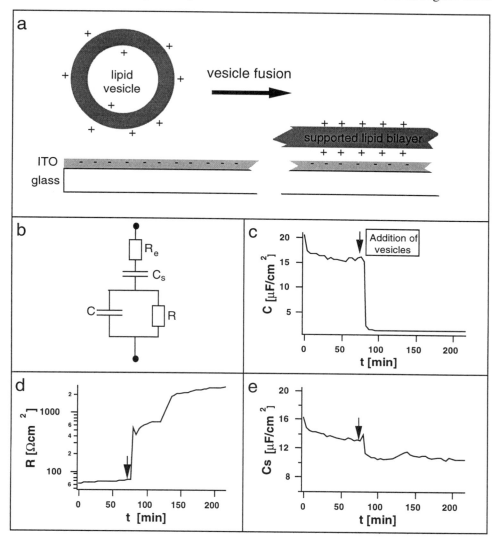

Figure 12.1. (a) Fusion of vesicles (cholesterol/dimyristoyl-L-phosphatidylcholine [DMPC]/dioctadecyldimethylamonium bromide, 5/4/1, mol/mol) onto an indium-tin oxide (ITO)–coated glass surface observed by frequency-dependent impedance analysis (0.1–10,000 Hz). (b) A four-parameter model was used to fit the measured impedance data. After vesicle fusion and subsequent formation of a planar bilayer, the capacitance C (c) and resistance R (d) are changed. The large increase in R demonstrates membrane function and the high resistance of the lipid bilayer. As a consequence of the serial lipid bilayer formation, the capacitance C decreases significantly. (e) As a control, the capacity C_s, which can be interpreted as electrode capacitance, shows only a small change. The electrolyte resistance R_e (not shown) remains constant.

Simultaneously, the membrane properties (resistance, capacitance) are followed by impedance spectroscopy.

Other substrates, such as gold, silver, or different types of glass, are useful only with a modified surface. Gold and silver are important substrates for electrochemical or plasmon measurements (Brink et al., 1994; Stelzle et al., 1993), whereas glass is important in optical methods. A gold surface can be coated with thiols, and glass with silanes. The functionality, surface charge, and hydrophobicity of the head groups can be varied. If the alkane chain is long (n≥8), a well-defined brushlike structure is formed (Figure 12.2). First, lipid monolayers can be transferred directly onto self-assembled hydrophobic alkane chains that simulate the first lipid layer of the membrane. Second, charged lipid vesicles can adsorb and fuse onto the oppositely charged carboxy- or amino-functionalized monolayers in a manner similar to membrane formation on the ITO surface. Third, calcium ions can induce vesicle fusion of negatively charged lipids to carboxy-thiol film (Stelzle et al., 1993). As discussed in Section 12.1, the system can be stabilized by the forces between the membrane and the modified surface. Stability can also be established via chemical binding of the membrane to the support. For this approach, the membrane and solid surface need to be functionalized. Two examples can be given of this procedure.

The first example exploits the high affinity binding of the biotin-(strept)avidin system. Vesicles doped with biotinylated lipids spontaneously form a stable membrane on a streptavidin-coated gold surface. The experiments show that there is no interaction between the protein film and nonbiotinylated vesicles, and that small amounts of incorporated biotin lipids are sufficient to induce adhesion. This suggests that by varying the surface density of biotin and streptavidin, adhesion can be controlled. By surface plasmon spectroscopy, and also by impedance spectroscopy (which allows a very sensitive detection of membrane defects), we can demonstrate that this membrane has properties similar to those of biological and free model membranes (Figure 12.3).

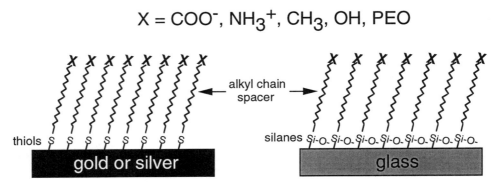

Figure 12.2. Modification of gold and SiO$_2$ surfaces by self-assembly of thiols and silanes. Functional groups X, e.g., carboxy, amino, alkyl, hydroxy, and polyethylene oxid (PEO), are separated from the solid support by a homoalkane chain of variable length (spacer).

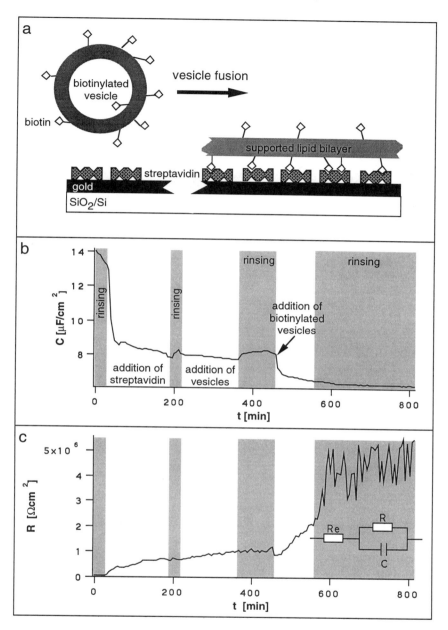

Figure 12.3. (a) Schematic view of vesicle fusion onto a functionalized surface. Egg–phosphatidylcholine (PC) vesicles were doped with biotinylated lipid (2 mol%) and prepared in a buffer solution (10 mM Hepes, 150 mM NaCl, pH = 7.5). Streptavidin was adsorbed to a freshly deposited gold layer on glass in aqueous solution. During steptavidin adsorption and vesicle fusion, the capacitance C (b) and resistance R (c) are followed by frequency-dependent impedance spectroscopy (0.1–10,000 Hz). A three-parameter model is applied to fit the data. The data demonstrate that biotinylated vesicles interact specifically with the modified solid surface in contrast to pure egg–PC vesicles. The high resistance R after the addition of biotinylated vesicles indicates the formation of a tight supported lipid bilayer.

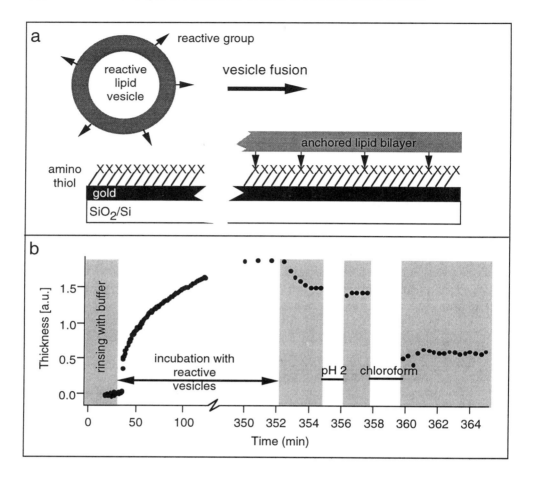

In the second example, the gold surface is coated with a amino-functionalized thiol film. The vesicles contain a small amount of an active ester lipid that reacts with amino-functionalized surfaces in aqueous solution, fixing the membrane covalently to the solid support. Membrane formation, thickness, and stability against low pH or organic solvents can be monitored by near infrared surface plasmon spectroscopy (NIR–SPS) (Brink et al., 1994) (Figure 12.4).

12.3 Polymer-supported membranes, a special form of supported membranes

Polymer-supported membranes are separated from their solid support by a thin (100–1,000 Å) hydrophilic polymer network. Because of this polymer cushion, biocompatible membranes can be designed on solid, eventually nanofabricated surfaces. The polymer layer is intended to act as a distance holder between the

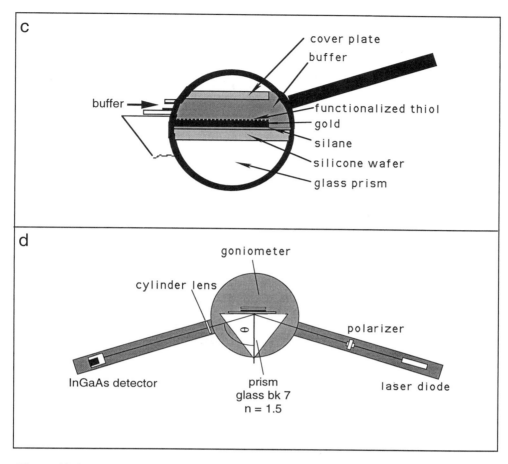

Figure 12.4. (a) Fusion of DMPC vesicles doped with a reactive lipid containing an active ester headgroup (10 Mol%) to an amino-functionalized thiol film on a gold layer. Vesicle fusion and bilayer formation was followed by near-infrared surface plasmon spectroscopy (NIR–SPS). (b) The change in resonance angle is plotted in arbitrary units. Addition of reactive lipid vesicles to a modified surface resulted in an increase of optical thickness and, consequently, in a proportional change of the resonance angle. After several hours of incubation, the surface was rinsed with pure buffer, removing excess lipid. Rinsing was performed in buffer solution (10 mM phosphate buffer, 150 mM NaCl, pH 5). Rinsing with acid (pH 2) did not alter the thickness, whereas washing with chloroform resulted in removal of the lipid layer. Note that covalently linked lipid is still attached to the surface. Schematic views of the flow chamber (c) and apparatus (d) are given. NIR light passes through the silicon base of the microchamber onto the back of the gold layer. It is absorbed at angles of incidence characteristic for the optical thickness of the specimen surface and is otherwise reflected back through the silicon plate. (Brink et al., 1994)

solid support and the bilayer to reduce the strong, mostly attractive, van der Waals interaction between the rigid surface and the fluid bilayer, thus improving the dynamic properties of lipid bilayer. Restoring the dynamic properties of the membrane causes an annealing of defects and results in a defect-free membrane of high specific resistance (a tight, supported bilayer has a resistance of 10^5 Ωcm^2) (Stelzle and Sackmann, 1989). Furthermore, the soft, nonrigid polymer interface allows the incorporation of membrane-spanning proteins into the supported membrane in a native and functional way, because those hydrophilic parts of the protein protruding to the substrate can be kept in a fluid aqueous environment; they are embedded in a hydrophilic polymer network (Kühner et al., 1994). This hydrophilicity is very important for the functional reconstitution of transmembrane proteins. Diffusion measurements of a lipopeptide embedded into a polymer-supported bilayer provide strong evidence that the protein significantly penetrates into the polymer network. Despite this effect, the lipopeptide also shows a remarkable lateral diffusion (Kühner et al., 1994). Without the polymer, the protein can denature at the solid surface and lose its function. The incorporation of the protein into the membrane can cause defects in the membrane itself (Figure 12.5). A further important point is that the polymer cushion is an isolated compartment between solid support and membrane and is analogous to the inner phase of a cell ("phantom cell").

12.3.1 Design of polymer-supported membranes

The design of polymer-supported membranes can be divided into two parts: (1) the design and fixation of the polymer support itself onto a solid substrate and (2) the deposition and fixation of a membrane onto the polymer cushion.

There are two main possibilities for depositing a thin polymer film onto a solid substrate. First, we will discuss the simple case of polymer adsorption onto the substrate. Here we can decide between random adsorption and self-organized adsorption. Adsorption of polymers (for example, polyethylene oxide) (Killmann et al., 1988) leads to diffuse polymer cushions whose thickness depends on substrate, polymer, and polymer concentration in the solution. Proteins as a special class of polymers can also adsorb onto solid surfaces, as described in Section 12.2. Another possible way to design polymer cushions by adsorption is the self-assembly technique. Here the adsorption process is completely controlled by self-assembly of the polymers onto the substrate. These amphiphilic polymers are spread at the air–water interface and then deposited by the LB technique. Examples of such polymers are lipids with polymeric head groups (Needham et al., 1992) or multilayer polymers based on phthalocyaninate polysiloxane or cellulose that is substituted with alkoxy chains (Wegner, 1992). Other candidates for self-assembly are block copolymers consisting of one adsorbing part and one part that stays in solution.

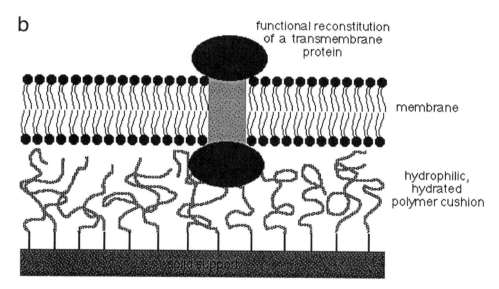

Figure 12.5. Reconstitution of integral membrane proteins into a supported membrane. (a) Denaturation of an incorporated transmembrane protein by contact with the rigid solid surface and the rise of membrane defects around the protein. (b) Polymer cushion as a soft and hydrophilic support for the nonperturbed reconstitution of a protein in a defect-free membrane. The hydrophilic parts of the protein protrude into the polymeric network. On the solid-support side, these protruding parts of the protein are embedded into the polymer cushion.

Grafted polymers are a completely different way to design and immobilize thin polymers onto substrates. Here we will limit our discussion to two important substrates, Si/SiO$_2$ and glass–gold systems, and concentrate on a few examples. The first step in the design is the functionalization of the substrate with a coupling agent. These coupling agents are reactive silanes (chloro-, methoxy-, and ethoxy-silane) for glass or SiO$_2$ surfaces, and thiols for gold surfaces, respectively (see Section 12.2). Silanes in particular offer the opportunity to combine biologically relevant materials with nanofabricated silicon structures via the approximately 10-Å-thick natural silicon oxide layer. The coupling agents carry a second functional group separated from the silane or thiol group by a short alkyl chain, which is easily accessible for coupling reactions. The macromolecules, such as polyethylene oxide, dextran, polyacrylamide, or agarose, can be attached to the surface after synthesis. Alternatively, the polymeric cushion for membrane support can be obtained by polymerization of monomers onto reactive surface groups (Kühner et al., 1994).

The deposition of lipid mono- or bilayers onto hydrophilic polymers is analogous to the deposition directly onto hydrophilic substrates by the LB technique, the Langmuir–Schäfer technique, or vesicle fusion. But the van der Waals interaction between the solid substrate and the lipid layer is now reduced by the interlaying quasi-fluid polymer cushion. This reduction of van der Waals forces leads to the problem of a lower stability against detachment of the membrane from the support. The strategies for stabilization are the same as for the deposition of membranes on pure solid substrates: (1) electrostatic or chemical fixation of the lipids, (2) lipids or polypeptides that span the whole membrane, and (3) polymerization of the lipids to create a rigid network that stabilizes large parts of the fluid membrane. A very special technique for the design of polymer-supported membranes was established in Ringsdorf's group (Spinke et al., 1992). Here, a polymer-supported lipid layer is formed by self-assembly of a multifunctional hydrophilic polymer containing disulfide anchor groups and hydrophobic alkyl chains. A bilayer is then formed by vesicle fusion.

12.3.2 Characterization of polymer-supported membranes

Over the past years, a broad spectrum of surface-sensitive methods has been developed for the characterization of thin films; some of these powerful techniques are very useful for investigating membranes on solid surfaces. The homogeneity of polymer films and deposited supported membranes can be studied by the microfluorescence technique, which yields a lateral resolution of 1 μm and indicates phase separations and inhomogeneities in the lipid layer. The elaborated techniques of fluorescence recovery after photobleaching (FRAP) (Axelrod et al., 1976; Kühner et al., 1994) and steady state photo bleaching (Dietrich and Tampé, 1995; Peters et al., 1981) provide dynamic information about the diffusion of labeled molecules within the polymer network as well as in the membrane. This allows the characterization of both sides of a supported bilayer with respect to mobility, diffusion hindrance, and homogeneity (Kühner et al., 1994). Some new methods in scanning force microscopy yield data about surface viscosity, surface friction, and surface elasticity on high lateral resolutions of about 10 nm (Radmacher et al., 1992) and allow a microscopic characterization of thin polymer films. Neutron reflectometry can determine density profiles of polymer cushions (Field et al., 1992). Spatially resolved ellipsometry and laterally resolved surface plasmon spectroscopy yield information about the film thickness homogeneity of thin polymer cushions and lipid films with a thickness resolution of 1 Å. The hydrophilicity of the substrate can be investigated by measuring contact angles or by measuring the swelling of the polymer cushion under controlled conditions in the range between 60% and nearly 100% relative humidity. This method was previously suggested and described for solution films and adsorbed thin polymer and polymer-solution films (Elender and Sackmann, 1994). The measurement of wetting and dewetting of thin hydrated polymer cushions by lipid monolayers, as well as the study of the stability of asymmetric soap films, are possible ways to investigate interactions between substrate and lipid layer (Frey and Sackmann, 1992). Another way to investigate small forces (down to 10^{-15} pN) between substrates and lipid layers is the study of distance fluctuations by microinterferometry (RICM) (Rädler and Sackmann, 1992). In addition, the surface force apparatus is very well suited for the measurement of weak forces between different surfaces (Israelachvili, 1992).

Frequency-dependent electrical measurements can detect the occurrence of small defects and the subsequent high specific resistance of supported bilayers, as well as the adsorption of proteins (Stelzle and Sackmann, 1989; Stelzle et al., 1993). A fairly new method for the local enrichment of charged particles in two-dimensional systems is microelectrophoresis in supported membranes (Dietrich and Tampé, 1995; Stelzle et al., 1992). By the local enrichment of receptor proteins, microelectrophoresis leads to signal amplification and an improvement of signal-to-noise ratio for the detection of specific binding reactions. New work in this field has shown that the great problem of counterworking electroosmosis is highly reduced in polymer-supported membranes by the polymer network. In consequence, by microelectrophoresis the charge determination of membranes molecules is possible (Dietrich and Tampé, 1995) (Figure 12.6).

The stability of lipid layers on thin polymer cushions (and on any other substrate) can be detected by microfluorescence. It should also be possible to detect detachment of the membranes by all other methods available, such as ellipsometry, surface plasmon spectroscopy, and impedance spectroscopy, to name only

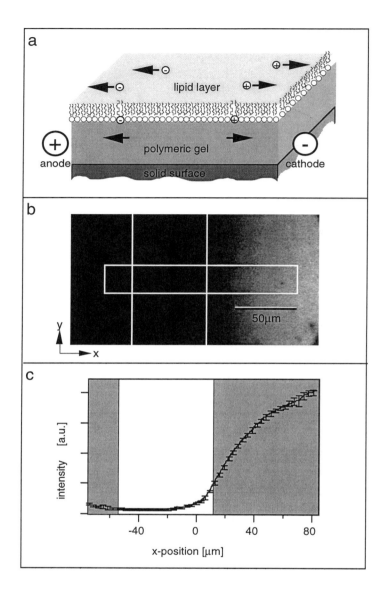

Figure 12.6. Two-dimensional microelectrophoresis and bleach profiles of a DMPC monolayer doped with 2 mol% of a negatively charged, fluorescent membrane probe. The monolayer is deposited on an agarose substrate at room temperature and at 20 mN/m. The pictures were recorded after the slit was removed. Without an electric field, the intensity profile is symmetric. In the presence of an electric field, the symmetry disappears. Due to the electric field (anode is on the left side), negatively charged dyes are drifting from the left side into the illuminated region, where a part of the fluorophore is bleached. This leads to a decrease of fluorescence toward the left side. By analysis of the bleach profile, the charge of the mem-

some of the methods. In the next section, a broader overview is given: It is certainly far from complete, but it shows the broad range of techniques available for the investigation of membrane properties in such systems.

12.4 Methods

Some methods for investigating supported membranes are well known from other areas of surface science and nanotechnology. The use of radiolabeled molecules to measure mass changes and relative amounts of surface material is also well known from other fields. Many well-established methods of contemporary surface science, such as x-ray photoelectron spectroscopy, or secondary ion-mass spectroscopy, that require ultrahigh vacuum during the measurements have been successfully applied to the research of supported biomembranes. Like those methods, electron microscopy calls for conditions far from the natural environment of lipids, but it has given good insight into the properties of supported biomembranes. Preservation of the morphology of organic and biological materials is the major clue to studying supported biomembranes with electron microscopy and other vacuum techniques.

Surface science instrumentation was developed for relatively robust systems. Many of the properties of supported biomembranes, however, call for methods that take softness, fluidity, mixing, and phase seperating events into account. Many of those investigation methods use fluorescent labeling of the biomolecules. In some cases fluorescent labeling serves the same goal as radiolabeling, namely the detection of trace amounts of substance. Unlike radiolabeling or other tracer methods, fluorescent probes are detected and imaged in real time. Therefore, dynamic processes can be visualized. From the analysis of fluorescence recovery after photo bleaching (FRAP), one can calculate the mobility of fluorescent components in the membrane and derive conclusions on the homogeneity and fluidity of the membrane. Another application of fluorescence takes advantage of the evanescent electrodynamic field accompanying any total reflection. Total internal reflection fluorescence (TIRF) performed from within the supporting body illuminates only molecules close to the supporting surface. Thereby, molecules in the volume above the surface can be discriminated from molecules bound to the surface.

Recently another labeling technique has become increasingly successful. In nuclear magnetic resonance (NMR), the biocomponent is labeled without changes in its chemical composition by replacement of hydrogen atoms with deuterium. Similarly, other atom species can be used to label molecules by isotope exchange. Broad-band NMR measures diffusion and other dynamic parameters in time constants of 10^{-6} to 10^{-9} sec. The orientation parameter can be studied as well and gives additional information, for example, on phase transitions.

An important trend in biosurface science and technology is the quest for real-time, label-free instrumentation. Ellipsometry, an optical technique for measuring film thicknesses, can achieve a thickness resolution of < 1 Å. In ellipsometry, light of known polarization and angle of incidence is reflected from a specimen. The reflectivity and change of polarization are measured. From Fresnel coefficients and with respect to multiple interferences, thicknesses and/or complex indices of refractivity can be determined for different layers of the specimen. Application of imaging ellipsometry has given insight into wetting and dewetting processes in the construction of supported monolayers (Frey and Sackmann, 1992). More recently it has been used to study lipid layers on polymer supports (Elender and Sackmann, 1994). Computerized image processing has given light microscopy new importance. Microinterferometry (RICM) is one example in which light microscopy could be applied to quantify dynamic properties of supported membranes and their assembly.

As demonstrated in Section 12.2, membrane function (resistance and capacity) can be investigated by electrochemical techniques such as impedance spectroscopy and cyclic voltammetry. These methods have proved to be highly sensitive to defects in the membrane coverage of the supporting surface (electrode). Most of these techniques, however, are not suited for easy-to-use laboratory devices. Engineers and scientists have tried to use other principles to detect surface adsorption, such as quartz microbalance or surface plasmon spectroscopy; the latter has been made commercially available (Chaiken et al., 1992).

Recently, we introduced near-infrared surface plasmon spectroscopy (NIR–SPS) (see Figure 12.4). This spectrometer is based on nanofabricated devices that work at a wave length of 1,300 nm, where both silicon and water exhibit a transmission gap. NIR–SPS also exhibits other advantages: Because the dielectric constants of metals show a distinct wave length dependence, surface plasmons with sharper resonance and higher field enhancement close to the surface can be observed (Brink et al., 1994).

12.5 Biofunctionalization of lipid interfaces

Natural membranes contain a wide variety of covalently and noncovalently attached macromolecules, such as carbohydrates and proteins that are integral to the function of the membranes. Therefore, biofunctionalization of artificial membranes (that is, the reconstitution of isolated membrane components, the covalent coupling of biological macromolecules, and the generation of functionalized lipid surfaces using ligand–receptor interactions) will be essential for the simulation of biological properties.

In contrast to adsorption, the covalent coupling of biomolecules to lipid interfaces offers a number of advantages. This system is independent of pH, charge, and salt concentration; it also offers the possibility of immobilizing receptors, ligands, or carbohydrates in a well-defined way. Important applications of this principle include the synthesis of glycolipids and hapten-bound lipids, which makes possible the study of recognition processes at membranes. Not only can small molecules be covalently coupled to lipids, but proteins such as antibody fragments can be coupled as well (Martin et al., 1981). These protein–lipid conjugates have been used as models for lipoproteins or for membrane-bound immunoglobulins. A third class of important synthetic lipid-conjugates are lipopeptides. In particular, immunogenic lipopeptides are useful instruments in the investigation of early steps in the immune response (Metzger et al., 1991).

Because of its high affinity, the biotin–(strept)avidin system is well-established. Today, biotin-lipids of variing chemical composition and spacer length are available, and this system is frequently used for studying receptor–ligand interactions on lipid interfaces (Blankenburg et al. 1989). Because biotinylated DNA, proteins, and peptides are now commercially available, streptavidin can be used as a docking interface to design an architecture for molecular layers on lipid surfaces (Mueller et al., 1993).

In molecular biology, immobilized metal affinity chromatography (IMAC) is a frequently applied method for the identification, one-step purification, and characterization of gene products expressed as fusion proteins (Hochuli et al., 1988; Porath et al., 1975). In principle, the recombinant protein contains an additional short histidine sequence fused to the C- or N-terminus of the protein. This histidine–tag has a high affinity to immobilized metal complexes such as nickel N-nitrilotriacetic acid (NTA).

By synthesis of a novel class of chelator lipids, we have combined this principle with the fundamental properties of lipids (Schmitt et al. 1994). These lipids carrying a chelator head group are separated by a flexible hydrophilic spacer molecule and form supermolecular assemblies. Films of these chelator lipids are highly metal ion–sensitive (Schmitt et al., 1994). After complex formation, imidazol, histidine, or histidine-tagged biomolecules can be immobilized and oriented at the lipid interface in a functional manner. Due to the phase behavior of various lipid mixtures, the chelator lipids and therefore the histidine-tagged biomolecules can be structured laterally, forming two-dimensional protein arrays (Dietrich et al., 1995; Gritsch et al., 1995). In contrast to the biotin–streptavidin system, reversibility is simply achieved by complexing Ni^{2+} with a stronger chelator, by a competitor, or by lowering the pH. In comparison to the biotin–streptavidin system, this system has several advantages: (1) It is not limited to (strept)avidin as the docking element; (2) chemical biotinylation of macromoleucles is difficult to control, whereas the histidine tags in fusion proteins are always located in a defined position; (3) a vast variety of histidine-tagged proteins are expressed and purified; therefore, no additional synthetic effort like biotinylation of protein is

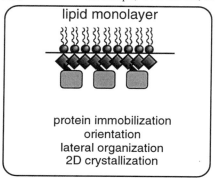

lipid monolayer

protein immobilization
orientation
lateral organization
2D crystallization

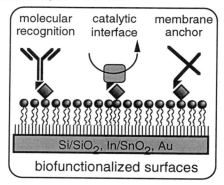

molecular   catalytic   membrane
recognition interface   anchor

Si/SiO$_2$, In/SnO$_2$, Au

biofunctionalized surfaces

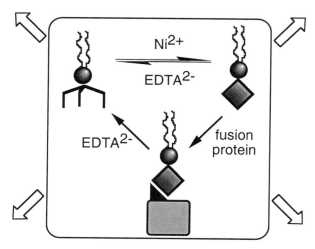

Ni^{2+}

EDTA^{2-}

EDTA^{2-}

fusion
protein

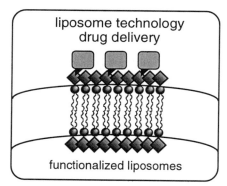

liposome technology
drug delivery

functionalized liposomes

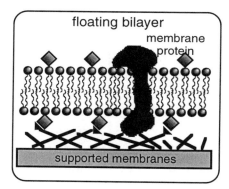

floating bilayer

membrane
protein

supported membranes

Figure 12.7. Schematic view of the complex formation and specific binding of histidine-tagged biomolecules to the novel class of chelator lipids (center). Applications of the chelator lipid in life and material sciences are illustrated. Beside the biofunctionalization of lipid interface, this concept can be applied to anchor lipid bilayers on a solid or polymer support in a well-defined way.

necessary; and, (4) histidine-tagged proteins can be reversibly immobilized, oriented, and organized at the lipid interface. The principle of these chelator lipids and their possible application are illustrated in Figure 12.7.

Biofunctionalized interfaces are of major interest in the material and life sciences, for example, for catalytic and switching interfaces, as models for molecular and cellular recognition processes, and in liposome technology, drug delivery, imaging techniques, and sensor applications. In combination with the large variety of fusion proteins already available, this novel concept represents a powerful and flexible technique for specific and reversible orientation, organization, and two-dimensional crystallization of biomolecules at self-assembled interfaces.

Acknowledgments

The authors like to thank Dr. Gunnar Brink and Dr. Gilberto Weissmüller for providing unpublished results on SPS and impedance spectroscopy, and Dr. Erich Sackmann for fruitful discussions.

References

Axelrod, D., Koppel, D. E., Schlessinger, J., Elson, E., and Webb, W. W. (1976) Mobility measurement by analysis of fluorescence photobleaching recovery kinetics. *Biophys. J.* 16:1055-1069.

Blankenburg, R., Meller, P., Ringsdorf, H., and Salesse C. (1989) Interaction between biotin lipids and streptavidin in monolayers: formation of oriented two-dimensional protein domains induced by surface recognition. *Biochemistry* 28: 8214-8221.

Brink, G., Schmitt, L., Tampé, R., and Sackmann, E. (1994) Self assembly of covalently anchored phospholipid supported membranes by use of DODA-Suc-NHS-lipids. *Biochim. Biophys. Acta* 1196:227-230.

Chaiken, I., Rosé, S., and Karlsson, R. (1992) Analysis of macromolecular interactions using immobilized ligands. *Anal. Biochem.* 201:197-210.

Derjaguin, B. V. and Landau, L. (1941) *Acta Physcochim. URSS* 14:633-662.

Dietrich, C., Schmitt, L., and Tampé, R. (1995) Molecular organization of histidine-tagged biomolecules at self-assembled lipid interfaces using a novel class of chelator lipids. *Proc. Natl. Acad. Sci. USA* 92: 9014–9018.

Dietrich, C. and Tampé, R. (1995) Charge determination of membrane molecules in polymer-supported lipid layers. *Biochim. Biophs. Acta* 1238:183–191.

Elender, G. and Sackmann, E. (1994) Wetting and dewetting of Si/SiO_2-wafers by free and lipid-monolayer covered aqueous solutions under controlled humidity. *J. Phys II France* 4:455-479.

Frey, W. and Sackmann, E. (1992) Solitary waves in asymmetric soap films. *Langmuir* 8:3150-3154.

Fromherz, P., Ofenhausser, A., Vetter, T., and Weis, J. (1991) A neutron silicon junction a retzius cell of the leech on an insulted-gate field-effect transistor. *Science* 262:1707-1708.

Gritsch, S., Neumaier, K., Schmitt, L., and Tampé R. (1995) Engineered fusion molecules at chelator lipid interfaces imaged by reflection interference contrast microscopy (RICM). *Biosensors Bioelectronics,* in press

Helfrich, W. Z. (1978) Steric interaction of fluid membranes in multilayer systems. *Z Naturforsch.* 33A:305-315.

Hochuli, E., Bannwarth, W., Döbeli, H., Gentz, R., and Stüber, D. (1988) Genetic approach to facilitate purification of recombinant proteins with a novel metal chelate absorbent. *Bio/Technology* 6:1321-1325.

Israelachvili, J. N. (1992) *Intermolecular and surface forces.* London: Academic Press.

Killmann, E., Maier, H., and Baker, J. A. (1988) Hydrodynamic layer thickness of various adsorbed polymers on precipitated silica and polystyrene latex. *Colloids Surf.* 31:51-71.

Kühner, M., Tampé, R., and Sackmann, E. (1994) Lipid Mono- and Bilayer Supported on Polymer Films - Composite Polymer-lipid Films on Solid Substrates. *Biophys. J.* 67:217-226.

Mahanty, J. and Ninham, B. W. (1976) *Dispersion Forces.* London: Academic Press.

Martin, F. J., Hubbell, W. L., and Papahadlopoulus, D. (1981) Immunospecfic targeting of lipsomes to cells: A novel and efficient method for covalent attachment of FAB´ fragments via disulfide bounds. *Biochemistry* 20:4229-4238.

McConnell, H. M., Owicki, J. C., Parce, J. W., Miller, D. L., Baxter, G. T., Wada, H. G., and Pitchford S. (1992) The cytosensor microphysiometer: biological applications of silicon technology. *Science* 257:1906-1912.

McConnell, H. M., Watts, T. M., Weiss, R. M., and Brian, A. A. (1986) Supported planar membranes in studies of cell-cell recognition in the immune system. *Biochim. Biophys. Acta* 864:95-106.

Metzger, J., Weismüller, K.-H., Schaude, R., Bessler, W. G., and Jung, G. (1991) Synthesis of novel immunologically active tripalmitoyl-S-glycerylcysteinyl lipopeptides as useful intermediates for immunogen preparations. *Int. J. Peptide Protein Res.* 37:46-57.

Mueller, W., Ringsdorf, H., Rump, E., Wildburg, G., Zhang, X., Angelmeier, L, Knoll, W., Liley, M., and Spinke, J. (1993) Attempts to mimic docking processes of the immune system: redognition-induced formation of protein multilayers. *Science* 262:1706-1708.

Needham,.D., McKintosh, T. J., and Lasic, D. D. (1992) Repulsive interactions and mechanical stability of polymer-grafted lipid membranes. *Biochim. Biophys. Acta* 1108: 40-48.

Peters, R., Brünger, A., and Schulten, K. (1981) Continuous fluorescence photolysis: a sensitive method for study of diffusion processes in single cells. *Proc. Natl. Acad. Sci. USA* 78:962-966.

Porath, J., Carlsson, J., Olsson, I., and Belfrage, G. (1975) Metal chelate affinity chromatography, a new approach to protein purification. *Nature* 258:598-599.

Radmacher, M., Tillmann, R. W., Fritz, M., and Gaub, H. E. (1992) From molecules to cells- imaging soft samples with the AFM. *Science* 257:1900-1905.

Rädler, J. and Sackmann, E. (1992) On the measurement of weak repulsive and frictional colloidal forces by reflection interference contrast microscopy. *Langmuir* 8:848-853.

Schmitt, L., Dietrich, C., and Tampé, R. (1994) Synthesis and characterization of chelator-lipids for reversible immobilization of engineered proteins at self-assembled lipid interfaces. *J. Am. Chem. Soc.* 116:8485-8491.

Singer, S. J. and Nicholson, G. L. (1972) The fluid mosaic model of the structure of cell membranes. *Science* 175:720-731.

Spinke, J., Yang, J., Wolf, H., Liley, M., Ringsdorf, H., and Knoll, W. (1992) Polymer-supported bilayer on a solid substrate. *Biophys. J.* 63:1667-1671.

Stelzle, M., Miehlich, R., and Sackmann, E. (1992) Two dimensional microelectrophoresis in supported lipid bilayers. *Biophys. J.* 63:1346-1354.

Stelzle, M. and Sackmann, E. (1989) Sensitive detection of protein adsorption to supported bilayers by frequency dependent capacitance measurements and microelectrophoresis. *Biochim. Biophys. Acta* 981:135-142.

Stelzle, M., Weissmüller, G., and Sackmann, E. (1993) On the application of supported lipid bilayers as receptive layer for biosensors with electrical detection. *J. Phys. Chem.* 97:2974-2981.

Tamm, L. K. and McConnell, H. M. (1985) Supported phospholipid bilayers. *Biophys. J.* 47:105-113.

Ulman, A. (1991) An introduction to ultrathin films: from Langmuir-Blodgett to self-assembly. London: Academic Press.

Verwey, E. J. W. and Overbeck, J. T. G. (1948) Theory of stability of lyophilic colloid. Amsterdam: Elsevier.

Wegner, G. (1992) Ultrathin films of polymers: architecture, characterization and properties. *Thin Solid Films* 216:105-116.

13

Molecular assembly technology for biosensors

M. AIZAWA, K. NISHIGUCHI, M. IMAMURA, E. KOBATAKE,
T. HARUYAMA, and Y. IKARIYAMA

Department of Bioengineering
Tokyo Institute of Technology
Nagatsuta, Midori-ku
Yokohama 227, Japan

13.1 Introduction

Because the size of many protein molecules is in the nanometer range, nanofabrication has long been anticipated for protein-molecular electronic devices that could be operated by reflecting the properties of individual protein molecules (Aizawa, 1994). Through the capabilities of macroscale fabrication, protein-molecular electronic devices such as biosensors have been designed on the basis of the collective properties that are derived from averaging the individual molecular properties (Aizawa, 1991). We have devoted our research program to the development of nanofabrication technology for functional protein molecules on solid surfaces, as well as sensing technology for an individual protein molecule in native form. This chapter deals with our current progress with organized protein arrays on solid surfaces and molecular sensing of the individual protein molecules on the protein arrays.

There have been extensive efforts in obtaining ordered arrays of protein molecules on solid surfaces; however, such efforts have concentrated on the fabrication of proteins such as bacteriorhodopsin, which is intrinsically easy to spread over a water surface by the creation of Langmuir–Blodgett (LB) films. Most functional proteins are, however, too soluble in water to fulfill the requirements for LB film fabrication. Another fabrication technology, self-assembly, has aroused keen interest in applications where an ordered structure of protein arrays on solid surfaces can be created. Some proteins in solution are self-assembled on metal surfaces such as mercury and gold, due to affinity of their specific groups. Such properties strongly suggest that a combination of LB film technology and self-assembling technology should be feasible in the fabrication of functional protein arrays on solid surfaces with retention of their activities.

In the first part of this chapter, two types of self-assembling processes are described. One method is a potential-assisted self-assembly of redox enzymes on the surface of an electrode. This potential-assisted self-assembly is followed by electrochemical deposition of a conducting polymer that functions as a molecular interface to enable the redox enzyme to communicate with the electrode. The other process is a self-assembly of mediator-modified redox enzymes on the surface of a porous gold electrode.

The last part of this chapter is concerned the the fabrication of antibody arrays on solid surfaces using LB film technology coupled with self-assembly technology. Antibody molecules have been considered; it is thought that they could be aligned in ordered form on solid surfaces, because the corresponding antigen molecules can be selectively bound to the surface by molecular recognition of the substrate-bound antibody. Generally, it is difficult to obtain antibody arrays on solid surfaces using LB film methods because of the antibodies' solubility in water. In our previous paper (Owaku et al., 1993), we reported that protein A (Prot A), which has a specific binding affinity to the Fc region of immunoglobulin G (IgG), was found to be spread over the water surface to form a monolayer. This has led to the establishment of a novel method of creating an ordered antibody array on solid surfaces.

In recent years, biosensing technology has made marked progress with the implementation of biological selectivity into electronic and optoelectronic devices. Enzymes as well as antibodies have been important proteins in the development of molecular recognition in biosensing. Most biosensing devices implement these protein molecules on solid surfaces and rely on their ability to sense changes in physicochemical properties after the corresponding molecular recognition. Because these physicochemical changes are derived from the averaging properties of the surfaces, such biosensing technologies have suffered from perturbations of various artifacts. One of the ultimate goals of biosensing technology is to develop a new technology to quantitatively select individual molecules at specific sites.

Scanning probe microscopy (SPM) seems promising in molecular enumeration, although several problems remain unsolved in obtaining sharp molecular images of proteins. Atomic force microscopy (AFM) has successfully been applied to quantitate individual antigen molecules that are aligned on the ordered structure of the corresponding antibody protein array.

13.2 Potential-assisted self-assembly and molecular interfacing of redox enzymes on the electrode surface

One of the key technologies required for fabricating biomolecular electronic devices concerns the molecular assembly of electronic proteins (such as redox enzymes) in monolayers on electrode surfaces. Furthermore, the molecularly assembled electronic proteins are required to communicate with the electrode (Aizawa, 1989; Aizawa et al., 1989; Cardosi et al., 1991; Gleria and Hill, 1992). Individual protein molecules on electrode surfaces should be electronically accessed through the electrode. To fulfill these requirements, two fabrication processes have been proposed by us. One is a potential-assisted self-assembly of redox enzymes on the electrode surface, which is followed by an electrochemical fabrication of a monolayer-scale conducting polymer on the electrode surface for molecular interfacing. The other is a self-assembly of mediator-modified redox enzymes on a porous gold electrode surface through the thiol–gold interaction.

The potential-assisted self-assembly is carried out in an electrolytic cell equipped with three electrodes: a platinum or gold electrode (working electrode) on which a protein monomolecular layer is formed, a platinum counter electrode, and an Ag/AgCl reference electrode. The potential of the working electrode is precisely controlled with a potentiostat with reference to the Ag/AgCl electrode. The protein solution should be prepared with the isoelectric point in mind, because proteins are negatively charged in pH ranges above the isoelectric point.

Fructose dehydrogenase (FDH) is a redox enzyme that has pyrrole-quinoline quinone (PQQ) as a prosthetic group. Upon enzymatic oxidation of D-fructose, the prosthetic group (PQQ) is reduced to $PQQH_2$, and an electron acceptor reoxidizes $PQQH_2$ to PQQ with the liberation of two electrons. FDH is a requisite element of a biosensor for fructose because it can selectively recognize D-fructose as a result of electron transfer from the D-fructose to an electron acceptor in solution. Due to steric hindrance, however, it is difficult for FDH to make the electron transfer from fructose directly to an electrode in place of an electron acceptor in solution. Fructose dehydrogenase is, therefore, one of the typical redox enzymes that have demanding conditions for molecular assembly resulting in electronic communication on the electrode surface.

A monolayer of FDH has been formed both on platinum electrode and on gold electrode surfaces by potential-assisted self-assembly, as schematically illustrated in Figure 13.1 (Khan et al., 1992). FDH was dissolved in phosphate buffer (pH 6.0) to make its net charge negative. FDH molecules instantly adsorb on the electrode surface, primarily due to electrostatic interaction. Under controlled electrode potential, FDH adsorption increased with time and reached a steady state, as shown in Figure 13.2. In the potential range from 0 to +0.5 V, the adsorption rate of FDH sharply increased with the electrode potential. FDH molecules may be self-assembled on electrode surfaces in such a manner that the negatively charged site of the FDH molecule faces the positively charged surface of the electrode. Enzyme assay clearly showed that electrode-bound FDH retained its enzyme activity without appreciable inactivation.

Adsorption isotherms were obtained at various electrode potentials for potential-assisted self-assembly of FDH on the electrode surface. From the isotherms, the amount of self-assembled FDH can be precisely regulated by electrode potential and potential-controlled time. One can easily obtain a monolayer of electrode-bound protein either as full surface coverage or as less surface coverage with its biological function.

In the next step, a molecular wire of molecular interface was prepared for the electrode-bound FDH by potential-assisted self-assembly. Polypyrrole was used as a molecular wire of molecular interface for the electrode-bound FDH and was synthesized by electrochemical oxidative polymerization of pyrrole.

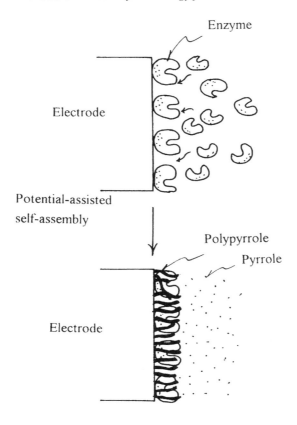

Enzyme

Electrode

Potential-assisted
self-assembly

Polypyrrole

Pyrrole

Electrode

Electrochemical polymerization

Figure 13.1. Fabrication process of potential-assisted self-assembly and electrochemical polymerization.

Electrochemical oxidative polymerization of the pyrrole was performed on the FDH-adsorbed electrode in a solution containing 0.1 M pyrrole and 0.1 M KC1 under anaerobic conditions at a potential of 0.7 V. The thickness of the polypyrrole membrane was controlled by polymerization electricity. The electrochemical polymerization was stopped when the monolayer of FDH on the electrode surface was presumably covered by the polypyrrole membrane. The total electricity of electrochemical polymerization was controlled at 4 mC. The molecular-interfaced FDH was thus prepared on the electrode surface.

Electronic communication between electrode-bound FDH and an electrode has been confirmed by differential pulse voltammetry. Differential pulse voltammetry of the molecular-interfaced FDH was conducted in a pH 4.5 buffered solution. A pair of anodic and cathodic peaks was observed for the molecular-interfaced FDH, as shown in Figure 13.3. The peaks are attributed to the electrochemical oxidation and reduction of the PQQ enzyme at redox potentials of 0.08 and 0.07 V vs. Ag/AgCl,

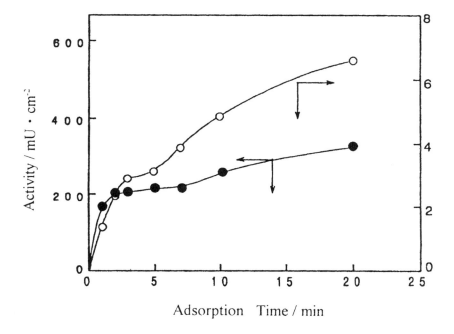

Figure 13.2. Time-course of potential-assisted self-assembly of FDH on an electrode surface.

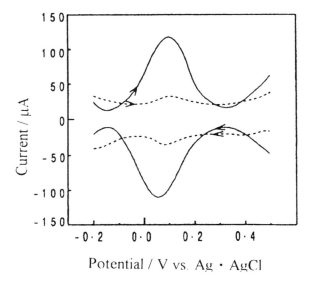

Figure 13.3. Differential pulse voltammogram of the polypyrrole-interfaced fructose dehydrogenase on an electrode surface.

respectively. In addition, the anodic and cathodic peak shapes and peak currents of the molecular-interfaced FDH were identical, which suggests reversibility of the electron transfer process. On the other hand, FDH exhibited no appreciable peaks in differential pulse voltammetry on the electrode surface without the polypyrrole molecular interface. These results indicate that polypyrrole works as an effective molecular interface for electronic communication between FDH and an electrode.

In addition to FDH, potential-assisted self-assembly has successfully been applied to several redox enzymes, including glucose oxidase and alcohol dehydrogenase. The self-assembled redox enzymes have also been molecularly interfaced with the electrode surface via a conducting polymer.

13.3 Self-assembly of mediator-modified redox enzymes on the porous gold electrode surface

In contrast to the molecular wire of molecular interfaces, electron mediators are covalently bound to a redox enzyme in such a manner that an electron-tunneling pathway is formed within the enzyme molecule. Therefore, enzyme-bound mediators work as a molecular interface between an enzyme and an electrode. Degani et al. (1987) proposed that the intermolecular electron pathway of ferrocene molecules was covalently bound to glucose oxidase. Few fabrication methods, however, have been developed to form a monolayer of mediator-modified enzymes on such electrode surfaces. We have succeeded in developing a novel electron transfer pathway of mediator-modified enzyme by self-assembly in a porous gold-black electrode.

Glucose oxidase (from *Asperigilus niger*) and ferrocene carboxyaldehyde were covalently conjugated by the Schiff base reaction, which was followed by $NaBH_4$ reduction. The conjugates were dialyzed against phosphate buffer with three changes of buffer and assayed for their protein and ion contents. Porous gold-black was electrodeposited on a micro-gold electrode by cathodic electrolysis with chloroauric acid and lead acetate. Aminoethane thiol was self-assembled on a smooth gold disk electrode (5 mm in diameter) and a gold-black electrode (100 μm in diameter). Ferrocene-modified glucose oxidase was covalently linked to either a modified plain gold or gold-black electrode by glutaraldehyde, as shown in Figure 13.4.

The ferrocene–glucose oxidase conjugates were characterized by molar ratios of ferrocene to enzyme in the range from 6 to 11. All the oxidase conjugates retained enzyme activity. Figure 13.5 shows the cyclic voltammograms of ferrocene–glucose oxidase in solution on a smooth gold electrode with and without glucose. A pair of redox peaks (solid line) indicates a reversible electron transfer of ferrocene-modified glucose oxidase on the electrode. The dashed line, indicating an increase in catalytic current, shows that the enzymatic oxidation of the conjugate is efficiently coupled with the electrochemical oxidation of modified enzyme. Cyclic voltammetry was also carried out for ferrocene–glucose oxi-

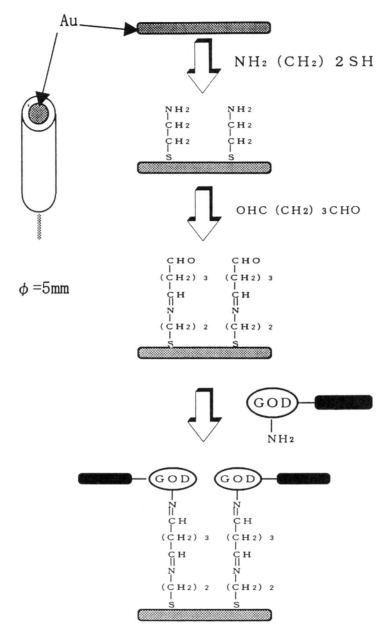

Figure 13.4. Self-assembled ferrocene-modified GOD on a gold surface through thiol groups.

dase conjugate in self-assembled form on the smooth gold electrode. Self-assembled ferrocene–glucose oxidase conjugate on the gold disk electrode showed reversible electron transfer. The anodic peak currents in the cyclic voltammograms were independent of the molar ratio of ferrocene to enzyme in the range

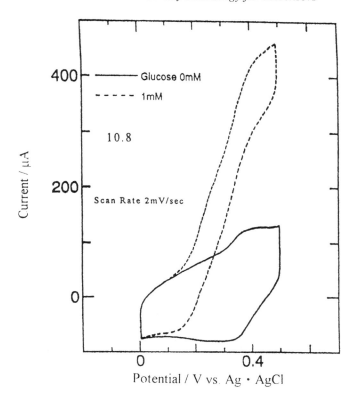

Figure 13.5. Cyclic voltammograms of ferrocene-modified GOD with and without glucose.

from 6 to 11. It is noted that the anodic peak current prominently increases with an increase in the molar ratio of ferrocene to glucose oxidase, whereas the amount of enzyme self-assembled on the electrode surface is fixed. This indicates that each modified ferrocene may contribute to electron transfer between the enzyme and the electrode in the case of the porous gold-black electrode. The ferrocene-modified enzyme could form multi-electron transfer paths on the porous gold-black electrode.

Substrate concentration dependence of response current of the gold-black electrode was compared with that of the gold disk electrode. The ferrocene-modified glucose oxidase used in this measurement had 11 ferrocenes per glucose oxidase. The electrode potential was controlled at 0.4 V vs. Ag/AgCl. The response current was recorded when the output reached a steady state. The response current was enhanced when ferrocene-modified glucose oxidase was self-assembled on a porous gold-black electrode.

The porous matrix of the gold-black electrode has enabled ferrocene-modified glucose oxidase to perform the smooth electron transfer by means of easy access between self-assembled molecules and electrode surface (schematically postulated in Figure 13.6).

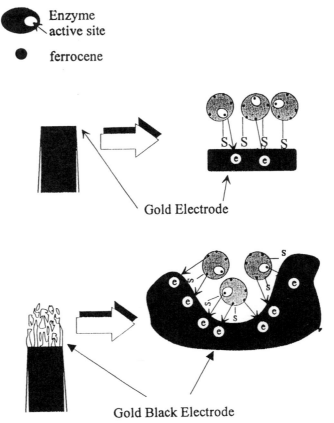

Figure 13.6. Postulated schemes of electron transfer of ferrocene-modified glucose oxidase on smooth gold and porous gold-black electrodes.

13.4 Self-assembled antibody protein array on protein A monolayer

Biosensors may be classified into two categories: biocatalytic biosensors and bioaffinity biosensors. Biocatalytic biosensors contain a biocatalyst, such as an enzyme to recognize the analyte selectively. Bioaffinity biosensors may involve an antibody, a binding protein, or a receptor protein that forms stable complexes with the corresponding ligand. An immunosensor, in which an antibody is used as the receptor, may represent a bioaffinity biosensor.

Advanced biotechnology and monoclonal antibody production have provided strong support for bioaffinity biosensors, and various new principles for electrochemical and optical immunosensors have been proposed. Concentrated efforts have been sharply focused on the development of homogeneous immunosensors, which are based on a single-step measuring procedure. Examples include an optical immunosensor based on surface-plasmon resonance (SPR), an optical fiber immunosensor based on flourescence determination using an evanescent wave, and an optical fiber electrode immunosensor based on electrochemical luminescence determination. These immunosensors are characterized by a single step of determination and high selectivity as well as high sensitivity. The response of these immunosensors, however, result from averaging the physicochemical properties of the antibody-bound solid surface. We have succeeded in fabricating an ordered array of antibody molecules on a solid surface and in quantitating individual antigen molecules that are complexed with the antibody array.

Protein A is a cell-wall protein from *Staphylococcus aureus* and has a molecular weight of 42 kDa. Since protein A binds specifically to the Fc region of IgG from various animals, it has been widely used in immunoassays and affinity chromatography. We found that protein A could be spread over the water surface to form a monolayer membrane using LB methods (Figure 13.7). On the basis of this finding, an antibody array on a solid surface can be obtained by the following two steps: The first step is fabrication of an ordered protein A array on the solid surface by the LB method; the second step is self-assembling of antibody molecules on the protein A array by biospecific affinity between protein A and the Fc region of IgG.

A Fromhertz-type of LB trough was used for fabrication of protein A arrays on a highly oriented pyrolytic graphite (HOPG) plate (15 x 15 x 2 mm). Protein A was dissolved in ultrapure water to make a 0.1×10^{-6} g/m solution. With a micropipet, 0.2 ml of protein A solution was dropped onto 150 cm² of the air–water interface of the compartment that contained ultrapure water as subphase. The protein A layer was compressed at a rate of 10 mm²/sec with a barrier. Compression was stopped at a surface pressure of 11 mN/m and the monomolecular layer of protein A was transferred to an adjacent compartment containing 0.5% glutaraldehyde solution at a rate of 10 mm²/sec. The protein A layer was incubated for 1 h to be crosslinked by glutaraldehyde, which was followed by transfer to a compartment containing ultrapure water for rinsing. The protein A molecular membrane was then transferred onto the surface of an HOPG plate by the horizontal method. The molecular imaging of the preparation was obtained by AFM in solution.

For preparation of an antibody protein array, a monolayer of protein A that was compressed at a surface pressure of 11 mN/m was transferred to a compart-

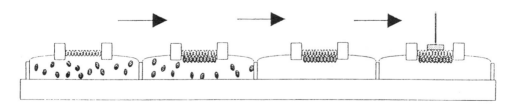

Figure 13.7. Fabrication process of self-assembled antibody array on a protein A layer.

ment containing anti-ferritin antibody in 10 mM phosphate buffer (pH 7.0). The
antibody molecules were self-assembled onto the protein A layer. The protein
A/antibody molecular membrane was transferred to a compartment containing
ultrapure water for rinsing and was transferred onto the surface of an HOPG plate
by the horizontal method. AFM measurements were made in 10 mM phosphate
buffer (pH 7.0) solution. AFM imaging of the protein A array deposited on the
HOPG plate showed an ordered alignment of protein molecules when the measurement was made in phosphate buffer at a controlled force of 4×10^{-11} N and a
scanning rate of 0.6 Hz. However, an ordered structure was not observed unless
protein A molecules were not crosslinked by glutaraldehyde.

The antibody array that was self-assembled on the protein A array was also
visualized in molecular alignment by AFM. The AFM measurement was conducted at a controlled force of 1.8×10^{-11} N and a scanning rate of 0.5 Hz.
Molecular size of the antibody was estimated as 7 nm in diameter.

The antibody array was soaked in different concentrations of ferritin solutions
for 1 h and was assayed for AFM imaging in solution. AFM imaging of the ferritin molecules, which were recognized and fixed by the antibody array, is presented in Figure 13.8. Ferritin concentration was 10 ng/ml. Individual ferritin
molecules on the antibody array can be selectively quantitated by AFM.

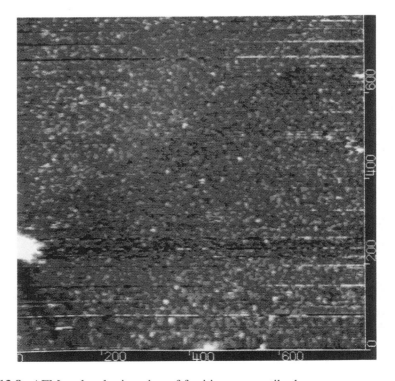

Figure 13.8. AFM molecular imaging of ferritin on an antibody array.

13.5 Conclusion

No general method has been developed for the fabrication of functional protein arrays on solid surfaces while retaining their activities. A combination of LB film technology and self-assembling technology seems most feasible in a wider range of protein molecules. An antibody array has been fabricated as an example by depositing a Protein A monolayer on the solid surface using the LB film technology, which is followed by bioaffinity-based self-assembly of antibody molecules on the surface of the Protein A layer.

References

Aizawa, M. (1994) Molecular interfacing for protein molecular devices and Neurodevices. *IEEE Engineering in Medicine and Biology* Feb/Mar 1994:94–102.

Aizawa, M. (1991) Principles and applications of electrochemical and optical biosensors. *Anal. Chim. Acta* 250:249–256.

Aizawa, M. (1989) Protein molecular assemblies and molecular interface for bioelectronic devices. In *Molecular Electronics-Science Technology* ed. A. Aviram, pp. 301–308. New York:Engineering Foundation.

Aizawa, M., Yauki, S., and Shinohara, H. (1989) Biomolecular interface. In *Molecular Electronics* ed. F. T. Hong, pp. 269–276. New York: Plenum Press.

Cardosi, M. F. and Turner, A. P. F. (1991) Mediated electrochemistry: A practical approach to biosensing. In *Advances in Biosensors* ed. A. P. F. Turner, pp. 125–170. London:JAI Press.

Degani, Y. and Heller. A. (1987) Direct electrochemical communication between chemically modified enzymes and metal electrodes. *J. Phys. Chem.* 91:6–12.

Gleria, K. D. and Hill, A. O. (1992) New developments in bioelectrochemistry. In *Advances in Biosensors* ed. A. P. F. Turner, pp. 53–78. London: JAI Press.

Khan, G. F., Kobatake, E., Shinohara, H., Ikariyama, Y., and Aizawa, M. (1992) Molecular interface for an activity controlled enzyme electrode and its application for the determination of fructose. *Anal.Chem.* 64:1254–1258.

Owaku, K., Goto, M., Ikariyama, Y., and Aizawa, M. (1993) Optical immunosensing for IgG. *Sensors and Actuators B* 13-14:723–724.

14

Self-organized ordered growth of III-V semiconductor quantum wires

KLAUS H. PLOOG

Paul-Drude-Institut für Festkörperelektronik
D-10117 Berlin
Germany

14.1 Introduction

Spontaneous atomic ordering on the group III or group V sublattice has been observed in epitaxial layers of ternary and quaternary III-V semiconductor alloys grown by molecular beam epitaxy (MBE) or metalorganic vapor-phase epitaxy (MOVPE) (Gomyo et al., 1987, 1988; Ihm et al., 1987; Kuan et al., 1985; Murgatroyd et al., 1990; Shahid et al.,1987; Suzuki et al., 1988, 1991). Formation of ordered phases is not only of scientific significance but also of technological importance, because of the strong correlation between the actual band-gap energy and the degree of ordering (Kurimoto and Hamada, 1989; Suzuki et al., 1988). Several of the ordered phases, of which the CuPt type is the most prominent, can be described as alternating arrangements of {111} planes of two different atoms on the group III or V sublattice. Due to the cubic symmetry of the disordered phases, the formation of four variants is expected in the ordered phases. However, only a few variants were observed in layers grown in the [100] direction. Although the asymmetry of the atomic structure on the (100) face of the zincblende structure limits the number of variants, the microscopic mechanisms that induce the formation of a specific phase at the substrate–epilayer interface are not yet well understood. Surface reconstruction plays an essential role in the formation of ordered phases through self-assembly (self-organization). First experimental evidence for the correlation between distinct surface reconstructions and different types of ordered phases was recently observed for growth of $Al_{0.48}In_{0.52}As$ on (100) InP substrate (Gomyo et al., 1994).

In this chapter we discuss another phenomenon of ordered growth of III-V semiconductors by means of self-assembly (self-organization) that provides a new concept for the direct formation of nanometer-scale quantum wire (QWR) and quantum dot (QD) structures by epitaxial growth. The basis for this kind of ordered growth is the existence of distinct steps and terraces on the growing surfaces. Such well-ordered corrugations can be created in situ during epitaxy particularly on non-(100)-oriented III-V surfaces (Nötzel et al., 1992b) because these surfaces have a strong tendency to lower their free energy by transformation of the original flat surface into a regular hill-and-valley structure (Chadi, 1984). The steps and terraces cause the impinging atoms to nucleate, become localized, and form distinct patterns ("supramolecular" entities) through self-assembly (self-organization). The distinct phase shift of the surface corrugation, observed during heteroepitaxy (for example, GaAs/AlAs or $Ga_{0.47}In_{0.53}As/InP$) (Nötzel et al., 1991), is understood in terms of a spontaneous self-organization of the system on a macroscopic scale. The evolution of the surface corrugations and the self-organized growth can be monitored in real time by means of reflection high-energy electron diffraction (RHEED) (Däweritz and Ploog, 1994), and the resulting QWR and QD structures exhibit electronic properties that are characteristic for low-dimensional semiconductor systems (Nötzel et al., 1992a, 1993a).

The described results demonstrate that even in epitaxial growth, the selective manipulation of matter to atomically engineered structures with unique properties has reached an unprecedented degree of control. Hence the previous distinctions between the disciplines of molecular electronics, semiconductor (metal) nanostructures, and molecular manipulation are more and more blurred, and synergetic activities now need to be developed.

14.2 Concept of self-organized ordered epitaxial growth

The concept of self-organized ordered epitaxial growth is based on the existence of distinct steps and terraces on the growing crystal surfaces, which are used to induce the self-organized ordered growth of III-V semiconductor heterostructures. The periodic arrays of steps, facets, and/or bunches on various non-(100)-oriented III-V surfaces often represent the equilibrium configuration defined by surface energy rather than a growth-induced phenomenon. This implies that, at appropriate growth conditions, at least a local equilibrium is established if the migration length of the species governing the growth is in the order of the lateral periodicity given by the equilibrium structure. Under these conditions the system, comprising many units and subject to constraints, is able to organize itself in various spatial, temporal, or spatiotemporal activities.

The formation of doping wires in GaAs by preferential attachment of impurity atoms at misorientation steps of vicinal (100) GaAs relies on the proper combination of lattice step growth and atomic-plane ("delta") doping (Däweritz et al. 1993; Schubert, 1990; Wood et al., 1980). The direction of misorientation and the growth conditions must be chosen to ensure that the terraces are smooth and the step edges as straight as possible. In addition, sufficient dopant migration lengths on the terraces and negligible distortions of the growth front are required. The stability of the growth front is important as the step height is only one lattice plane ($\triangleq$ monolayer, ML), that is, 2.83 Å, and the wirelike dopant incorporation can induce local strain effects. Another important parameter is the stability of the growth front during over-growth of the aligned dopant atoms with the matrix material.

When the step height is increased, the growth-front stability is improved and stronger ordering effects occur. The generation of macrosteps requires the exis-tence of very stable facets. On (311)A GaAs the macrosteps (facets) form well-ordered parallel channels along [$\bar{2}$33] of about 10 Å depth and 30 Å periodicity (Nötzel et al., 1991). The spontaneous self-organization of the system on a macro-scopic scale manifests itself by a distinct phase shift of the surface corrugation during heteroepitaxy, that is, AlAs on GaAs, and GaAs on AlAs. The distinct pat-tern ("supramolecular" entities) formed through self-assembly (self-organization) consists of alternating wide and narrow GaAs regions in the AlAs matrix running along [$\bar{2}$33]. The resulting GaAs QWR structure is free from any structural defects due to the self-organization mechanism operating during growth.

The stability of the surface and interface corrugation on (311)A GaAs and other III-V semiconductors allows for only a limited tuning of the lateral period-icity and step height by adjusting the strain. A much wider range of tunability is obtained through the accumulation of steps (step "bunching") during epitaxial growth (Nötzel et al., 1993b). We demonstrate this concept for (210) GaAs, where periodic arrays of steps with a height of 20 Å and a periodicity of more than 200 Å can be created by step bunching, which can be directly visualized by atomic force microscopy (AFM).

14.3 Equilibrium morphology of non-(100)-oriented surfaces

The surface topography of III-V semiconductors during MBE growth and after growth interruption is usually the result of a competition between the growth-kinet-ics–related morphology and the equilibrium morphology that is given by the ori-entation dependence of the surface free energy. For many years, MBE has been considered to be a process highly off equilibrium. The growth conditions in the present study, however, are characterized by a high surface mobility of the rate-determining species. The processes, leading to a growthfront morphology that depends on growth conditions, proceed on different length and time scales. A local equilibrium controlled by processes on a short time scale, such as fast surface migration, is reached within a time that is short compared to the relaxation time of the global system to the complete equilibrium (Mbye and Massies, 1990). It is the breakup of some nonequilibrium surfaces into a hill-and-valley structure to lower their energy (Herring, 1951), under the constraint of a constant surface orientation, that is used here for the distinct synthesis of QWR structures. Other constraints, such as surface stoichiometry (reconstruction) and strain, may thus also become important, and the equilibrium shape can be modified by kinetic effects.

Figure 14.1 shows the three types of periodic surface corrugations that were exploited for the present study. Periodic arrays of ordered steps with 1 ML height

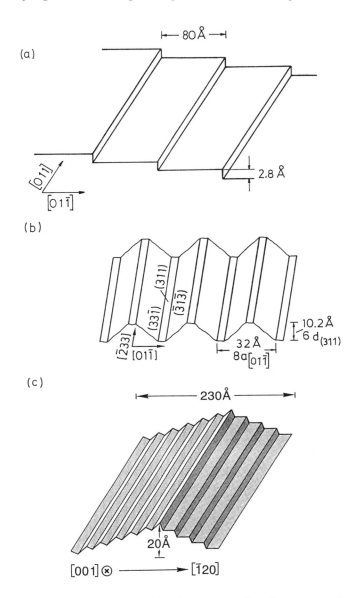

Figure 14.1. Schematic illustration of in situ–generated surface corrugations on (a) (100) GaAs 2° misoriented toward <011>, (b) (311) GaAs, and (c) (210) GaAs after step bunching, which are exploited in this study.

and flat terraces with a width of $w = \sqrt{2}\tan\phi$, where a is the lattice constant, are easily generated on the low-index (100) GaAs surface by a slight misorientation of angle ϕ toward [011] (Figure 14.1a). The ordered step arrangement on vicinal (100) GaAs was first used by Petroff et al. (1984) to fabricate QWR structures in "lateral" or "tilted" GaAs/Al$_x$Ga$_{1-x}$As superlattices directly by epitaxial growth. However, in a more detailed study of the step-flow growth on vicinal (100) planes

– growth that is essential for the formation of ordered lateral superlattices – Horikoshi et al. (1990) showed that there is a distinct anisotropy of the growth front for steps created by misorientation toward [011] or [01$\bar{1}$], due to the given [01$\bar{1}$] orientation of the arsenic dimers on the (2 x 4) reconstructed surfaces (terraces). (For clarity it should be noted that steps running along [011], that is, misorientation in the [01$\bar{1}$] direction, are also called "As-terminated" steps, and steps running along [01$\bar{1}$], that is, misorientation in the [011] direction, are called "Ga-terminated" steps, according to the atoms forming the edges of the terraces.) Scanning tunneling microscopy (STM) data by Pashley et al. (1991) confirmed that As-terminated steps on the vicinal (100) GaAs surface are ragged (although they display the better step ordering), whereas Ga-terminated steps form rather straight ledge profiles. This anisotropy of the growth fronts on the vicinal (100) substrates makes the fabrication of ordered lateral GaAs/Al$_x$Ga$_{1-x}$As superlattices extremely difficult (Sundaram et al., 1991), and it must be taken into account for any ordered incorporation of impurity atoms.

The (311)A GaAs surface breaks up into an ordered array of two types of {331} facets to lower the surface energy (Figure 14.1b). A careful analysis of the RHEED patterns taken along [01$\bar{1}$] and along [$\bar{2}$33] (Däweritz and Ploog, 1994; Nötzel et al., 1992a) reveals that the surface is composed of (311) terraces of 4 Å width (= $a_{[110]}$) and two sets of (33$\bar{1}$) and ($\bar{3}$13) facets corresponding to upward and downward steps of 10.2 Å height. The {331} facets are actually composed of stripes with alternating (110) and (111) surface meshes, hence corresponding to low-index planes with low surface energy (Nötzel et al., 1992b). The breaking up of the (311) surface is in agreement with the energetic benefit expected from a rearrangement of the surface atoms in a hill-and-valley structure. The key parameter for the energetically most-favorable configuration is the ratio $\Gamma = \gamma_{\{331\}}/\gamma_{\{311\}}$, where $\gamma_{\{hkl\}}$ is the free energy of the {hkl} surface. Theoretical studies have indeed shown $\Gamma < 1$ in the case of Si surfaces (Chadi, 1984).

The equilibrium nature of the surface corrugation on (311) GaAs is further supported by the following observations. First, the splitting in the RHEED pattern appears after the removal of the oxide from the substrate surface at 580 °C. Second, the stepped (311) surface is stable below 680 °C down to room temperature. However, above 480 °C the stepped surface is observed only with an As$_4$ flux of about 10^{15}/cm^2 sec. When closing the As$_4$ shutter, the splitting in the RHEED pattern vanishes and only integral-order streaks remain, indicating a flattening of the surface. When As$_4$ is again supplied, the surface structure immediately reappears. Third, when GaAs is deposited below 400 °C, the splitting in the RHEED pattern also vanishes and only integral-order streaks remain, due to a filling of the channels with Ga of low surface mobility. However, when the substrate is heated, the splitting reappears during growth.

In order to increase the periodicity and step height of the interface corrugation to a range comparable to the exciton Bohr radius (around 100 Å in GaAs), an accumulation of the steps on non-(100)-oriented substrates during epitaxy is necessary. This is possible, for example, by one-dimensional step bunching on (210) GaAs leading to mesoscopic step arrays with a periodicity of more than 200 Å (Nötzel et al., 1993b). The width of the mesoscopic steps on (210) GaAs is much smaller than the saturated terrace widths of supersteps (0.1–0.3 μm) generated during growth on vicinal surfaces (Colas et al., 1989; Cox et al., 1989; Fukui and Saito, 1990; Hasegawa et al., 1992). In the case of supersteps, the terrace width is kinetically limited by the Ga adatom diffusion length. Hence, the size of the mesoscopic step array formed on (210) GaAs is probably defined by energy barriers limiting the height and the periodicity. These energy barriers have their origin in the interplay between terrace width and reconstruction during the formation of the vicinal (stepped) facet planes (Bartolini et al., 1989).

The characteristic corrugation of the (311) GaAs surface is observed also for various III_AIII_B-V layers grown on (311) GaAs and (311) InP substrates (Tournié et al., 1994). For lattice-matched materials, as well as for systems with high compressive or tensile strain, qualitatively the same diffraction patterns are obtained, with the characteristic spot splitting in the $[\bar{2}33]$ azimuth and the pronounced streaking in the $[\bar{1}10]$ azimuth due to the development of ordered two-level systems with {331} facets as upward and downward steps. However, the lateral periodicity and step height (normalized to the lattice units) depend on the material system and change continuously when going from the lattice-matched ternary materials toward the binary extremes. The results of a systematic study on lattice-matched, strained, and relaxed systems are summarized in Table 14.1.

For lattice-matched $Ga_{0.47}In_{0.53}As$ on InP, we found a lateral periodicity of 43 Å and a step height of 9 Å. In units of lattice distances this corresponds $10a_{[110]}^{InP}$ and $5d_{(311)}^{InP}$, respectively. The same relation holds for $Al_{0.48}In_{0.52}As$, which is also lattice matched to InP. The surface structure of both materials is stable in the temperature range 300–500 °C, and it is clearly observed on epilayers only a few nanometers thick. The RHEED pattern of a 6-Å-thick tensile-strained ($\varepsilon = 3.2\%$) InAs layer deposited on a (311) $Ga_{0.47}In_{0.53}As$ buffer layer indicates a lateral periodicity of ≈ 26 Å ($\approx 6a_{[110]}^{InP}$) and a step height of ≈ 6 Å ($\approx 3d_{[311]}^{InP}$). Similar values are obtained for compressive-strained GaAs layers ($\varepsilon = -3.9\%$) on (311) $Ga_{0.47}In_{0.53}As$ or $Al_{0.48}In_{0.52}As$ buffer layers (Table 14.1). However, one monolayer (ML) of highly strained InAs ($\varepsilon = 7.2\%$) deposited on (311) GaAs flattens the surface completely. On the other hand, a relaxed InAs layer as thick as 0.5 μm on a (311) GaAs buffer layer exhibits a surface structure that is identical to the structure of the underlying buffer layer.

The lateral periodicity and step height of the corrugated (311) surface are tunable by selecting the proper epilayer/substrate combination and/or overlayer strain; this demonstrates the equilibrium nature of this surface. For the (311) GaAs surface, the key parameter is the ratio $\Gamma = \gamma_{331}/\gamma_{311}$. An additional strain-energy term, including an important shear-strain component, must be taken into

Table 14.1. *Experimental values of the lateral periodicity* P *(given in Å and in units of* $a_{[110]}^{sub}$*) and of the step height* h *(given in Å and in units of* $d_{[311]}^{sub}$*) as determined from RHEED intensity profiles for several* III$_A$ III$_B$–V *systems. The error bar is ±2 Å for* h *and ±5 Å for* P.

Epilayer/Substrate	Strain (%)	P (Å/$a_{[110]}^{sub}$)	h (Å/$d_{[311]}^{sub}$)
Lattice-matched systems:			
GaAs/GaAs	0	32/8	10/6
Ga$_{0.47}$In$_{0.53}$As/InP	0	43/10	9/5
Al$_{0.48}$In$_{0.52}$As/InP	0	42/10	9/5
Strained systems:			
InAs/Ga$_{0.47}$In$_{0.53}$As	3.2	26/6	6/3
InAs/InP	3.2	25/6	5/3
GaAs/Ga$_{0.47}$In$_{0.53}$As	-3.9	25/6	5/3
InAs/GaAs	7.2	0	0
Relaxed systems:			
InAs/Ga$_{0.47}$In$_{0.53}$As	0	39/9	7.5/4
GaAs/Ga$_{0.47}$In$_{0.53}$As	0	37/9	8.5/5
InAs/GaAs	0	34/8	9/5

account (DeCaro and Tapfer, 1993). Consequently, the global energy balance depends not only on the orientation but also on the peculiarities of each material system, which then adopts its own topography.

The random occupation of the cation lattice sites in ternary alloys implies that local alloy fluctuations are unavoidable. This results in high local strain fields and bond distortions that change the local energy balance. The detrimental influence of local strain fields on the ordering of the surface manifests itself in a more diffuse RHEED pattern, with a much shallower minimum in the intensity profile across the streaks at out-of-phase conditions, as compared to the growth of (311) GaAs/AlAs structures.

14.4 Fabrication of doping wires in GaAs by self-organization of dopant impurities along step edges of vicinal surfaces

The engineering of doping wires in semiconductors (Bauer and van Gorkum, 1990) by epitaxy based on the preferential incorporation of dopant impurities along step edges requires (1) well-ordered vicinal surfaces, (2) sufficient migration lengths of the dopant species on the terraces, and (3) negligible distortion of the growth front during overgrowth with the matrix material. We have recently realized doping wire arrays in our laboratory (Däweritz et al., 1993, 1994; Ramsteiner et al., 1994) by employing vicinal (100) GaAs starting surfaces tilted by 2° toward the [011] direction [$\triangleq$ (111) Ga plane]. Both the evolution of steps and the surface chemical reactions were monitored in real time by RHEED, with the electron beam parallel to the step edges and an incidence angle close to the first out-of-phase condition for diffraction. In this geometry the terraces are not shadowed by steps, and the specular beam intensity is sensitive to adatom density, to changes in step edge roughness and strain-induced effects, and to nucleation processes on the terraces (Lagally et al., 1988). Hence, any Si-induced changes in surface reconstruction can be traced from the very beginning.

In addition to the preference of the vicinal surface tilted toward the (111) Ga plane (because of its straight step edges), three key parameters should be optimized to make the growth experiments successful. First, preparation of the regularly stepped surface with smooth terraces requires a substrate temperature above the transition from a two-dimensional nucleation and step propagation mechanism to the desired step flow mechanism at T_{crit}, which can be monitored by the change from an oscillating to a constant RHEED intensity (Neave et al., 1985). The value of T_{crit} ranged from 500 to 630 °C, depending on the angle and orientation of the misorientation (Däweritz et al., 1993). Second, the terraces on the vicinal surface have to be smoothened by successive deposition of several monolayers of GaAs above T_{crit} followed by growth interruptions of several minutes to eliminate undesired holes and islands. The existence of well-ordered step arrays is evidenced by the clear splitting of the RHEED streaks into inclined slashes (Lagally et al., 1988). Third, the Si deposited above T_{crit} is supplied *not* continuously but in pulses of a certain duration interrupted by annealing periods. This interruption allows the Si atoms to migrate on the terraces and arrange themselves along the step edges. In addition, overgrowth of the ordered Si distribution with GaAs is performed at a temperature below T_{crit} in order to minimize adverse effects on the morphology of the growth front.

Figure 14.2 shows the RHEED intensity behavior during Si deposition in pulses of 90 sec duration and 180 sec interruption using a flux of 1×10^{11} cm^{-2}/sec ($\approx 1.7 \times 10^{-4}$ ML/sec) and a schematic illustration of the Si incorporation along a step edge. The RHEED intensity (Figure 14.2a) decreases nonlinearly with Si coverage and exhibits a certain recovery when the Si flux is interrupted. When a coverage of 0.041 ML is reached, the intensity increases drastically; at the same time a (3×2) reconstruction develops, as evidenced by half-order spots in the [011]H and third-order spots in the [01$\bar{1}$] azimuth. These findings are consistent with the model of the Si-covered vicinal surface shown in Figures 14.2b and 14.3. The twofold periodicity in the [011] direction is due to the formation of areas with dimerized Si atoms on Ga sites (at the given growth conditions Si is incorporated only as donor, as confirmed by independent measurements on reference samples). The threefold periodicity in the [01$\bar{1}$] direction is produced by an ordered array of

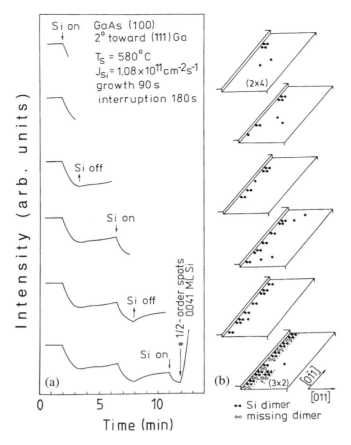

Figure 14.2. (a) RHEED intensity recorded in the [110] azimuth during interrupted Si deposition at 580°C on vicinal (100) GaAs tilted by 2° toward (111) Ga; (b) model for incorporation of Si as (3 x 2) units along the step edge.

Si dimers in (3 x 2) units which consist of two Si dimers and one missing dimer per unit mesh. The fractional-order spots in the RHEED pattern thus appear after a certain completion of (3 x 2) units along the step edge and the attachment of (3 x 2) units in a second row. In the case of ideal wirelike incorporation of Si dimers, the first stripe of (3 x 2) units should be completed at an Si coverage of 4/3 θ_{crit} ($\triangleq 0.066$ ML), where the critical coverage θ_{crit} is the number of Ga sites at the step edges, which is given by the misorientation angle. In the experiments, Däweritz et al. (1993) found a lower value of 0.041 ML. This discrepancy can be understood from the neglect of step-edge roughness in the ideal model. In addition, a second stripe of (3 x 2) units starts to develop in the actual experiments before the first one is completed. It should be mentioned here that a single Si layer forming an ideal (3 x 2) structure is completed after deposition of 0.67 ML Si (with 1 ML equivalent to 6.25×10^{14} atoms cm^{-2}). In reference experiments, the

Si on GaAs (100)
misoriented toward (111)Ga

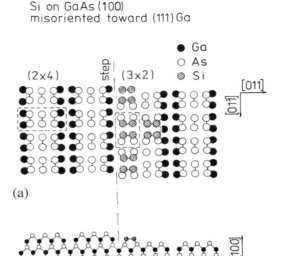

(a)

(b)

Figure 14.3. Schematic model of (2 × 4) reconstructed vicinal (100) GaAs misoriented toward (111) Ga with Si atoms attached in (3 × 2) symmetry along the step edge in (a) top view and (b) side view.

transition from the (3 × 2) to a distorted (1 × 3) structure was indeed observed at coverages of about 0.6 ML Si. This can be explained by the 90° rotation of the dangling bond direction when changing from the Ga to the As plane, and it hence indicates the incorporation of Si atoms in the second layer above a critical concentration

The conclusion of a wirelike incorporation of the Si atoms at step edges is thus based on the following observations (Figure 14. 2a). First, the RHEED intensity during Si deposition decreases nonlinearly. Second, the Si-induced reconstruction and the final intensity increase appear at a coverage that corresponds to the number of sites at the step edge. The attachment of migrating Si atoms as dimers at the step edges generates kinks. The kink density increases sublinearly with coverage, because with increasing density a rising number of dimers become direct neighbors. This accounts for the observed sublinear intensity decrease. In addition, strain effects are important. The completion of a stripe of (3 × 2) units along the step edge to a high degree and the attachment of (3 × 2) units in a second stripe lead to the threefold and twofold periodicities observed in the [01$\bar{1}$] and the [011] directions, respectively. The abrupt change in reconstruction and strain accompanying this process account for the final intensity increase.

The RHEED intensity behavior during Si deposition becomes very different when the terraces on the vicinal (100) GaAs surface are not smooth or when the terrace width w becomes larger than the Si migration length (that is, at small mis-orientation angle ϕ or for Si deposition without flux interruption). In these cases the RHEED intensity decreases linearly, and the intensity minimum and the appearance of fractional-order spots are correlated neither with the amount and the direction of misorientation nor with the critical coverage. The linear intensity decrease indicates that the Si adatoms form a lattice gas of increasing density. Clustering of the Si atoms on the terraces can then lead to the formation of islands with a (3 x 2) structure, evidenced by the occurrence of fractional-order spots in the respective azimuths.

In principle, a wirelike incorporation of (3 x 2) Si units along step edges would also be possible on vicinal (100) surfaces tilted toward the [01$\bar{1}$] direction ($\triangleq$111) As plane). However, the As-terminated steps on these surfaces are very ragged (Pashley et al., 1991), and they therefore make a long-range ordering of the wire-like incorporation of impurity atoms more difficult. On the other hand, the critical terrace width for wirelike Si incorporation is larger for misorientation toward (1$\bar{1}$1) As than toward (111) Ga, because the preferential path for Ga (as well as for Si adatom diffusion) is along the [011] direction, that is, normal to the As-terminated steps for the tilt toward the (1$\bar{1}$1) As plane but parallel to the Ga-terminated steps for the misorientation toward the (111) Ga plane (Shiraishi, 1992).

After the wirelike incorporation of the impurity atoms along step edges, this ordered arrangement must be stabilized during overgrowth with the GaAs host material when the impurity atoms are rebonded to become part of a three-dimensional solid. The studies of Däweritz et al. (1993, 1994) have shown that several tens of lattice planes of GaAs must be deposited at temperatures below T_{crit} so that the growth-front morphology of the stepped surface is not adversely affected. Under these conditions, the Si segregation is also reduced.

The conclusion of a self-organization of Si incorporation along step edges drawn from the RHEED studies is confirmed by recent results obtained by Raman scattering (Ramsteiner et al., 1994). The wirelike Si incorporation that occurs in the sample misoriented 2° toward (111) Ga induces a distinct polarization asymmetry in the Raman scattering intensity of collective intersubband plasmon–phonon modes arising from the doping layer.

14.5 GaAs quantum wires in AlAs matrix

14.5.1 Self-organized growth on macrosteps and facets

As discussed in Section 14.3, the (311) A GaAs surface develops well-ordered macrosteps and facets during MBE growth with lateral periodicities and step heights extending over several lattice constants (Figure 14.1b). The reciprocal lattice of the corrugated surface can be directly imaged by means of RHEED (see Figures 14.4 and 14.5). For the present study, it is important to note that a unique self-organization mechanism leads to a phase change of this corrugation during heterogeneous growth of AlAs on GaAs and vice versa. This phase change allows for the fabrication of arrays of alternating narrow and wide regions of GaAs in an AlAs matrix running along [$\bar{2}$33] and being 32 Å apart. To obtain full information from the RHEED pattern on the evolution of the surface topography during growth and after growth interruption, in principle one needs to measure the intensity profile of distinct streaks as a function of time. However, the specular beam intensity is also a good measure of surface roughness, that is, step and kink density. Under carefully selected diffraction and recording conditions (that is, out-of-phase condition, maximum of the intensity profile), the intensity is directly proportional to the smoothness or ordering of the surface (Däweritz and Ploog, 1994). The growth mechanism on vicinal and corrugated surfaces changes with increasing mobility of the group III cations (for example, at elevated substrate temperature) from a two-dimensional nucleation and step propagation mechanism (indicated by RHEED intensity oscillations) to a step flow mechanism. In this case, the adatoms are directly incorporated at the step edges and the surface topography does not change appreciably with time. As a consequence, the RHEED intensity remains constant after an initial rise or decrease (Neave et al., 1985).

The RHEED patterns of the (311)A surface taken along the [01$\bar{1}$] direction (Figure 14.4a) and along the [$\bar{2}$33] direction (Figure 14.4b) directly reveal the

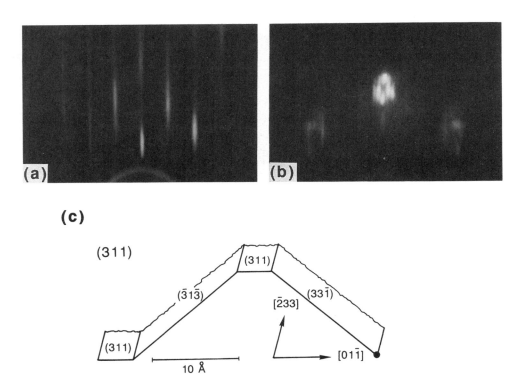

Figure 14.4. RHEED pattern of the (311) GaAs surface taken (a) along [01$\bar{1}$] and (b) along [$\bar{2}$33]. (c) Schematic of the stepped surface.

breaking up of the flat (311) surface into a well-ordered array of upward and downward steps oriented along the $[\bar{2}33]$ direction, as illustrated in Figure 14.4c. With the electron beam along the $[01\bar{1}]$ direction, the diffraction pattern shows a pronounced streaking, which indicates a high density of steps along the perpendicular $[\bar{2}33]$ direction. Taking the $[\bar{2}33]$ azimuth parallel to the steps, the streaks are found to be split into sharp satellites or unsplit, depending on the scattering vector $k_\perp$ that is, the position along the streaks. Furthermore, the intensity maximum of the satellites corresponds to an intensity minimum of the (00) streak for constant $k_\perp$ values and vice versa. The RHEED pattern recorded in this direction thus directly images the reciprocal lattice of a two-level system oriented along $[\bar{2}33]$. Figure 14.5 presents the corresponding intensity profiles measured as a function of $k_\parallel$, that is, across the streaks for different k-values at maximum (Figure 14.5c) and minimum (Figure 14.5d) intensity of the (00) streak, and as a function of $k_\perp$ along the (00) streak (Figure 14.5e). The scale is taken from the separation of the zero- and first-order diffraction streaks of the (100) GaAs reference sample in the $[01\bar{1}]$ azimuth. The separation of the satellites in Figure 14.5(d) gives the lateral periodicity of 32 Å ($\triangleq 8a_{[1\bar{1}0]}^{\text{GaAs}}$)for the stepped surface, where a is an in-plane lattice constant, in this case in the $[\bar{1}10]$ direction. In this profile, the intensity of the (00) streak is canceled compared to the intensity of the profile shown in Figure 14.5c, which was taken at a $k_\perp$ value where the satellite intensity has a minimum. This intensity

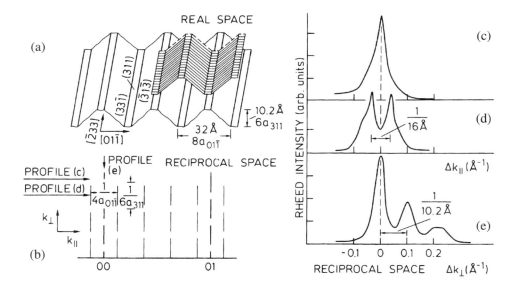

Figure 14.5. (a) Schematic illustration of the stepped (311) GaAs surface. The upper surface (shaded) illustrates the phase change of the surface corrugation during heterogeneous growth of GaAs on AlAs and vice versa. (b) Schematic of the observed RHEED pattern taken along [233]. (c–e) RHEED intensity profiles for the (311) GaAs surface along [233] measured as a function of the scattering vectors $k_\perp$ and $k_\parallel$, respectively.

distribution evidences the high degree of ordering and the presence of an almost perfect two-level system. From the splitting of the profile measured along the (00) streak (Figure 14.5e), the step height is deduced to be 10.2 Å.

The {331} facets forming the upward and downward macrosteps of the corrugated surface are composed of stripes with alternating (110) and (111) surface meshes. Actually, the alternating (110) and (111) surface configurations are not one, but two, unit meshes wide (Nötzel et al., 1992b). This finding underlines the importance of the energy-minimizing (2 x 2) surface reconstruction of (111) planes (Tong et al., 1984) for stabilization of the surface structures on a nanometer scale. The lateral extension of the {331} facets in the <116> directions on the corrugated (311)A GaAs surface, and hence the height of the corrugation, corresponds to the minimum facet size conserving the structure of the extended (331) surface.

During growth of (311) GaAs/AlAs multilayer structures, the RHEED intensity dynamics show a pronounced oscillation at the onset of GaAs and AlAs growth, respectively (Figure 14.6) (Nötzel et al., 1991). This oscillation corresponds to the deposition of three (311) monolayers, that is, lattice planes, as deduced from the

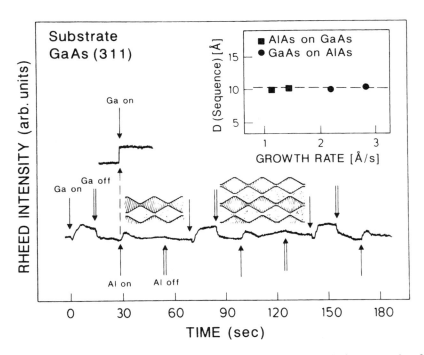

Figure 14.6. RHEED intensity dynamics taken along [233] during growth of (311) GaAs/AlAs multilayer structures. The upper curve shows the RHEED intensity during deposition of GaAs on GaAs. The inset shows the deposited layer thickness for different growth rates of GaAs on AlAs and vice versa until stable growth conditions are reached. The schematic cross section images the GaAs/AlAs multilayer structure resulting from the phase change of the surface corrugation during the heterogeneous deposition of the first monolayers of GaAs on AlAs and AlAs on GaAs.

growth rates. During deposition of the next three monolayers, the intensity approaches the value found in the RHEED pattern of the stable stepped surface during growth. The whole sequence corresponds to the deposition of six (311) lattice planes, that is, 10.2 Å, as shown in the inset of Figure 14.6 for different growth rates of GaAs and AlAs. This sequence results from a phase change of the surface corrugation during the heterogeneous deposition of GaAs on AlAs and vice versa. The phase change includes quasi-filling of the corrugation during the first deposition of the first three monolayers and rearrangement of the stepped surface during the deposition of the next three monolayers. After completion of the phase change, the growth continues layer by layer with conservation of the surface corrugation, as indicated by the constant RHEED intensity. The stepped character of the {331} facets explains the morphological stability of the growing surface (Däweritz, 1993) and the RHEED intensity behavior, which is analogous to that for a vicinal surface under step flow conditions (Neave et al., 1985). As for vicinal surfaces, an abrupt change is also observed in the RHEED intensity from the nongrowing GaAs(AlAs) surface to the growing GaAs(AlAs) surface within the deposition of one monolayer (see upper curve of Figure 14.6). The direction of this change, which is opposite to that usually found for growth on vicinal surfaces, is explained by the specific diffraction and recording conditions. We assume that the distinct phase of the surface corrugation at the onset of GaAs growth on AlAs (and vice versa) is caused by strain and/or surface reconstruction. A different distribution of strain on the facets and terraces and/or differences in surface reconstruction between GaAs and AlAs will result in nonequivalent growth conditions, hence favoring growth either on the facets or on the terraces.

Although the microscopic growth mechanism during the interface formation of GaAs/AlAs multilayers is not yet understood in detail, the formation of well-ordered alternating narrow and wide regions of GaAs and AlAs oriented along [$\bar{2}$33] is well established by high resolution transmission electron microscopy (HREM) (Nötzel et al., 1991, 1992a). The HREM images confirm that the unique arrangement of alternating narrow and wide regions of GaAs and AlAs indeed forms an as-grown GaAs QWR structure in an AlAs matrix. Detailed x-ray diffraction measurements further reveal that the structural perfection of GaAs/AlAs multilayer structures grown on (311) substrates is as excellent as that of (100) reference samples (Nötzel et al., 1991, 1992a). In addition, the average GaAs and AlAs layer thicknesses (neglecting the specific interface corrugation) are found to be comparable to those of the reference samples; that is, the overall sticking coefficient of Ga and Al is unity for the [311] and [100] orientations.

The presence of alternating narrow and wide regions of GaAs in the AlAs matrix gives rise to additional lateral size quantization, which results in unusual optical properties (Nötzel et al., 1992a, 1993a). The photoluminescence (PL) and the heavy-hole (hh) and light-hole (lh) exciton resonances in photoluminescence excitation (PLE) are always shifted to *lower* energies compared to (100) multiple

Table 14.2. *Dependence of the measured and calculated redshift of the hh-exciton resonance and the increment of the effective GaAs-layer thickness of (311) GaAs/AlAs QWR structures on the average GaAs layer thickness, as compared to (100) reference samples.*

	(311) GaAs/AlAs		
Average GaAs-layer thickness in Å	66	56	43
Measured redshift of the hh-exciton resonance in meV	4	5.6	7.4
Calculated redshift of the hh-exciton resonance from the valence-band anisotropy in meV	4.0	5.6	7.4
Increment of the effective GaAs layer thickness corresponding to the measured redshift in Å	0	3.8	4.6

quantum well (MQW) reference samples grown side by side. The anisotropy of the valence band that results in larger hh-masses along non-[100] crystallographic directions cannot account for the observed redshift. Table 14.2 shows that with decreasing GaAs-layer thickness the observed redshift of the hh-exciton resonance increases compared to that calculated from the valence-band anisotropy. This increase of the redshift with decreasing GaAs thickness can even be observed at room temperature (Figure 14.7), and it is accompanied by a strong enhancement of the integrated luminescence intensity compared to the (100) reference sample, which exceeds one order of magnitude for a 43-Å (311) GaAs QWR structure. This behavior, which does not degrade up to 400 K, arises from the additional lateral confinement and the strong localization of excitons in the QWR structures.

The additional lateral confinement by the interface corrugation results in a strong enhancement of the exciton continuum energies in the (311) samples, which are resolved even in the PL spectra (Nötzel et al., 1993a). The two-dimensional hh-exciton continuum energy of the reference (100) MQW structure amounts to hh = 12 meV, in agreement with the investigations of Miller and Kleinman (1985), whereas the one-dimensional hh-exciton continuum energy of the (311) quantum-wire structure amounts to hh = 28 meV. The measured one-dimensional hh-exciton continuum energy and the model of Degani and Hipolito (1987) can be used to estimate the lateral extension of the hh-exciton in the QWR structure. In this model, the exciton binding energy in square-shaped QWR is calculated as a function of the vertical and lateral extensions of the wires. The given vertical extension of 66 Å in our structure and 29 meV observed for hh yields a lateral extension of about 80 Å. This lateral exciton extension exceeds the 32 Å geometrical width of the channel region of our QWR structure, indicating a strong lateral coupling between the wire regions. The lateral coupling in the present structure, where the vertical potential is much stronger than the lateral one, is further revealed by the smooth low-energy onset of the hh- and lh-exciton continuum energies. In a coupled QWR array, one consequence of the lateral coupling is that the peaked one-dimensional density of states becomes blurred, whereas the corresponding maximum of the hh- and lh-continuum in PLE is normally attributed to the energy position of the unperturbed one-dimensional density-of-states peak.

The increased stability of the laterally confined excitons in (311) QWR structures is manifested not only by the enhancement of the excito continuum energies but also by the strong exciton–phonon interaction (Hopfield, 1959), as revealed by the drastic enhancement of the intensity of LO-phonon sidebands of the exciton luminescence (Nötzel et al., 1993a) and by the observation of hot-exciton relaxation (Permogorov, 1975). When the detection wavelength is set to the high-energy side of the PL line, strong LO- and TA-phono–related lines are resolved in the PLE spectra of the (311) samples (Figure 14.7) because of the high probability that the later-

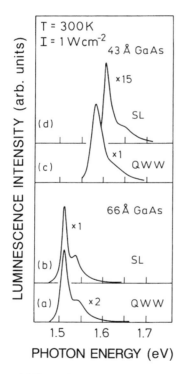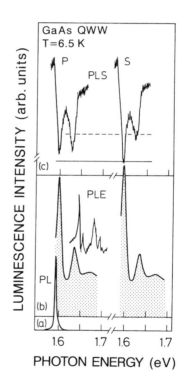

Figure 14.7. Left panel: Room-temperature luminescence of (a) (311) 66 Å GaAs/62 Å AlAs QWR, (b) (100) 66 Å GaAs/62 Å AlAs MQW, (c) (311) 43 Å GaAs/47 Å AlAs QWR, and (d) (100) 43 Å GaAs/47 Å AlAs MQW. Right panel: Low-temperature PL (a), PLE (b), and PLS (photoluminescence suppression) (c) spectra of (311) GaAs/AlAs multilayer structure with GaAs-layer thickness of 66 Å and AlAs-layer thickness of 60 Å. *P* corresponds to light polarized perpendicular to the wire axis, and *S* to light polarized parallel to the wire axis. The additional lines in (b) are detected on the high-energy side of the PL line and correspond to LO- and TA-phonon–related lines.

ally confined excitons created above the band gap will relax as a whole. The energy threshold for damping of the phonon lines increases for thinner average GaAs-layer thicknesses, reflecting the increased influence of the interface corrugation; thus it can be used to estimate the one-dimensional confinement energy for excitons. The striking result of this estimate is that the one-dimensional exciton confinement energy reaches values up to 90 meV for 43-Å (311) GaAs QWR structures.

The PL and PLE spectra, as well as the optical absorption of the (311) samples, exhibit a pronounced polarization anisotropy of the excitonic resonances (Figure 14.7). The lh-exciton resonance is more pronounced in the spectra taken with the light polarized parallel to the [01$\bar{1}$] direction, whereas the hh-resonance is more pronounced for the respective perpendicular polarization, in agreement with the asymmetric interface corrugation in the (311) samples. The corresponding polarization behavior is observed also in the PL spectra. As would be expected for quantum wells, no optical anisotropy is observed in the (100) reference samples.

14.5.2 Self-organized growth on mesoscopic step arrays created by step bunching

In Section 14.5.1 it was shown that the lateral extension of hh-excitons exceeds the 32-Å geometrical width of the GaAs channel region in the (311) quantum-wire structures. To increase the lateral periodicity to a value comparable to the exciton Bohr radius, the accumulation of steps to periodic mesoscopic step arrays must be achieved. This concept has been applied to the (210) GaAs surface (Nötzel et al., 1993b), where the step bunching can be monitored in real time by RHEED, as depicted in Figure 14.8. The RHEED pattern taken along [001] during growth of GaAs below 560 °C indicates a regular step array with a lateral periodicity of about 13 Å ($\hat{=} a_{[120]}$) and a height of about 3 Å ($\hat{=} 2d_{(210)}$). During GaAs growth above 560 °C, an accumulation of the microscopic steps that are related to the existence of (110) and (100) facets manifests itself by a continuous transition of the RHEED pattern from Figure 14.8a to 14.8b. The appearance of tilted, nonperiodically arranged streaks shows the presence of mesoscopic facets. The tilt angle of the streaks relative to the surface normal gives the tilt angle of the facet planes to about 6° off the singular (100) and (110) planes. During growth of AlAs on the facets, however, the RHEED pattern of Figure 14.8a returns, indicating a "flattening" of the surface. The change toward microscopic steps during growth of AlAs is governed by the lower migration length of Al compared to Ga (Hata et al., 1990; Joyce, 1990). The high uniformity of the mesoscopic step array and the homogeneous coverage of the surface after GaAs growth above 560 °C is indicated by the high intensity of the tilted streaks in Figure 14.8b and by the extinction of the streaks corresponding to the microscopically stepped surface in Figure 14.8a. The homogeneity is further evidenced by the pronounced streaking of the RHEED pattern taken along the perpendicu-

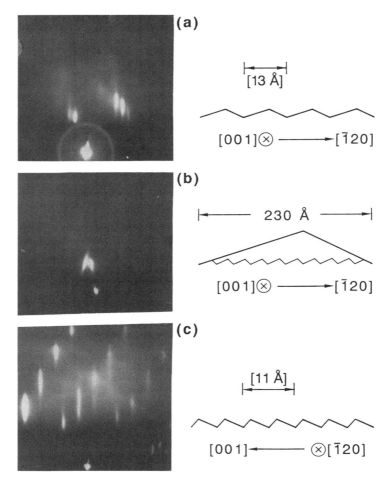

Figure 14.8. RHEED patterns and schematic diagrams of the surface corrugations of the (210) GaAs surface (a) taken along [001] after GaAs growth at 500 °C, (b) taken along [001] after GaAs growth at 600 °C, and (c) taken along [$\bar{1}$20].

lar [$\bar{1}$20] azimuth (Figure 14.8c), which is not affected by the step accumulation observed along [001]. This latter RHEED pattern arises from the periodic step array of the microscopic surface configuration of the (210) plane with a lateral periodicity of 5.6 Å ($\triangleq a_{[100]}$) and a step height of 2.6 Å.

The topography of the mesoscopic surface corrugation after GaAs growth at 600 °C is directly imaged by AFM. Figure 14.9 shows an AFM topographical image of a 2-μm-thick GaAs epilayer. We have obtained similar AFM topographical images from GaAs layers with thicknesses less than 100 Å; this reveals the morphological stability of the mesoscopic surface corrugation during growth, which has been also deduced from the RHEED dynamics. The single-line scan perpendicular to the steps shown in Figure 14.9b can be used to determine that the

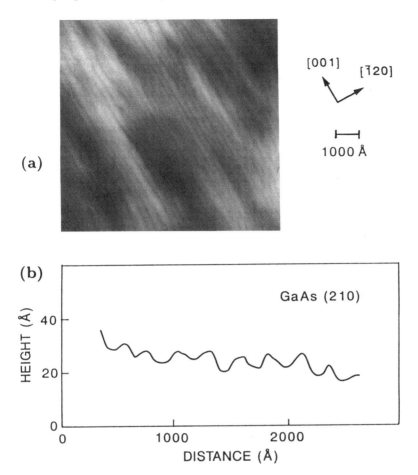

Figure 14.9. (a) Atomic force microscopy (AFM) topographical image of (210) GaAs grown at 600 °C and (b) single-line scan perpendicular to the grooves, showing the high periodicity of the surface corrugation.

lateral periodicity is 230 Å. The height of the steps is not imaged accurately by the AFM single-line scan (about 10 Å), because the oxidation of the sample after removal from the MBE growth chamber smoothens the surface corrugation. Although the oxidation process may introduce fluctuations in the shape of the step array, the AFM micrograph directly illustrates the high uniformity and almost perfect periodicity of the mesoscopic step array.

The shape of the mesoscopic steps has been directly imaged by HREM, viewing the [001] cross section of a 48-Å GaAs/50-Å AlAs multilayer structure grown at 600 °C (Nötzel et al., 1993b). The tilt angles of the facet planes with respect to the macroscopic (210) surface observed in the HREM image agree with the RHEED pattern in Figure 14.8b, with the streaks perpendicular to the mesoscopic facets. The observed height of the steps (21–23 Å) is in agreement with estimates from the RHEED intensity dynamics. The different growth behaviors of GaAs and AlAs on the (210) surface above and below 560 °C offer a new flexibility in engineering GaAs/AlAs interfaces. It allows the formation of GaAs/AlAs multilayer structures with microscopically stepped interfaces at low growth temperatures (no accumulation of steps) and with alternating microscopically and mesoscopically stepped interfaces at elevated growth temperatures (accumulation of steps during GaAs growth). The influence of these different interface structures on the optical properties of GaAs/AlAs multilayer structures manifests itself in the room-temperature luminescence of 48 Å GaAs/50 Å AlAs samples grown at 500 °C and 600 °C (Nötzel et al., 1993b). The formation of mesoscopic steps during GaAs growth at 600 °C results in a 26-meV redshift of the luminescence with respect to that of the (100) reference sample. This value strongly exceeds the 11 meV redshift for the samples grown at 500 °C, which is attributed to the presence of microscopic steps. The most important result, however, is the integrated luminescence intensity of the (210) samples grown at 600 °C, which is strongly enhanced compared to that of the (100) reference structure. Again, this behavior is assigned to a reduced diffusion of carriers in samples with mesoscopic corrugation of the interfaces. In these samples, carrier diffusion is free only along the steps, which diminishes the probability of encountering nonradiative recombination centers.

14.6 Conclusion

We have discussed new phenomena of ordered growth of III-V semiconductors by means of self-assembly (self-organization) that leads directly to the formation of nanometer-scale QWR structures by epitaxial growth. An important aspect of this self-assembly of atoms into "supramolecular" entities is the structural perfection and the extremely low defect density of the resulting QWR structures. The technological advances – and the unprecedented degree of control – in the fabrication of these artificially constructed materials systems will certainly lead to significant advances in the understanding of the basic physics of these systems and to new electronic and optical applications. Revolutionary nanoscale electronics that operate on the single-electron level will thus become a reality.

References

Bartolini, T. A., Ercolessi, F., and Tosatti, E. (1989) "Magic" vicinal surfaces stabilized by reconstruction. *Phys. Rev. Lett.* 63:872–875.
Bauer, G. E. W. and van Gorkum, A. A. (1990) Electronic properties of doping quantum wires. In *Science and Engineering of One and Zero Dimensional. Semiconductors*, Sotomajor Torres, pp. 133–138. New York: Plenum Press.

Chadi, D. J. (1984) Theoretical study of the atomic structure of silicon (211), (311), and (331) surfaces. *Phys. Rev. B* 29:785–792.

Colas, E., Kapon, E., Simhony, S., Cox, H. M., Bhat, R., Kash, K., and Liu, P. S. D. (1989) Generation of macroscopic steps on patterned (100) vicinal GaAs surfaces. *Appl. Phys. Lett.* 55:867–869.

Cox, H. M., Liu, P. S., Yi-Yan, A., Kash, K., Seto, M., and Bastos, P. (1989) Formation of laterally propagating supersteps of InP/InGaAs on vicinal wafers. *Appl. Phys. Lett.* 55:472–474.

Däweritz, L. (1993) RHEED studies of steps, islanding and faceting on singular, vicinal and high-index surfaces. *J. Cryst. Growth* 127:949–955.

Däweritz, L., Hagenstein, K., and Schützendübe, P. (1993) Si incorporation during molecular beam epitaxy growth of GaAs and preferential attachment of Si atoms at *J. Vac. Sci. Technol. A* 11:1802–1806.

Däweritz, L.and Kostial, H. (1994) Self-organized in-plane incorporation of Si atoms in GaAs by molecular beam epitaxy. *Appl. Phys. A* 58:81–86.

Däweritz, L. and Ploog, K. (1994) Contribution of reflection high-energy electron diffraction to nanometre tailoring of surfaces and interfaces by molecular beam epitaxy. *Semicond. Sci. Technol.* 9:123–136.

De Caro, L. and Tapfer, L. (1993) Elastic lattice deformation of semiconductor heterostructures grown on arabitrarily oriented substrate surfaces. *Phys. Rev. B 48*:2298–2303.

Degani, M. H. and Hipolito, O. (1987) Exciton binding energy in quantum well wires. *Phys. Rev. B* 35:9345–9348.

Fukui, T. and Saito, H. (1990) Natural supersteps formed on GaAs vicinal surface by metalorganic chemical vapor deposition. *Jpn. J. Appl. Phys.* 29:L483–L485.

Gomyo, A., Suzuki, T. Kobayashi, K., Kawata, S., Hino, I., and Yuasa, T. (1987) Evidence for the existence of an ordered state in $Ga_{0.5}In_{0.5}P$ grown by metalorganicvapor phase epitaxy and its relation to band-gap energy. *Appl. Phys. Lett.* 50:673–675.

Gomyo, A., Suzuki, T., and Iijima, S. (1988) Observation of strong ordering in $Ga_xIn_{1x}P$ alloy semiconductors. *Phys. Rev. Lett.* 60:2645–2648.

Gomyo, A., Makita, K., Hino, I., and Suzuki, T. (1994) Observation of new ordered phase in $Al_xIn_{1-x}As$ alloy and relation between ordering structure and surface reconstruction during molecular-beam-epitaxial growth. *Phys. Rev. Lett.* 72:673–676.

Hasegawa, S., Kimura, K., Sato, M., Maehashi, K., and Nakashima, H. (1992) Step structures during MBE growth of GaAs and AlGaAs films on vicinal GaAs (110) surfaces inclined toward (111)B. *Surf. Sci.* 267:5–7.

Hata, M., Isu, T., Watanabe, A., and Katayama, Y. (1990) Distribution of growth rates on patterned surfaces measured by scanning microprobe reflection high-energy electron diffraction. *J. Vac. Sci. Technol. B* 8:692–696.

Herring, C. (1951) Some theorems on the free energies of crystal surfaces. *Phys. Rev.* 82:87–93.

Hopfield, J. J. (1959) A theory of edge-emission phenomena in CdS, ZnS and ZnO. *J. Phys. Chem. Solids* 10:110–119.

Horikoshi, Y., Yamaguchi, H., Briones, F., and Kawashima, M. (1990) Growth process of III-V compound semiconductors by migration-enhanced epitaxy. *J. Cryst. Growth* 105:326–338.

Ihm, Y. E., Otsuka, N., Klem, J., and Morkoc, H. (1987) Ordering in $GaAs_{1-x}Sb_x$ grown by molecular beam epitaxy. *Appl. Phys. Lett.* 51:2013–2015.

Joyce, B. A. (1990) The evaluation of growth dynamics in MBE using electron diffraction. *J. Cryst. Growth* 99:9–17.

Kuan, T. S., Kuech, T. F., Wang, W. I., and Wilkie, E. L. (1985) Long-range order in $Al_xGa_{1-x}As$. *Phys. Rev. Lett.* 54:201–204.

Kurimoto, T. and Hamada, N. (1989) Electronic structure of the $(GaP)_1/(InP)_1$ strained-layer superlattice. *Phys. Rev. B* 40:3889–3895.

Lagally, M. G., Savage, D. E., and Tringides, M. C. (1988) Diffraction from disordered surfaces: An overview. In *Reflection High Energy Electron Diffraction and Reflection Electron Imaging of Surfaces,* ed. P. K. Larsen and P. J. Dobson, pp. 139–172. New York: Plenum Press.

Mbaye, A. A. and Massies, J. (1990) Transient dynamics of the growth front surface morphology in molecular beam epitaxy: A deterministic viewpoint. *Europhys. Lett.* 11:769–774.

Miller, R. C. and Kleinman, D. A. (1985) Excitons in GaAs quantum wells. *J. Lumin.* 30:520–540.

Murgatroyd, I. J., Norman, A. G., and Booker, G. R. (1990) Observation of {111} ordering and [110] modulation in molecular beam epitaxial $GaAs_{1-y}Sb_y$ layers: Possible relationship to surface reconstruction during layer growth. *J. Appl. Phys.* 67:2310–2319.

Neave, J. H., Dobson, P. J., Joyce, B. A., and Zhang, J. (1985) Reflection high-energy electron diffraction oscillations from vicinal surfaces – A new approach to surface diffusion measurements. *Appl. Phys. Lett.* 47:100–102.

Nötzel, R., Ledentsov, N. N., Däweritz, L., Hohenstein, M., and Ploog, K. (1991) Direct synthesis of corrugated superlattices on non-(100)-oriented surfaces. *Phys. Rev. Lett.* 67:3812–3815.

Nötzel, R., Ledentsov, N. N., Däweritz, L., and Ploog, K. (1992a) Semiconductor quantum wire structures directly grown on high-index surfaces. *Phys. Rev. B* 45:3507–3515.

Nötzel, R., Däweritz, L., and Ploog, K. (1992b) Topography of high- and low-index GaAs surfaces. *Phys. Rev. B* 46:4736–4743.

Nötzel, R., Ledentsov, N. N., and Ploog, K. (1993a) Confined excitons in corrugated GaAs/AlAs superlattices. *Phys. Rev. B* 47:1299–1304.

Nötzel, R., Eissler, D., Hohenstein, M., and Ploog, K. (1993b) Periodic mesoscopic step arrays by step bunching on high-index GaAs surfaces. *J. Appl. Phys.* 74:431–435.

Pashley, M. D., Haberern, K., and Gaines, J. M. (1991) Scanning tunneling microscopy comparison of GaAs(001) vicinal surfaces grown by molecular beam epitaxy. *Appl. Phys. Lett.* 58:406–408.

Permogorov, S. (1975) Hot excitons in semiconductors. *Phys. Status Solidi* B 68:9–42.

Petroff, P. M., Gossard, A. C., and Wiegmann, W. (1984) Structure of AlAs-GaAs interfaces grown on (100)vicinal surfaces by molecular beam epitaxy. *Appl. Phys. Lett.* 45:620–622.

Ramsteiner, M., Wagner, J., Jungk, G., Behr, D., Däweritz, L., and Hey, R. (1994) Raman spectroscopic study of the wirelike incorporation of Si dopant atoms on GaAs(001) vicinal surfaces. *Appl. Phys. Lett.* 64:490–492.

Schubert, E. F. (1990) Delta doping of III-V compound semiconductors: Fundamentals and device applications. *J. Vac. Sci. Technol.* A 8:2980–2996.

Shahid, M. A., Mahajan, S., Laughlin, D. E., and Cox, H. M. (1987) Atomic ordering in $Ga_{0.47}In_{0.53}As$ and $Ga_xIn_{1-x}As_yP_{1-y}$ semiconductors. *Phys. Rev. Lett.* 58:2567–2570.

Shiraishi, K. (1992) Ga adatom diffusion on an As-stabilized GaAs(001) surface via missing As dimer rows. First principles calculations. *Appl. Phys. Lett.* 60:1363–1365.

Sundaram, M., Chalmers, S. A., Hopkins, P. F., and Gossard, A. C. (1991) New quantum structures. *Science* 254:1326–1335.

Suzuki, T., Gomyo, A., Iijima, S., Kobayashi, K., Kawata, S., Hino, I., and Yuasa, T. (1988) Band-gap energy anomaly and sublattice ordering in GaInP and AlGaInP grown by metalorganic vapor phase epitaxy. *Jpn. J. Appl. Phys.* 27:2098–2106.

Suzuki, T. and Gomyo, A. (1991) Re-examination of the formation mechanism of CuPt-type natural superlattices in alloy semiconductors. *J. Cryst. Growth* 111:353–359, and references therein.

Tong, S. Y., Xu, G., and Mei, W. N. (1984) Vacancy-buckling model for the (2x2)GaAs(111) surface. *Phys. Rev. Lett.* 52:1693–1696.

Tournié, E., Nötzel, R., and Ploog, K. H. (1994) Tunable generation of nanometer-scale corrugations on high-index III-V semiconductor surfaces. *Phys. Rev. B* 49:11053–11059.

Wood, C. E. C., Metze, G. M., Berry, J. D., and Eastman, L. F. (1980) Complex free-carrier profile synthesis by "atomic-plane" doping of MBE GaAs. *J. Appl. Phys.* 51:383–387.

15

Formation of a simple model brain on microfabricated electrode arrays

AKIO KAWANA

NTT Basic Research Laboratories
3-1, Morinosato Wakamiya
Atsugi-shi, Kanagawa 243-01
Japan

15.1 Introduction

The remarkable capacity of the brain for information processing remains poorly understood, largely because of the vast number of neurons and the overwhelming complexity of their interconnections. The main purpose of this study has been to understand the operation of the single neuron and of neural networks in the brain by investigating simplified, artificially cultured neurons and neural networks formed from biological neural cells.

15.2 Cultured neurons and neural networks

Cultured systems of vertebrate neurons provide a useful methodology for studying the physiology of neurons and neural networks in a simplified, controlled environment. Recent progress in the technique of cell culture has made it possible to routinely culture neurons from the brain and to form synaptic connections among them. However, neurons cultured in the usual way present a number of difficulties for studying the details of signal transmission and processing in single neurons and in neural networks.

Usually a neuron receives signals from many other neurons through synapses on neurites. The received signals propagate on the neurite to the soma, where they sum the inputs to form the neuron's output. Because the signals are believed to undergo significant processing during propagation, an understanding of the signal processing and transmission in neurites is of central importance in studying the function of single neurons. One of the main difficulties in this endeavor is the complexity of the neurite's morphology. Figure 15.1 shows a mouse sensory neuron cultured on a conventional dish. A profusion of neurites grows randomly from the cell somas, and their morphology is too complicated

258

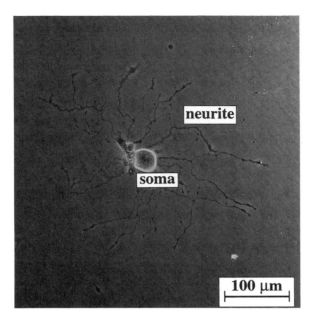

Figure 15.1. A mouse sensory neuron cultured on a normal dish. Many neurites grow randomly from the cell soma.

to allow a meaningful investigation of their physiology. To achieve an understanding of signal transmission and processing in neurites, their morphology must be simplified by controlling the direction of outgrowth. In order to study the function of neurites, it is also necessary to introduce signals into a neurite from intended and localized points, since it is very difficult to form synaptic connections at controlled positions.

In order to understand the function of neural networks, it is necessary to have a knowledge of the neural connections as well as a method to measure simultaneously the electrical activity from many neurons. Figure 15.2 shows a network formed from chick brain neurons. Although the number of neurons in the networks is rather small, the network of connections is already complex, to the degree that individual connections between particular neurons cannot be recognized. Clearly, it would be extremely useful if we could construct real neural networks in an analogous way to that in which integrated electronic circuits are built. For this purpose, guidance of the neurite outgrowth and placement of neurons at intended positions are necessary. A microelectrode is usually used to observe electrical activity of neurons, but it cannot measure the activity from many neurons simultaneously. A multisite recording system for electrical activity should therefore be developed. Our approach to these technical challenges will be described.

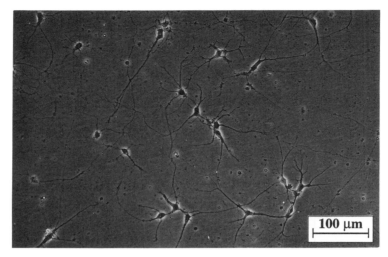

Figure 15.2. Chick brain neurons cultured on a normal dish. They formed simple neural networks, but it is difficult to recognize the connections among them.

15.3 Signal transmission and processing in neurites

15.3.1 Guiding the neurite outgrowth direction

It is well known that neurites recognize the three-dimensional geometrical configuration of the surface of substrates (Dow et al., 1987) and differences in the surface chemistry (Hammarback et al., 1985; Kleinfeld et al., 1988). In order to control the neurite outgrowth direction, we fabricated microgrooves and patterns of metal oxide, which are expected to be more adhesive than silica glass, on silica glass substrates, using conventional photolithography (Hirono et al., 1988; Torimitsu and Kawana, 1990).

Figure 15.3 shows a schematic diagram of the fabrication of substrates with microgrooves or metal oxide patterns. The surface of silica glass was spin-coated with photoresist. A mask with a designed pattern was laid down on the surface, which was exposed through the mask by ultraviolet light. The designed patterns of the photoresist were obtained after developing. The microgrooves were fabricated using reactive ion etching (RIE), and metal oxide patterns were formed by vapor deposition of metal oxide. Figure 15.4 shows a scanning electron micrograph of the microgroove and its cross section measured with a thickness meter. Using RIE, the edge of the silica glass was etched sharply so that we could obtain clear boundaries of grooves as shown in the figure. Figures 15.5(a), (b), and (c) show a mouse sensory neuron cultured on a silica glass substrate, respectively, with microgrooves of 10 μm width and 1 μm depth; 2 μm width and 1 μm depth; and no microgrooves. In the figure, white lines and dark lines are microgrooves and microsteps, respectively. The neurites cultured on substrates with microgrooves

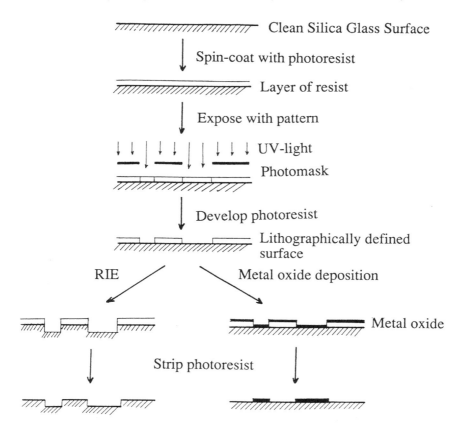

Clean Silica Glass Surface

Spin-coat with photoresist

Layer of resist

Expose with pattern

UV-light

Photomask

Develop photoresist

Lithographically defined surface

RIE   Metal oxide deposition

Metal oxide

Strip photoresist

Figure 15.3. Schematic diagram of the fabrication process of substrates with micro-gooves or metal oxide patterns. The process consists of three steps: patterned photoresist formation; etching of the surface, or deposition of metal oxide on it; and stripping off the remaining photoresist.

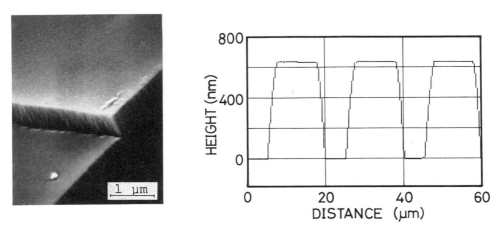

Figure 15.4. A scanning electron micrograph of a microgroove and its cross section measured with a thickness meter.

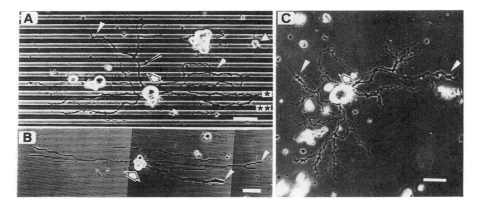

Figure 15.5. Mouse sensory neurons cultured on silica glass substrates with microgrooves. ★ and ★ ★ in (A) indicate the microgroove and the microstep, respectively. ∇ and ⊅ are growth cone and soma, respectively. Scale bar is 50 μm. (A) The width and depth of microgrooves are 10 μm and 1μm, respectively. (B) The width and depth of microgrooves are 2 μm and 1 μm, respectively. (C) Flat silica glass.

grow along microgrooves and almost exclusively within them. Deeper microgrooves show a more efficient guiding ability. The movement of the growth cones was observed by time-lapse video to be confined in the microgrooves by their walls. These results suggest that the growth cones were physically confined in the microgrooves and produced neurites lying mainly in the microgrooves.

Figure 15.6 shows mouse sensory neurons cultured on a substrate with a metal oxide pattern. The black lines in the figure are indium oxide strips of 10 μm width

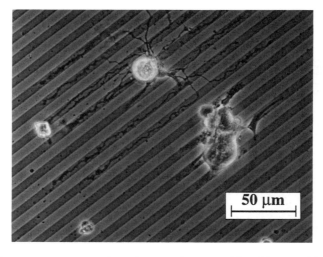

Figure 15.6. A sensory neuron cultured on a substrate with a 10-μm-wide indium oxide pattern. The thickness of the indium oxide is 50 nm. The black lines and the white lines are the indium oxide and the silica glass substrate, respectively.

and 50 nm thickness. The neurites mainly grow on the metal oxide that was deposited on the silica glass substrate. Figure 15.7 shows the movement of a growth cone on the substrate, in a sequence of frames taken from a time-lapse video. The growth cone on the metal oxide spread when it came to the border between the metal oxide and the silica glass. The growth cone seems to try to find its way at the border. After filopodia were extended onto the silica glass, they retracted back to the metal oxide, and the growth cone grew again on the metal oxide. These processes resulted in the neurite's growing on the metal oxide in a zigzag manner, as seen in the figure. This phenomenon is different from the confinement in microgrooves because of the extreme thinness of the pattern, and it probably corresponds to a difference in adhesiveness of the metal oxide. In Table 15.1, the ability of different classes of metal oxides to guide outgrowth is shown. Aluminum and indium oxide guided neurites very well, and 80% of neurites were on the metal oxide. These oxides may have a suitable electrical surface charge that is highly correlated with adhesiveness. For our

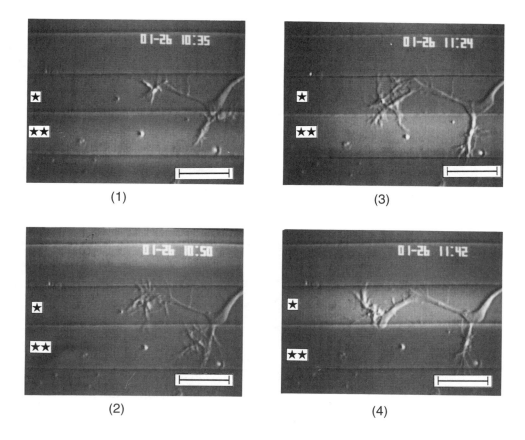

Figure 15.7. The outgrowth of a neurite on a substrate with an indium oxide pattern. ★ and ★★ indicate the indium oxide and the silica glass substrate, respectively. The scale bar is 10 μm. These photos were taken from a time-lapse video and the time is shown in each photo.

Table 15.1. *Relation between metal oxide electrical negativity values and the percentage (S.D.) of total neurite length growing on the metal oxide (Torimitsu and Kawana, 1990).*

Metal oxide	Electronegativity of metal	Neurites on metal oxide (%)
SnOx	1.72	71 ± 16
InOx	1.49	79 ± 10
AlOx	1.47	78 ± 8
TiOx	1.32	49 ± 19
YOx	1.11	40 ± 9

study, it is necessary to be able to guide the neurites in any specific direction. To illustrate this capability, a chick sensory neuron was cultured on a substrate with a hexagonal indium oxide pattern. The width and thickness of the metal oxide were 10 μm and 50 nm, respectively. The neurites grew along the hexagonal indium oxide pattern, as shown in Figure 15.8. This result indicates that the neurite outgrowth may be guided efficiently in multiple directions using this method.

15.3.2 Stimulating neurites at localized points

In order to stimulate neurites in restricted, well-specified location, we fabricated substrates with embedded electrode arrays. Several authors have already described

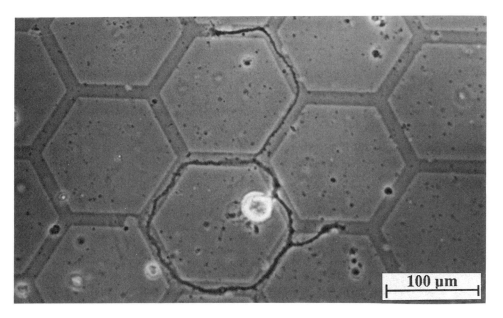

Figure 15.8. A chick sensory neuron cultured on a substrate with a hexagonal pattern of 10-μm-wide indium oxide (black lines) (Torimitsu and Kawana, 1990).

the stimulation of neurites by similar methods (Pine, 1980; Regehr et al., 1989). However, in previous work, the neurites grew randomly, and consequently they had only a low probability of forming adequate electrical coupling to the electrodes. We combined the electrode array with a guiding structure, as shown in Figure 15.9, to direct the neurites efficiently onto electrodes (Jimbo and Kawana, 1992). In this figure, indium tin oxide (ITO) is used for the electrode and aluminum oxide is used as passivation. An overlying polymethylmethacrylate (PMMA) layer guides the neurite on the electrode.

Figure 15.10 shows the process of fabrication of electrode array substrates. Silica glass coated with sputtered ITO was etched in HCl solution and the electrode pattern was formed. The surfaces of the ITO electrodes were then partially platinized electrochemically in order to reduce the interface impedance between the electrode and the electrolyte solution. After that, aluminum oxide was deposited onto the substrates as the passivation film except at the very tips of the electrode. Finally, a PMMA layer was spin-coated and patterned by RIE.

Figure 15.11 shows a cultured mouse sensory neuron on the substrate. It can be seen that the neurite was guided onto the electrode. An electrical pulse was applied to the substrate electrode pair through an isolating circuit. Constant current pulses were applied as a stimulating signal.Membrane potential and current in the cell body region were measured by the whole-cell patch-clamp technique (Hamill et al., 1981). Figure 15.12 shows the cell's response to a current stimulus to its neurite through the substrate electrode. The lower trace is the waveform of the stimulation current, which had an amplitude of 5 μA and a duration of 5 msec. Every three pulses, generation of an action potential was observed.This result indicates that a single pulse of that

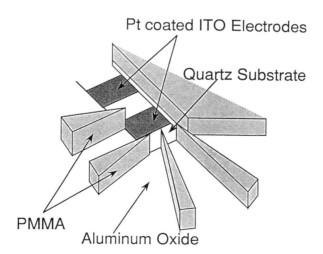

Figure 15.9. A schematic diagram of the surface microstructure of the electrode array substrate. The electrodes were coated with an aluminum oxide passivation film. A PMMA wall was constructed on the film to guide neurite outgrowth (Jimbo and Kawana, 1992).

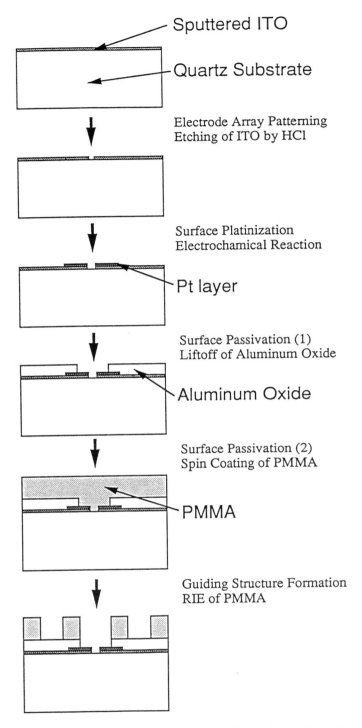

Figure 15.10. Schematic diagram of the fabrication of the planar electrode array. The process consists of the following three steps: electrode array formation, deposition of a passivation layer, and construction of the guiding structure (Jimbo and Kawana, 1992).

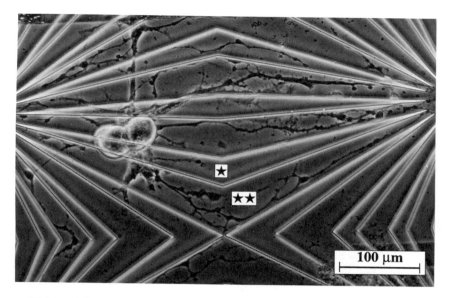

Figure 15.11. Cultured mouse sensory neurons on the planar electrode array substrate. Neurites grow along the guiding structure. ★ and ★★ are the aluminum oxide passivation layer and a PMMA guiding layer, respectively (Jimbo and Kawana, 1992).

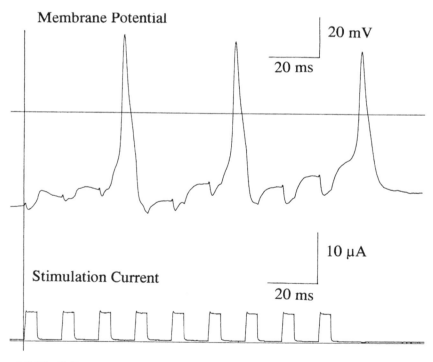

Figure 15.12. Cell response to current stimulation on its neurite. An action potential was generated with every three stimulation pulses (Jimbo and Kawana, 1992).

amplitude produces only subthreshold depolarization, whereas a train of three pulses triggers the action potential as a result of temporal summation.

15.4 Stimulating and monitoring the activity of neurons by electrode arrays

The ability to stimulate and monitor the activity of neurons and neural circuits non-invasively is essential for understanding their function. For this purpose, substrates with embedded electrodes have been used extensively since their development by Gross (1979). Planar electrode arrays with wells and conduits were fabricated on silica glass as previously described, except for the final coating procedure (Jimbo et al., 1993). The surface was coated with a 10-µm layer of polyimide instead of PMMA, because polyimide films can be deposited more thickly than PMMA films. Wells were fabricated at each electrode by etching the polyimide as well as the connective conduits. A photograph of the substrate with the electrode array is shown in Figure 15.13. The substrate had 16 wells, each of which was 150 µm square. In each well, a pair of electrodes was embedded, with each electrode measuring 20 x 100 µm. The 16 wells were connected by 20-µm-wide conduits.

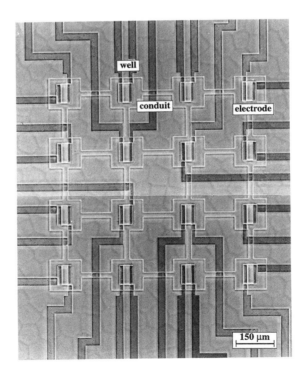

Figure 15.13. Phase contrast micrographs of the substrate surface. There are 16 square wells that are connected by narrow conduits. A pair of electrodes is in each well (Jimbo et al., 1993).

Networks using neurons from rat brain were formed on the substrate, and the spontaneous electrical activity of neurons was monitored through the embedded well electrodes. In nominally Mg-free solution, the neurons showed periodic action potential bursting (Jimbo et al., 1993; Robinson et al., 1993). The maximum amplitude of the extracellularly recorded bursting signal was around 100 μV, and the system noise was 10 μV. Simultaneous recording from four electrodes is shown in Figure 15.14. The positions of the recording sites are indicated in the figure. The electrical firing, which was synchronized among the electrodes, was observed at approximately 10-sec intervals, which varied somewhat from culture to culture. This result suggests that the neurons over a region of 1 mm were synaptically connected. Their firing pattern was found to correspond to intracellular Ca^{2+} rises by simultaneous recording with an optical method using Ca^{2+} indicator dyes (Jimbo et al., 1993).

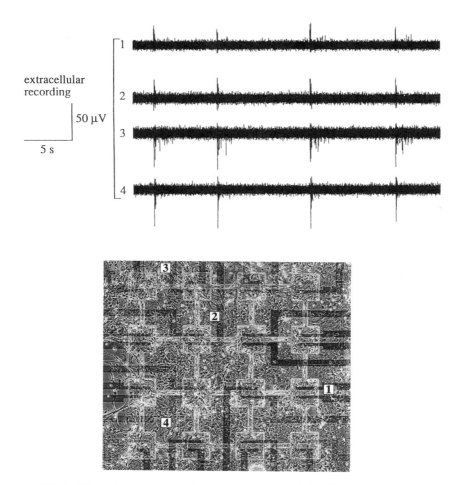

Figure 15.14. Multichannel recording of electrical activity. Four simultaneous extracellular signals are shown. The position of recording of each signal is indicated on the photograph (Jimbo et al., 1993).

The period of the firing could be controlled by stimulation through these electrodes (Jimbo et al., 1993). Extracellular stimulation was applied using substrate electrode pairs. Figure 15.15 shows the response of the network activity. The top trace shows the stimulation current. The detailed waveform of the current is shown in the inset. This capability is very useful in probing the function of networks and in measuring the effects of stimulation on the development of networks.

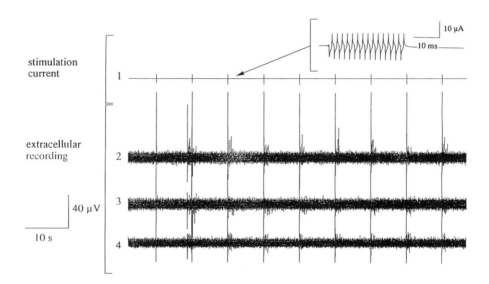

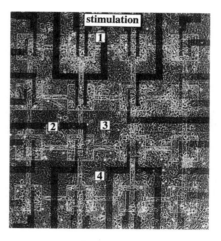

Figure 15.15. Response of network activity to extracellular stimulation. The stimulation current and three traces of extracellular recording are shown. The sites of stimulation and recording are indicated in the photograph. The waveform of the stimulating current is shown in the inset (Jimbo et al., 1993).

15.5 Simple artificial neural networks constructed by cultured neurons

Placement of neurons at arbitrarily specified positions is essential for forming artificially designed neural networks. This is a very difficult problem, however, because neurons are too small and fragile to handle with something like tweezers. A silica substrate with wells, plus a metal mask with holes at positions corresponding to the substrate wells, were employed to control the plating position of neurons (Jimbo et al., 1993). A schematic diagram of this method is shown in Figure 15.16. A silica substrate was used with 150 x 150 μm wells, connected by

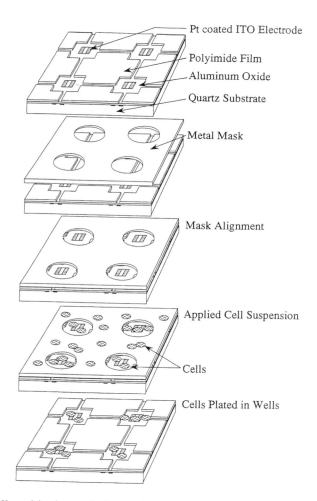

Pt coated ITO Electrode

Polyimide Film

Aluminum Oxide

Quartz Substrate

Metal Mask

Mask Alignment

Applied Cell Suspension

Cells

Cells Plated in Wells

Figure 15.16. Cell positioning technique using a metal mask. A stainless steel mask with 16 holes was aligned with the pattern of the electrode array substrate. A dense cell suspension was applied over it and the mask was removed after cell adhesion was complete (Jimbo et al., 1993).

50-μm-wide microgrooves of 10 μm depth and 200 μm length. A metal mask with 130-μm-diameter circular holes was laid over the substrate. A suspension of dissociated rat brain neurons was plated over the metal mask and incubated for 12 h. Then, culture medium was added and the metal mask was removed. The cell density in the wells was controlled by varying the density of the suspension or the diameter of the holes. When we used a metal mask with 150-μm-diameter holes and a silica substrate with 150 x 150 μm wells, the number of neurons in each well was about 20. Figures 15.17 and 15.18 show the process of simple neural network formation observed using time-lapse video. Two or three days after starting the culture, neurons within the same well formed synaptic connections, as shown in Figure 15.17. Following the formation of connection in the wells, the neurites

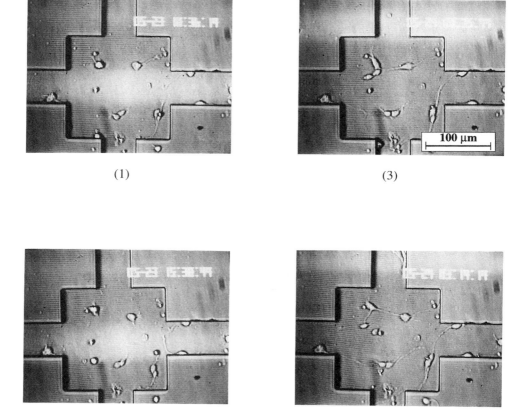

(1)   (3)

(2)   (4)

Figure 15.17. Formation of the neural network in the wells. Two to three days after the culture was started, neurons in the same well formed synaptic connections. These photos were taken from a time-lapse video. The time intervals between each photo are around 3 h. Scale bar is 100 μm.

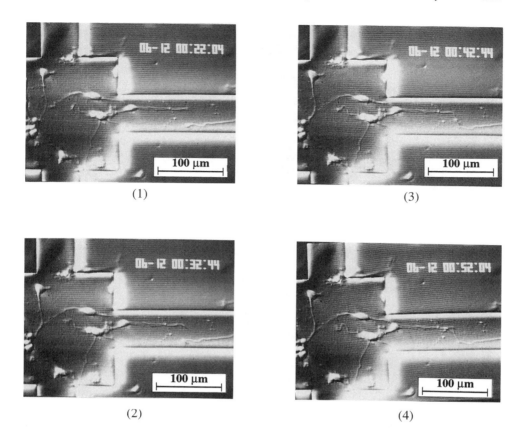

Figure 15.18. Formation of interwell connections. Following the formation of intrawell connection, some neurites grew out from the well and formed interwell connections. These photos are taken from a time-lapse video and the time of each photo is indicated.

grew out from the wells along the microgrooves to form interwell synaptic connections, as shown in Figure 15.18.

In order to confirm the formation of functional synaptic connections among the cells, we monitored the activity of the neurons using fluorescent dyes. Ca^{2+} indicator dyes were used for measuring cell activity because the Ca^{2+} level is an accurate index of neural activity. The result is shown in Figure 15.19. The left side is a photograph of the fluorescence of neurons, which are mainly in the wells. The right panel of the figure shows the fluorescence changes of the neurons. Each fluorescence trace corresponds to a numbered position on the left side, as indicated. As shown in Figure 15.19, periodic intracellular Ca^{2+} changes were observed and they were synchronized among the cells. The synchronization in the Ca^{2+} level reflects the exchange of electrical signals through the con-

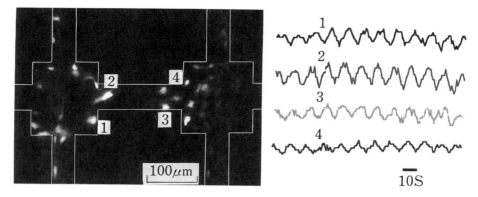

Figure 15.19. Intercellularly synchronized Ca²⁺ transients in simplified neural networks. (a) shows the spatial fluorescence distribution of the neurons; (b) shows the temporal changes in fluorescence of the neurons, as numbered in (a).

necting synapses between cells (Kudo and Ogura, 1986; Kuroda et al., 1992), which thus form functional networks on the silica glass substrate.

15.6 Conclusion

We have developed a novel method of cell culture for controlling the neurite outgrowth direction and cell plating position. The morphology of neurites was greatly simplified, and the neurites were guided onto electrodes embedded in the substrates using this method, which makes it possible to stimulate neurites in restricted, localized areas. This technique should be of great use in elucidating the details of signal transmission and processing.

We succeeded in stimulating and monitoring the electrical activity of networks through electrode arrays embedded in the substrates. Simple neural networks were also formed on the substrates using masks. This should be a powerful tool for understanding the operation of cultured neural networks and, ultimately, of the brain.

Acknowledgments

Part of this work was done in collaboration with Dr. Jun Fukuda of Tokyo University and Dr. Yoichiro Kuroda of the Tokyo Metropolitan Institute for Neuroscience. I would like to thank them sincerely and the other members of my laboratory at NTT.

References

Dow, J. A., Clark, P., Connolly, P., Curtis, A. S. G., and Wilkinson, C. D. W. (1987) Novel methods for guidance and monitoring of single cell and simple networks in culture. *J. Cell Sci.* 8:55–79.

Gross, G. W. (1979) Simultaneous single unit recording in vitro with a photoetched laser deinsulated gold multielectrode surface. *IEEE Trans. Biomed. Eng.* 26:273–279.

Hamill, O. P., Marty, A., Neher, E., Sakman, B. and Sigworth, F. J. (1981) Improved patch-clamp techniques for high-resolution current recording from cells and cell-free membrane patches. *Pflugers Arch.* 391:85–100.

Hammarback, J. A., Palm, S. L., Furcht, L. T. and Letourneau, P. C. (1985) Guidance of neurite outgrowth by pathways of substratum-adsorbed laminin. *J. Neurosci. Res.* 13:213–220.

Hirono, T., Torimitsu, K., Kawana, A. and Fukuda, J. (1988) Recognition of artificial microstructures by sensory nerve fibers in culture. *Brain Res.* 446:189–194.

Jimbo, Y. and Kawana, A. (1992) Electrical stimulation and recording from cultured neurons using a planar electrode array. *Bioelectrochem. Bioenergetics.* 29:193–204.

Jimbo, Y., Robinson, H. P. C. and Kawana, A. (1993) Simultaneous measurement of intracellular calcium and electrical activity from patterned neural networks in culture. *IEEE Trans. Biomed. Eng.* 40:804–810.

Kleinfield, D., Kahler, K. H. and Hockberger, P. E. (1988) Controlled outgrowth of dissociated neurons on patterned substrate. *J. Neurosci.* 8:4098–4120.

Kudo, Y. and Ogura, A. (1986) Glutamate-induced increase in intracellular Ca^{2+} concentration in isolated hippocampal neurones. *Br. J. Pharmacol.* 89:191–198.

Kuroda, A., Ichikawa, M., Muramoto, K., Kobayashi, K., Matsuda, Y., Ogura, A. and Kudo, Y. (1992) Block of synapse formation between cerebral cortical neurons by a protein kinase inhibitor. *Neurosci. Lett.* 135:255–258.

Pine, J. (1980) Recording action potentials from cultured neurons with extracellular microcircuit electrodes. *J. Neurosci. Methods* 2:19–31.

Regehr, W. G., Pine, J., Cohan, C. S., Mishke, M. D. and Tank, D. W. (1989) Sealing cultured invertebrate neurons to embedded dish electrodes facilitates long-term stimulation and recording. *J. Neurosci. Methods* 30:91–106.

Robinson, H. P. C., Kawahara, M., Jimbo, Y., Torimitsu, K., Kuroda, Y. and Kawana, A. (1993) Periodic synchronized bursting and intracellular calcium transients elicited by low magnesium in cultured cortical neurons. *J. Neurophys.* 70:1606–1616.

Torimitsu, K. and Kawana, A. (1990) Selective growth of sensory nerve fibers on metal oxide pattern in culture. *Dev. Brain Res.* 51:128–131.

16

Cellular engineering: control of cell–substrate interactions

PHILIP E. HOCKBERGER,[1, 2] BARBARA LOM,[1]
ANITA SOEKARNO,[1] CARSON H. THOMAS,[3, 4] and KEVIN E. HEALY[3, 4]

Institute for Neuroscience[1] and Departments of Physiology[2] and Biomaterials[3]
Northwestern University Medical and Dental Schools
303 E. Chicago Ave., Chicago, IL 60611
and
Department of Biomedical Engineering[4]
Robert R. McCormick School of Engineering and Applied Science
Evanston, IL 60201

16.1 Introduction

There is substantial intellectual and commercial interest in products that employ biocompatible materials (biomaterials). These materials include metals, ceramics, and polymers that are used in a wide variety of therapeutic and diagnostic devices. Recent estimates indicate that biomaterials are found in more than 5,000 different medical devices and almost 40,000 different pharmaceutical products with a collective annual sales approaching $100 billion (Peppas and Langer, 1994).

Products based on biomaterials are being developed at the molecular, cellular, and tissue levels. At the molecular level, biomaterials are used in biosensors (for example, pregnancy and diabetes tests), biochemical analysis (for example, chromatography, blood dialysis), and drug delivery systems (for example, degradable coatings, controlled-release vehicles). At the cellular level efforts are focused on controlling the attachment, and movement, of cells to discrete locations on devices for diagnostic and therapeutic purposes (for example, cell sorting, nerve and bone regeneration, cell-based sensors). Engineering at the tissue level has resulted in orthopedic and dental prostheses, artificial organs, degradable devices, and contact lenses. Tissue engineering alone affects approximately 20 million patients per year in the United States, and devices used in blood transfusions affect another 18 million per year (Langer and Vacanti, 1993). If we include all products that utilize biomaterials, then the number of people affected worldwide exceeds 100 million per year.

In spite of the enormous need for, and investment in, developing products based on biomaterials, there have been only modest advances in the science underlying their design and implementation. The development of materials relies more often on trial and error and serendipity than on fundamental science (Weiss,

*Address all correspondence to Philip E. Hockberger, Department of Physiology, M211, Northwestern University Medical School, 303 E. Chicago Ave., Chicago, IL 60611

1991). Although this opportunistic approach has yielded remarkable results, virtually every biomaterial on the market has encountered troublesome and sometimes serious side effects when placed in the complex environment of the body. The desire to eliminate side effects has resulted in a heightened awareness of the need for more basic research in this area.

In order to develop safer, more effective, and less costly products employing biomaterials, three major changes must occur in the way in which this research is performed. First, an interdisciplinary effort is required that joins researchers who have expertise in the chemical and physical sciences with those who have expertise in the life sciences and engineering. Second, graduate and postgraduate training programs in biomaterials need to be developed and expanded. Training programs have played a major role in the success and growth of other interdisciplinary fields, such as neuroscience and materials science. Third, funding for biomaterials research needs to be earmarked in order to avoid the inevitable problems of competing with traditional disciplines for support. Because the rational design of products using biomaterials could save billions of dollars in health care as well as in liability costs, increased support for basic and applied research in biomaterials in consistent with performing science in the national interest (cf. Mikulski, 1994).

16.2 Cellular engineering and biomaterials

The goal of cellular engineering is to develop new diagnostic and therapeutic devices by capitalizing on the inherent capabilities of biological cells. Cells are natural chemical sensors capable of producing high-affinity receptors for many circulating metabolites and toxins. In addition, many cells produce growth-promoting and healing molecules involved in tissue repair. By controlling the attachment and placement of specific cell types on devices, it should be possible to utilize these natural capabilities to create new diagnostic and therapeutic devices as well as to improve the performance of existing ones.

A fundamental starting point in cellular engineering is selecting an appropriate material for interfacing cells and devices. The ideal biomaterial has a well-defined chemical composition, possesses the requisite structural features (tensile strength, ductility, durability), and displays minimal corrosive and toxic properties when placed in contact with biological fluids or tissues (Williams, 1987). Our interest is in controlling the attachment and growth of nerve and bone cells on metal oxides. Silicon-based materials have several advantageous features for studying cell–substrate interactions. First, silicon, quartz, and glass are chemically defined, durable, and relatively nontoxic and noncorrosive within biological tissues. Second, a wide variety of silane derivatives of biological interest can be covalently attached to these materials (see Section 16.4). Third, these derivatives can be patterned on silicon-based substrates using photolithographic techniques (Section 16.5). This feature is useful for studying directed cell growth (Section 16.6) and proliferation (Singhvi,

Kumar, et al., 1994). Fourth, quartz and glass are translucent substrates that allow optical analysis of cell–substrate interactions (Section 16.7). Fifth, silicon-based materials could be useful for constructing neural networks in vitro (Hockberger et al., 1994; also see Chapter 15) and interfacing regenerating neurons with solid-state devices in vivo (Kovacs, 1994; also see Chapter 3). The chemistry used to control cell behavior may also be useful for influencing the placement of cells on other metal oxides, such as titanium and its alloys, that are used in orthopedic devices (Ducheyne and Healy, 1988).

In this chapter we describe our efforts aimed at controlling the attachment of mammalian nerve and bone cells on planar silicon-based substrates. We have developed methods for chemically modifying the substrates to enhance cell adhesion and promote cell guidance. The methods were adapted from microfabrication technologies, and the analysis of modified substrates was facilitated by recent advances in surface analytical techniques. We have benefited from progress in cellular and molecular biological studies of cell–cell and cell–extracellular matrix interactions, which has resulted in a better understanding of the important roles these interactions play in the development and repair processes of tissues. By combining the information and tools from these disparate disciplines, we are learning how to control cell placement on silicon-based substrates.

16.3 Cell adhesive molecules

In the past decade there has been significant progress made in understanding cell adhesion in the native environment. Most prominent has been the discovery of different types of adhesive molecules involved in attaching cells to the extracellular matrix and to neighboring cells. One type involves a class of membrane receptors called integrins, which bind to specific glycoproteins found in the extracellular matrix (Ruoslahti, 1991). Glycoproteins, such as collagen, laminin, and fibronectin, are distributed throughout the matrix and may form a scaffolding for cell anchorage and movement. Cells also adhere to neighboring cells using different receptors called cell adhesion molecules (CAMs; Edelman and Cunningham, 1990), cadherins (Takeuchi, 1991), and selectins (Lasky, 1992). Both cell–cell and cell–extracellular matrix interactions have been implicated in a wide variety of cellular functions related to cell movement, including morphogenesis and metastasis (Takeuchi, 1991), wound healing and inflammation (Lasky, 1992), and neurite outgrowth (Edelman and Cunningham, 1990; Jessel, 1988).

Several lines of evidence indicate that adhesion receptors recognize discrete peptide sequences of their respective ligands. The most compelling evidence derives from studies showing that peptides that bind to these receptors, or antibodies directed against the peptide-binding sites, disrupt cell adhesion and migration. For example, the peptide Arg-Gly-Asp (RGD), the putative cell-binding domain of fibronectin, can block cell adhesion and migration in vitro and in vivo (Ruoslahti and

Pierschbacher, 1987; Yamada et al., 1992). Other peptides have also been implicated in cell adhesion and migration, including Tyr-Ile-Gly-Ser-Arg (YIGSR) for laminin (Graf et al., 1987; Massia et al., 1993), Arg-Glu-Asp-Val (REDV) for fibronectin (Ruoslahti and Pierschbacher, 1987), His-Ala-Val (HAV) for N-cadherin (Chuah et al., 1991), and Lys-Tyr-Ser-Phe-Asn-Tyr-Asp-Gly-Ser-Glu (KYSFNY-DGSE) for N-CAM (Rao et al., 1992). Some cells recognize more than one peptide sequence on a matrix protein, for example, fibronectin contains RGD and REDV, and the same peptides are present on different matrix molecules, for example, RGD is in laminin and fibronectin (Ruoslahti and Pierschbacher, 1987).

A biophysical explanation of how ligand–receptor interactions facilitate cell adhesion is lacking. It is unclear, for example, how forces generated by interactions among adhesive molecules differ from those generated by other ligands and receptors. If we assume that adhesive molecules somehow generate larger intermolecular forces, then it still is not clear whether this would facilitate or impede cell locomotion. Given the small size (10 μm average), mass (5×10^{-10} grams), and large surface-to-volume ratio (6×10^3/cm) of cells, it is conceivable that adhesion is not due to specific ligand–receptor interactions but is instead governed by weak intermolecular forces, such as electrostatic, van der Waals, H-bonding, and/or surface tension.

Analysis of cells in vitro has demonstrated that electrostatic forces can influence cell–cell as well as cell–substrate adhesion. Katchalsky et al. (1959) were the first to systematically study this relationship. They found that polymers of basic amino acids (such as lysine, ornithine, arginine), but not of acidic or neutral amino acids, facilitated agglutination of red blood cells in solution. Ultrastructural analysis indicated that the distance between cell membranes at attachment sites was approximately the size of monomeric molecules. They hypothesized that agglutination was therefore due to binding of basic amino acids (which are positively charged at physiological pH) to negatively charged components of adjacent membranes (such as phosphatidyl serine, sialic acid). The results were extended by other researchers, who showed that many cell types will bind to surfaces treated with basic, but not acidic or neutral, amino acids (Letourneau, 1975; McKeehan and Ham, 1976; Yavin and Yavin, 1974). These studies, as well as more recent ones utilizing a wide range of organic molecules, have led to the realization that electrostatic force is an important factor in cell adhesion.

One way in which electrostatic interactions may influence adhesion is by facilitating ligand–receptor binding. Peptides involved in adhesion generally have at least one basic amino acid that may facilitate binding to negatively charged receptors. As previously mentioned, however, the relationship between ligand–receptor binding and adhesion is uncertain. In order to gain a better understanding of this relationship, we have analyzed cells on simple organic molecules as well as on extracellular matrix proteins. The chemical structures of the simple organic molecules (amines and alkanes) are shown in Figure 16.1, together with two basic amino acids (lysine and arginine) and three peptides (RGD, REDV, and YIGSR) that facilitate cell adhe-

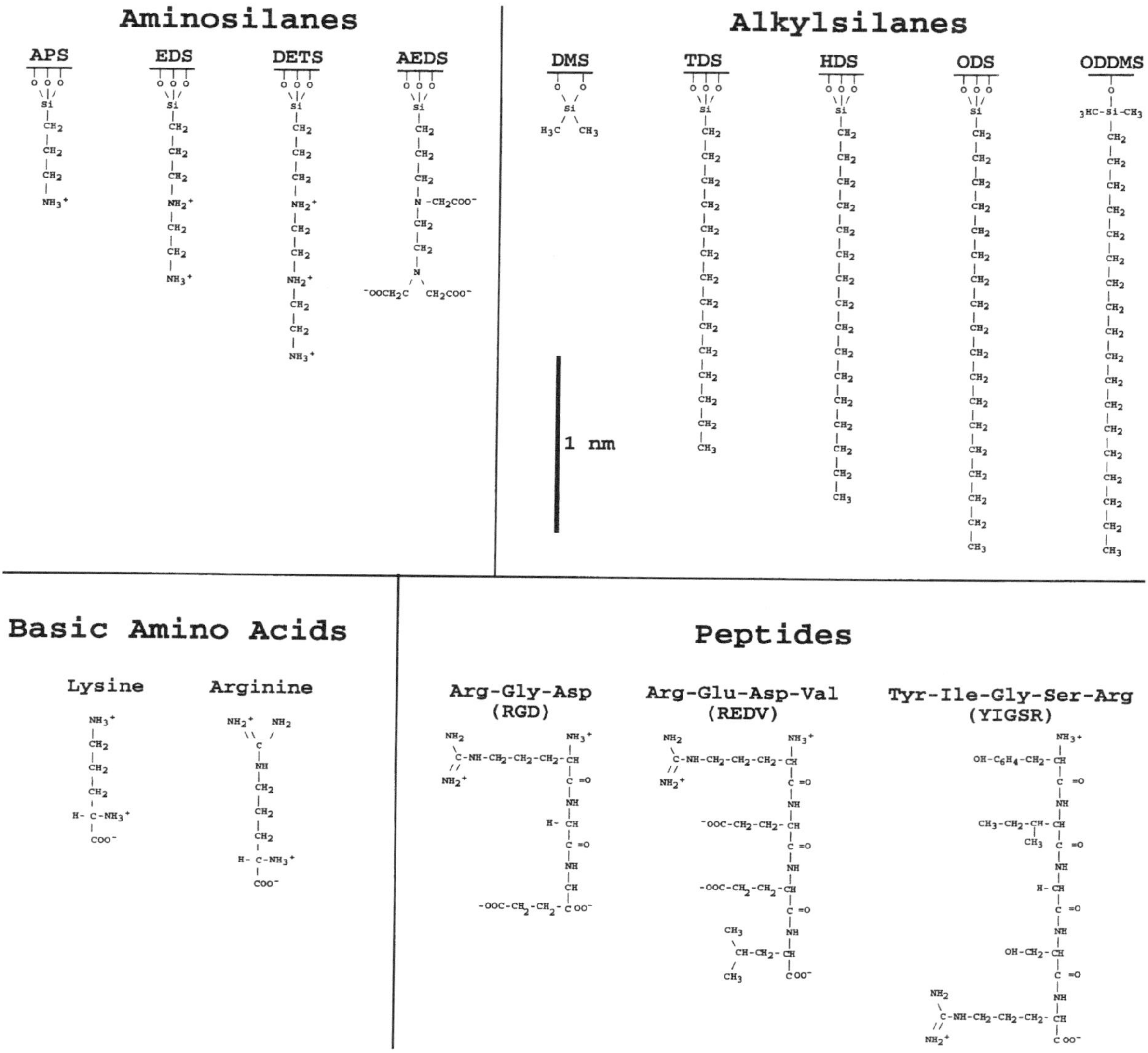

Aminosilanes

APS   EDS   DETS   AEDS

Alkylsilanes

DMS   TDS   HDS   ODS   ODDMS

1 nm

Basic Amino Acids

Lysine   Arginine

Peptides

Arg-Gly-Asp
(RGD)

Arg-Glu-Asp-Val
(REDV)

Tyr-Ile-Gly-Ser-Arg
(YIGSR)

Figure 16.1. Chemical structures of the amino- and alkylsilanes immobilized on silicon-based substrates, as well as two basic amino acids and three peptides that are known to facilitate cell–substrate adhesion. Ionic charges represent the ionization state of the molecular groups at physiological pH (7.4). The molecular lengths were estimated assuming a *trans*-extended chain projection with minimal tilting of the chains with respect to the surface normal (cf. Lom et al., 1993).

Abbreviations:
AEDS = acetylated EDS
APS = aminopropylsilane
DETS = diethylenetriamine propylsilane
DMS = dimethysilane
EDS = ethylenediamine propylsilane

HDS = hexadecylsilane
ODDMS = octadecyldimethylsilane
ODS = octadecylsilane
TDS = tetradecylsilane

sion. It is our belief that by investigating the interactions of cells with these molecules, we will gain insight into the adhesive properties of the natural ligands, which may lead to the development of better interfaces between cells and devices.

16.4 Immobilization of cell adhesive molecules on silicon-based substrates

To date, several classes of adhesive chemicals have been covalently attached to silicon-based substrates, including amines (Dulcey et al., 1991; Healy et al., 1994; Kleinfeld et al., 1988; Lom et al., 1993; Massia and Hubbell, 1992; Matsuzawa et al., 1993), amino acids (Massia and Hubbell, 1992), peptides (Britland, Perez-Arnaud et al., 1992; Fodor et al., 1991; Massia and Hubbell, 1990), and proteins (Britland, Perez-Arnaud et al., 1992, Britland, Clark et al., 1992). A common strategy has been to immobilize the compounds using silane-coupling chemistry. This approach involves creating a silane derivative of the compound of interest and covalently attaching it to the silicon substrate (Arkles, 1991; Pawlenko, 1986; Plueddemann, 1982). We have used this strategy to assess the influence of surface chemistry on cell adhesion to silicon-based substrates. In this section we describe our efforts to characterize the surface properties of these substrates.

There are many factors that can influence the deposition of silane derivatives on silicon-based substrates. One of the most critical is the chemical nature of the outermost layer (overlayer) of the silicon material. In ambient air, silicon is covered by a thermodynamically stable oxide film, and under proper preparation conditions it can approach near ideal SiO_2 stoichiometry. Under humidified conditions, the oxide becomes hydroxylated, which is essential for the immobilization of silane derivatives. Hydroxylation is fostered by cleaning the substrate in detergent, etching the surface in acid to remove residual organic and inorganic surface contaminants, rinsing it in ultrapure H_2O, and drying it under sterile conditions. Deposition is then accomplished using chlorinated derivatives of the alkylsilanes and methoxy derivatives of the aminosilanes (Figure 16.2).

The efficiency of molecular deposition (that is, thickness of the molecular overlayer, percent surface coverage, spatial uniformity) is dependent on the number of reactive sites on the substrate, the number of hydrolyzable groups on the derivative, pH of the reaction solution, temperature, and duration of drying. In order to assess the efficiency of our methods, we have performed three types of measurements on derivatized substrates: contact angle, ellipsometry, and x-ray photoelectron spectroscopy (XPS; also referred to as electron spectroscopy for chemical analysis, or ESCA). Contact-angle measurements indicate the wettability (or hydrophilicity) and chemical uniformity of the surface, and they thereby provide a rather simple test of the efficiency of silane deposition (Neumann and Good, 1979). Ellipsometry measures the thickness of the molecular overlayer, and

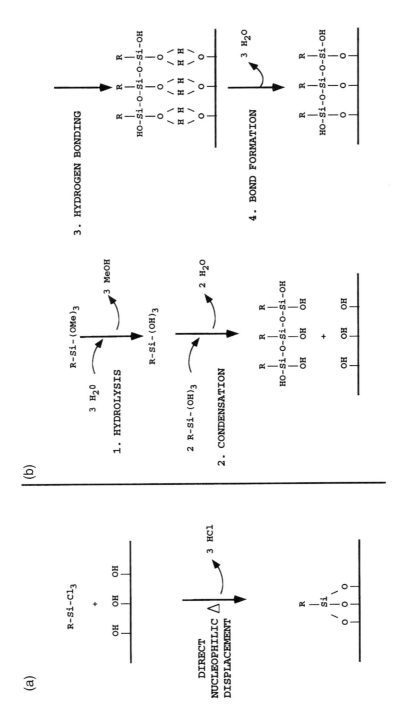

Figure 16.2. Chemical steps involved in covalently attaching amino- and alkylsilanes to silicon-based substrates. (a) Alkylsilanes were immobilized by direct nucleophilic displacement utilizing chlorinated derivatives dissolved in organic solvents at elevated temperatures. (b) Aminosilanes were immobilized by a four-step process using methoxy derivatives in ethanol at room temperature (modified after Arkles, 1991)

XPS can be used to verify the presence of specific atomic species (for example, nitrogen in aminosilanes) as well as to estimate the thickness and percent coverage of the overlayer (Fadley, 1976).

Table 16.1 shows the surface properties of borosilicate glass before and after immobilization of an aminosilane (ethylenediamine propylsilane, EDS) and two different alkylsilanes (dimethylsilane, DMS; octadecyldimethylsilane, ODDMS). The quasi-static advancing (θ_a) and receding (θ_r) contact angles increased on each of the derivatized substrates, indicating an increase in hydrophobicity concomitant with organosilane immobilization. The hysteresis (that is, the difference between θ_a and θ_r) in contact-angle measurements can be attributed to surface roughness, chemical heterogeneity, and chemical reactivity (Good and Neumann, 1979). In preliminary experiments using photon tunneling microscopy (Guerra et al., 1993), we have determined that the surface roughness of quartz and glass substrates was less than 10 nm. This indicates that chemical heterogeneity and reactivity were the dominant factors affecting hysteresis. For polar molecules like aminosilanes, reactivity with the solvent (water) can contribute significantly to the hysteresis. For nonpolar alkylsilanes, hysteresis is due primarily to heterogeneity in the surface coverage.

Estimates of the average thickness and percent coverage, calculated from angle-dependent XPS, indicated that the organosilane overlayers were of roughly nanometer thickness with less than perfect coverage (Table 16.1). Thickness of the overlayer ranged from 1 to 5 nm, and coverage ranged from 21% to 86% depending upon the silane. Similar results were obtained on quartz and silicon substrates, and the values obtained are similar to those reported by others under similar conditions (Table

Table 16.1. *Surface properties of acid-etched borosilicate glass before and after immobilization with silane derivatives (see Healy et al., 1994).*

	Borosilicate	EDS	DMS	ODDMS
θ_a [1]	$23 \pm 1°$	$33 \pm 4°$	$99 \pm 1°$	$103 \pm 2°$
θ_r [1]	$15 \pm 5°$	$17 \pm 5°$	$88 \pm 5°$	$90 \pm 1°$
Surface Coverage [2]	———	21%	46%	86%
Overlayer Thickness [2]	———	1.3 nm	0.4 nm	5.3 nm
Theoretical Thickness [3]	———	1.2 nm	0.5 nm	2.6 nm

[1] Measured by contact-angle analysis.
[2] Measured by x-ray photoelectron spectroscopy.
[3] Calculated using *trans*-extended molecular chain.

Abbreviations: EDS = ethylenediamine propylsilane, DMS = dimethylsilane, ODDMS = octadecyldimethysilane.

Table 16.2. *Surface properties of modified quartz and silicon substrates.*

	Quartz	EDS	EDS[1]	ODDMS[2]
$\theta_a{}^3$	< 10°	36 ± 1°	28–32°	110° (b)
$\theta_r{}^3$	< 10°	<10°	NM	100° (b)
Surface Coverage[4]	———	50%	NM	NM
Overlayer Thickness[4]	———	0.6 nm	NM	NM
Overlayer Thickness[5]	———	0.8 nm	0.8 nm	2.6 nm

[1] Dulcey et al. (1991).
[2] Wasserman et al. (1989).
[3] Measured by contact angle on quartz and silicon (b) substrates.
[4] Measured by x-ray photoelectron spectroscopy on quartz substrates.
[5] Measured by ellipsometry on silicon substrates.

Abbreviations: EDS = ethylenediamine propylsilane, ODS = octadecylsilane, NM = not measured.

16.2). These results, together with the contact-angle and ellipsometry data, indicated that EDS was immobilized on silicon-based substrates as a fractional hydrophilic monolayer film, whereas ODDMS and DMS were bound as hydrophobic multilayer films. In each case the surface roughness was two to three orders of magnitude below the limit of cell response to surface topography (Singhvi, Stephanopoulos, and Wang, 1994; also see Chapters 19 and 20). Thus, cell behavior on these chemically modified substrates is dominated by the surface chemistry.

16.5 Patterning cell adhesive molecules on silicon-based substrates

Surfaces patterned with biological molecules offer a highly controlled method for studying cell–material interactions. We have combined photolithographic, silane-coupling, and protein-adsorption procedures to pattern silicon-based substrates with aminosilanes, alkylsilanes, and proteins with micrometer spatial resolution (Healy et al., 1994; Lom et al., 1993). A diagram of the steps involved in patterning the silanes is shown in Figure 16.3. Photoresist and masking techniques were used to create a lithographically defined surface, and silane-coupling chemistry was used to attach different silanes to the exposed regions before and after removing the photoresist. The latter resulted in a chemically defined surface. The spatial distributions of the lithographically and chemically defined surfaces were confirmed using the endogenous fluorescence

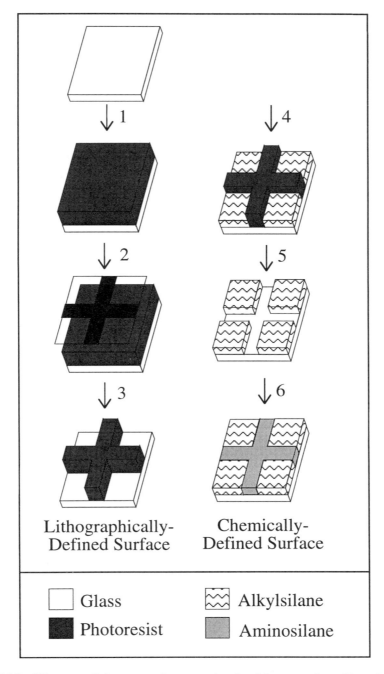

Lithographically-
Defined Surface

Chemically-
Defined Surface

☐ Glass   〰 Alkylsilane

■ Photoresist ▨ Aminosilane

Figure 16.3. Diagram of the processing steps involved in patterning silicon-based substrates with aminosilanes and alkylsilanes. (1) Acid-cleaned substrate was spin-coated with photoresist. (2) Photoresist was exposed to uv light through a photomask. (3) Exposed photoresist was removed with alkaline solution producing a lithographically defined surface. (4) Alkylsilane was attached to exposed regions. (5) Remaining photoresist was removed with acetone. (6) Aminosilane was attached to newly exposed regions, creating a chemically defined surface. Reproduced with permission from Lom et al., 1993.

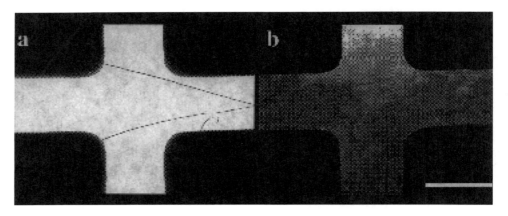

Figure 16.4. Fluorescent images of (a) lithographically defined and (b) chemically defined glass substrates prepared as described in Figure 16.3. Reproduced with permission from Lom et al., 1993. Calibration bar is 10 μm.

of the photoresist (Figure 16.4a) and fluorescent staining of the amines (Figure 16.4b), respectively. Staining of the chemically defined pattern indicated that amines were below the level of detection in the alkane regions. Cellular studies also confirmed the spatial integrity of the patterns (Sections 16.6 and 16.7).

Protein patterns on silicon-based substrates were created using two variations of the method just described. The first involves protein adsorption onto a lithographically defined surface, followed by photoresist removal. The substrate can then be linked via silane coupling to form a pattern of protein and amine (Figure 16.5a). Alternatively, we have formed protein–amine patterns by adsorbing protein onto a surface previously immobilized with amine and covered with patterned photoresist (Figure 16.5b). In both cases the spatial integrity of the pattern was confirmed using immunochemical staining of the proteins (for example, Figure 16.6). High-affinity binding of the antibody indicated that recognition of the protein was not significantly affected by adsorption of the protein to the surface nor by exposure of the adsorbed protein to solvents during the processing steps (that is, to ethanol and acetone). Furthermore, analysis of cell behavior revealed no noticeable differences between adsorbed and processed proteins (Soekarno and Hockberger, 1994).

16.6 Qualitative analysis of cell adhesion on chemically modified, silicon-based substrates

Our initial cellular studies focused on a qualitative evaluation of the effect of the surface charge on nerve-cell adhesion to silicon-based substrates (Kleinfeld et al., 1988). Cells from postnatal rat cerebellum or embryonic mouse spinal cord were enzymatically dissociated and plated onto either silicon or quartz substrates immobilized with

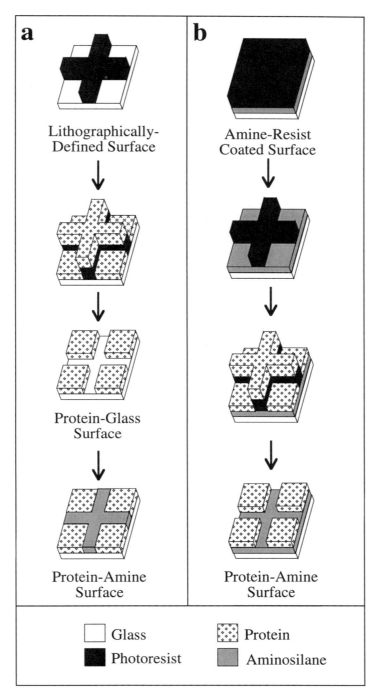

Figure 16.5. Diagram of two different methods for patterning proteins alongside amines on silicon-based substrates. (a) One method utilized protein adsorption onto a lithographically defined surface, followed by removal of photoresist and immobilization of aminosilane. (b) Alternatively, one can pattern photoresist on an amine-coated surface, followed by protein adsorption and removal of photoresist. Reproduced with permission from Lom et al., 1993.

Figure 16.6. Immunoperoxidase stain of laminin (gray areas) on an EDS-laminin patterned substrate. The pattern was visualized using bright-field microscopy, and the unstained regions were 10 μm wide. Reproduced with permission from Lom et al., 1993.

positively charged amines (EDS; aminopropylsilane, APS; diethyltriaminosilane, DETS), negatively charged amines (acetylated-EDS, AEDS), or uncharged alkanes (tetradecylsilane, TDS). Cells were plated and cultured in minimal essential medium (MEM) with or without 10% serum (horse or fetal calf). Serum-containing medium was required for survival of cells beyond a few hours after plating.

Within minutes after plating in either serum-containing or serum-free medium, both cell types adhered to each unpatterned substrate. The one exception was TDS, which was poorly adhesive when cells were plated in serum-containing medium. The poor adhesion may have been caused by deposition of adhesion-inhibiting molecules in serum on this hydrophobic surface (for example, albumin). After 2 days in culture, cells displayed characteristic differences between substrates. Cells were uniformly dispersed and displayed greater neuritic outgrowth on EDS and DETS compared with cells on APS, AEDS, or TDS. These differences were even more apparent after 10 days in culture, and results on EDS and DETS were comparable to those on polylysine.

The results are consistent with the earlier observations that positive surface charge enhances cell adhesion to substrates. We examined this relationship further using patterned substrates. Cells plated on patterns of EDS–TDS invariably attached

Table 16.3. *Cellular preference on EDS–DMS patterns developed over time in culture. Preference for EDS was accelerated by the presence of 10% fetal calf serum. The plus (+) indicates preference for EDS; the minus (-) indicates no preference between EDS and DMS.*

| Cell Type | Medium | Time after Plating | | | | |
		10 min	1 h	2 h	6 h	24 h
Rat cerebellum	MEM + serum	+	+	+	+	+
	MEM	-	+	+	+	+
Mouse neuroblastoma	MEM + serum	-	-	+	+	+
	MEM	-	-	-	-	+
Rat calveria bone	MEM + serum	+	+	+	+	+
	MEM	-	-	-	-	-
Human osteosarcoma	MEM + serum	-	-	-	-	+
	MEM	-	-	-	-	-

Abbreviation: MEM = minimal essential medium.

to the EDS regions within minutes, and in some cases (when the TDS spaces were large) cells remained there for up to 30 days in culture (Kleinfeld et al., 1988). Preference for positively charged surfaces was apparent for all cell types within each culture, and similar results have been reported for mouse and human neuroblastoma cells (Dulcey et al., 1991; Lom et al., 1993; Matsuzawa et al., 1993), osterosarcoma and primary bone cells (Healy et al., 1994, 1995), BHK fibroblasts (Healy et al., 1994), embryonic chick sensory ganglia, (Philip E. Hockberger, unpublished observations), and 3T3 fibroblasts (Soekarno et al., 1995).

An important concern when evaluating the performance of materials in culture is the possibility that components of the growth medium may alter the surface chemistry of the substrate. In fact, several reports have indicated that serum was required for cells to organize on patterned substrates (Kleinfeld et al., 1988; Lopez et al., 1993, Ranieri et al., 1993). Recent results in our laboratories may help to clarify this issue. We have found that many cell types will organize on patterned substrates in the absence of serum although serum accelerates the process. This is illustrated in Table 16.3, which shows results for two nerve- and two bone-cell types (one primary culture and one cell line for each) plated on EDS–DMS patterns. In serum-free medium, only the neuronal cultures developed a preference within 24 hrs. In serum-containing medium, both nerve- and bone- cell types displayed preferences, and the former developed preference in less time, indicating that serum influenced the response although it did not affect the choice. Though the effect of serum will be an important parameter when evaluating the performance of biomaterials in vivo, the remaining experiments described in this chapter were done in the absence of serum to avoid altering the surface chemistry.

Besides being helpful in analyzing the influence of surface charge, patterned substrates have been useful for comparing the relative adhesivity of different chemically modified substrates (Lom et al., 1993). We have evaluated this by examining cells on the following patterns: protein (or polylysine) and glass; protein (or polylysine) and amine; and amine and alkane. Preference for a surface was assigned when the majority of cells were located unambiguously on one region as determined by either phase-contrast microscopy (Figure 16.7a–c) or subsequent immunochemical staining (Figure 16.7d–f). The results for mouse neuroblastoma cells were determined after 24 h in culture, and they displayed the following hierarchy (starting with most preferred): laminin, fibronectin, collagen IV, polylysine > EDS, glass > DMS, HDS, ODS, ODDMS, albumin. A similar strategy was used to obtain the following hierarchy for postnatal cerebellar cells after 1 hr in culture: fibronectin, polylysine > EDS, glass > DMS, TDS, albumin, laminin. The fact that laminin was a nonpreferred substrate of cerebellar cells demonstrates the cell specificity of the hierarchy (Lom and Hockberger, 1995). More important, these results challenge the notion that chemicals are either adhesive or nonadhesive, and argue instead that adhesivity is a relative term. This interpretation is supported by our quantitative analysis of cell adhesion on chemically modified substrates (Section 16.7).

16.7 Quantitative analysis of cell adhesion on chemically modified, silicon-based substrates

In order to generate a more quantitative evaluation of adhesivity, we have used interference reflection microscopy (IRM) to analyze cell–substrate adhesion. This optical technique allows one to discriminate three types of attachment sites between cells and substrates: focal contact (10–15 nm distance between cell and substrate), close contact (20–50 nm distance), and no contact (>100 nm distance) (Verschueren, 1985). Focal contacts correspond to the black regions in IRM images, close contacts to gray regions, and regions of no contact are white. By comparing the relative amounts of these contacts on different surfaces, we have found quantitative differences in surface adhesivity.

For the IRM studies we have focused primarily on mouse neuroblastoma cells. The cell bodies and growth cones of these cells are large (20–50 μm in diameter) and therefore ideal for quantitative microscopic analyses. Furthermore, they can be cultured in MEM without serum (Soekarno et al., 1993). In fact, serum-free medium promotes neurite outgrowth (and, therefore, growth cone formation) in these cells. Figure 16.8 shows typical IRM profiles of neuroblastoma growth cones on six different chemically modified surfaces. The gray-scaled images were similar on amine (EDS), alkane (DMS), fibronectin, and albumin surfaces. The overall darkness of the images reflects closer contact of cells with these surfaces. In contrast, IRM profiles on laminin

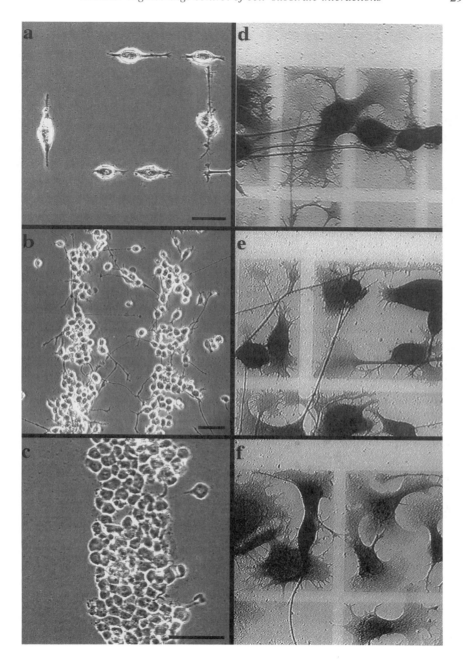

Figure 16.7. Attachment and neurite outgrowth of mouse neuroblastoma cells on chemically patterned glass substrates. Cells attached themselves to, and extended neurites preferentially on, EDS regions of (a) EDS–HDS, (b) EDS–DMS, and (c) EDS–BSA patterns. Cells preferred laminin on EDS–laminin patterns (d and e) and collagen on EDS–collagen IV patterns (f). Cells were photographed using phase-contrast microscopy (a–f); protein regions were visualized using immunostaining with laminin (d and e) and collagen antibodies (f). Calibration bars represent 50 μm (a) and 100 μm (b, c).Unstained regions in d–f were 10 μm wide. Reproduced with permission from Lom et al., 1993.

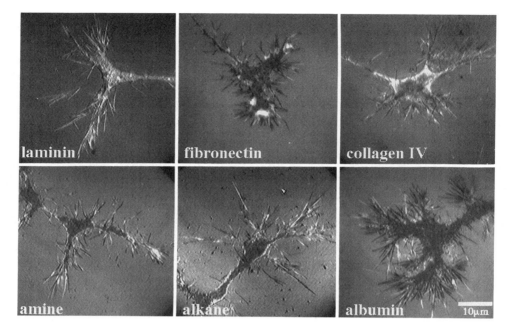

Figure 16.8. IRM images of neuroblastoma growth cones on chemically modified, unpatterned glass substrates. Gray-level values reflect proximity of the cell to the surface (black is nearest; white is farthest). See text for details. Reproduced with permission from Soekarno and Hockberger, 1994.

and collagen IV appeared brighter, indicating greater overall distance from the surface. Profiles beneath cell bodies and neurites were similar, indicating no apparent compartmentalization of the responses.

When we compared the IRM profiles with the preference hierarchy for the cells described in Section 16.6, we found no correlation between surface preference and adhesivity. That is, neuroblastoma cells preferred laminin and collagen IV over amine, alkane, and albumin even though there was closer contact with the latter three. The indistinguishable IRM profiles on fibronectin, amine, and alkane also belied the fact that cells preferred fibronectin over amine, and amine over alkane. These results indicated that preferred surfaces were neither more nor less adhesive than nonpreferred surfaces.

In order to get a more quantitative assessment of these results, we used digital image processing techniques to evaluate the IRM profiles on amine, alkane, and laminin surfaces (Soekarno and Hockberger, 1994). We calculated the percentage of cellular area devoted to each type of contact relative to the total ventral surface area of the cell on each substrate. As shown in Figure 16.9, the area corresponding to focal contacts was significantly greater on amine and alkane surfaces compared with laminin. Because there was more area of no contact on

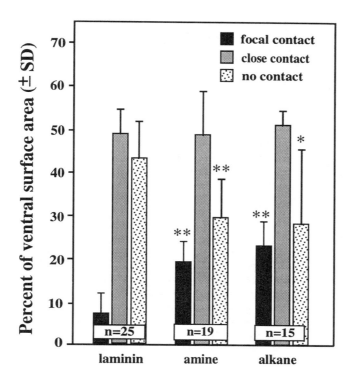

Figure 16.9. Comparison of the percent of ventral surface area of neuroblastoma cells devoted to focal contact, close contact, and regions of no contact when cultured on laminin, amine (EDS), or alkane (DMS) substrates. There was significantly more surface area devoted to focal contacts – and less area devoted to regions of no contact – on amines and alkanes compared with cells on laminin. Results within and between groups were evaluated statistically using analysis of variance and Welch's *t*-test for uncorrelated measure (* $p < 0.01$, ** $p < 0.0001$). Reproduced with permission from Soekarno and Hockberger, 1994.

laminin and no difference in the amount of close contact on any of the substrates, the brighter appearance of the IRM images on laminin was due to a combination of less area devoted to focal contact and greater area with no contact. Although this might indicate that laminin was preferred over amine because it was less adhesive, this rationale could not account for the preference of amine over alkane, because there was no statistical difference between the types of contacts on these surfaces.

    We further analyzed the relationship between chemical preference and adhesion on patterned substrates (Soekarno et al.,1993; Soekarno and Hockberger, 1994). An example is shown in Figure 16.10, which depicts a cell on an amine–alkane pattern, though similar results have been obtained on other patterns. The IRM images indicated that the cell body was attached on both amine and

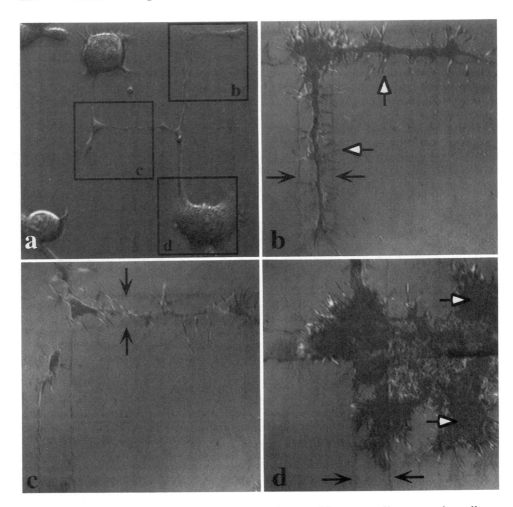

Figure 16.10. (a) Modulation-contrast image of a neuroblastoma cell on an amine–alkane patterned glass substrate. The pattern consisted of 10-μm-wide regions (between black arrows in b–d) separated by 50-μm² regions of alkane. The border between regions is visible in the IRM regions. (b–d) IRM images of the same cell showing regions boxed in (a) at a higher magnification. The neurite extended along the amine regions, whereas the cell body and filipodia were less discriminating (white arrows). Reproduced with permission from Soekarno and Hockberger, 1994.

alkane regions, even though its neurite was located centrally over the amine region. Filopodia extending from the cell body and from the neurite also attached to both chemistries. When we analyzed the types of contacts beneath cells on each surface, there was no statistical difference between laminin and amine (Figure 16.11a) or between amine and alkane (Figure 16.11b). That is, cells displayed the same percentage of focal, close, and no contacts on both substrates

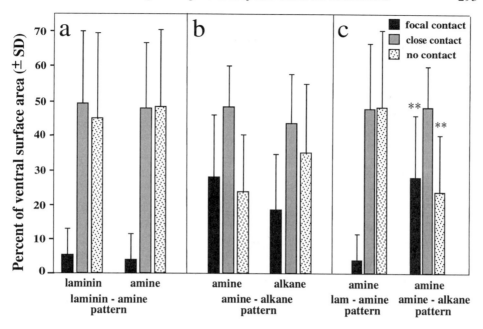

Figure 16.11. Group data comparing the percent of ventral surface area of cells devoted to different types of contact on patterns of laminin–amine (EDS) and amine (EDS)–alkane (DMS). No difference was found in the percent of each type of contact within a pattern (a and b). Comparisons between patterns, however, revealed fewer focal contacts and more area with no contact on amine when patterned alongside laminin than when patterned alongside alkane (c, ** p ≤ 0.0001). Reproduced with permission from Soekarno and Hockberger, 1994.

within a pattern. A surprising result was that cells made fewer focal contacts on amine regions when patterned alongside laminin than when patterned alongside alkane (Figure 16.11c) or when compared with unpatterned amine (Figure 16.9). These results indicate that differential adhesion is not responsible for the spatial organization of neuroblastoma cells on patterned substrates. An alternative possibility is that preference was influenced by a signal transduction mechanism (cf. Calof and Lander, 1991; Curtis et al., 1992; Lemmon et al., 1992).

One of the important lessons gleaned from these studies is that adhesivity does not necessarily correlate with the attractiveness of a chemical. In fact, our results suggest that the term "adhesive molecule" may be ambiguous, because cells can respond differently to the same molecule depending upon the surrounding environment. These results illustrate the problem with studies that test the relative adhesivity of a material in isolation, that is, on uniform substrates. As a further example of this, we have found that cerebellar cells prefer laminin over EDS if the laminin is mixed with polylysine (Lom and Hockberger, 1995). This type of experiment demonstrates the importance of controlling the presentation of a molecule, and it reinforces our belief that results in vitro may be quite different from what is found in vivo. In spite of this limitation, we are confident that continued analysis of cell–substrate interactions in vitro coupled with in vivo studies will provide further insights for designing new and better surfaces for interfacing cells and devices.

16.8 Summary

We have illustrated the utility of a multidisciplinary approach to a fundamental problem in cellular engineering, that is, interfacing cells with materials. We described the use of silane-coupling chemistry, photolithography, and protein-adsorption techniques to modify silicon-based substrates. The efficiency of silane immobilization on silicon, quartz, and borosilicate glass substrates was analyzed using surface chemistry analytical techniques (XPS, ellipsometry, and contact angle). This analysis indicated that amino- and alkylsilanes were deposited in films several orders of magnitude below the level needed for topographical detection by cells. Analysis of cell behavior on chemically modified surfaces (uniform and patterned) confirmed that adhesion was facilitated on positively charged surfaces. Furthermore, the organization of cells on patterned substrates revealed a hierarchy of chemical preferences that differed between cell types and developed over time in culture. Serum in the medium accelerated the adhesion process but did not affect the hierarchy in the cells we examined. The hierarchy was altered when two compounds were mixed together (such as polylysine and laminin), reminding us that results in vitro may differ from results in vivo. Quantitative analysis of adhesion profiles on substrates using interference reflection microscopy demonstrated that chemical preferences of cells did not correlate with adhesivity. These results challenge the notion that cells are guided by differential adhesion, and suggest instead that a signal-transduction mechanism may be involved.

Acknowledgments

We wish to thank the National Institutes of Health and the Whitaker Foundation for their support of this research. We thank John Guerra of the Polaroid Corporation (Cambridge, Mass.) for helping with the photon tunneling measurements.

References

Arkles, B. (1991) Silane coupling agent chemistry. In *Silicon Compounds: Register and Review*, ed. R. Anderson, G. L. Larson, and C. Smith, pp. 59–64. Piscataway, NJ: Hüls America.

Britland, S., Clark, P., Connolly, P., and Moores, G. (1992) Micropatterned substratum adhesiveness: a model for morphogenetic cues for controlling cell behavior. *Exp. Cell Res.* 198:124–129.

Britland, S., Perez-Arnaud, E., Clark, P., McGinn, B., Connolly, P., and Moores, G. (1992) Micropatterning proteins and synthetic peptides on solid supports: a novel application for microelectronics fabrication technology. *Biotech. Prog.* 8:155–160.

Calof, A. L. and Lander, A. D. (1991) Relationship between neuronal migration and cell-substratum adhesion: Laminin and merosin promote olfactory neuronal migration but are anti-adhesive. *J. Cell Biol.* 115:779–94.

Chuah, M. I., David, S., and Blaschuk, O. (1991) Differentiation and survival of rat olfactory epithelial neurons in dissociated cell culture. *Dev. Brain Res.* 60:123–32.

Curtis, A. S. G., McGraff, M., and Gasmi, L. (1992) Localized application of an activating signal to a cell: Experimental use of fibronectin bound to beads and the implication for mechanisms of adhesion. *J. Cell Sci.* 101:427–36.

Ducheyne, P. and Healy, K. E. (1988) Surface spectroscopy of calcium phosphate, ceramic, and titanium implant materials. In *Surface Characterization of Biomaterials*, ed. B. Ratner, pp. 175–192. Amsterdam: Elsevier.

Dulcey, C. S., Georger, J. H., Krauthamer, V., Stenger, D. A., Fare, T. L., and Calvert, J. M. (1991) Deep UV photochemistry of chemisorbed monolayers: patterned coplanar molecular assemblies. *Science* 252:551–554.

Edelman, G. M. and Cunningham, B. A. (1990) Place-dependent cell adhesion, process retraction, and spatial signalling in neural morphogenesis. *Cold Spring Harbor Symp. Quant. Biol.* 55:303–318.

Fadley, C. S. (1976) Solid state and surface analysis by means of angular-dependent X-ray photoelectron spectroscopy. *Prog. Solid State Chem.* 11:265–343.

Fodor, S. P. A., Read, J. L., Pirrung, M. C., Stryer, L., Lu, A. T., and Solas, D. (1991) Light-directed, spatially addressable parallel chemical synthesis. *Science* 251:767–773.

Good, R. J. and Neumann, A. W. (1979) Contact Angles. In *Surface and Colloid Science,* ed. R. J. Good and R. A. Stromberg, pp. 1–30. New York: Plenum.

Graf, J., Iwamoto, Y., Sasaki, M., Martin, G. R., Kleinman, H. K., Robey, F. A., and Yamada, Y. (1987) Identification of an amino acid sequence in laminin mediating cell attachment, chemotaxis, and receptor binding. *Cell* 48:989–96.

Guerra, J. M., Srinivasarao, M., and Stein, R. S. (1993) Photon tunneling microscopy of polymeric surfaces. *Science* 262:1395–1400.

Healy, K. E., Lom, B., and Hockberger, P. E. (1994) Spatial distribution of mammalian cells dictated by materials surface chemistry. *Biotech. Bioeng.* 43:792–800.

Healy, K. E., Thomas, C. H., Rezania, A., Kim, J. E., McKeown, P. J., Lom, B., and Hockberger, P. E. (1996) Kinetics of bone cell organization and mineralization on materials with patterned surface chemistry. *Biomaterials* 17:195–208.

Hockberger, P. E., Racker, D. K., and Houk, J. C. (1994) Culturing neural networks. In *Enabling Technologies for Cultured Neural Networks,* eds. D. A. Stenger and T. M. McKenna, pp. 35–49. New York: Academic Press.

Jessell, T. M. (1988) Adhesion molecules and the hierarchy of neural development. *Neuron* 1:1–13.

Katchalsky, A., Danon, D., Nevo, A., and DeVries, A. (1959) Interactions of basic polyelectrolytes with the red blood cell. II. Agglutination of red blood cells by polymeric bases. *Biochem. Biophys. Acta* 33:120–138.

Kleinfeld, D., Kahler, K. H., and Hockberger, P. E. (1988) Controlled outgrowth of dissociated neurons on patterned substrates. *J. Neurosci.* 8:4098–4120.

Kovacs, G. T. A. (1994) Introduction to the theory, design and modelling of thin film microelectrode for neural interfaces. In *Enabling Technologies for Cultured Neural Networks*, eds., D. A. Stenger and T. M. McKenna, pp. 121–165. New York: Academic Press.

Langer, R. and Vacanti, J. P. (1993) Tissue engineering. *Science* 260:920–926.

Lasky, L. A. (1992) Selectins: interpreters of cell-specific carbohydrate information during inflammation. *Science* 258:964–969.

Lemmon, V., Burden, S. M., Payne, H. R., Elmslie, G. J., and Hlavin, M. L. (1992) Neurite growth on different substrates: permissive vs. instructive influences the role of adhesive strengths. *J. Neurosci.* 12:818–26.

Letourneau, P. C. (1975) Cell-to-substratum adhesion and guidance of axonal elongation. *Dev. Biol.* 44:92–101.

Lom, B., Healy, K. E., and Hockberger, P. E. (1993) A versatile technique for patterning biomaterials onto glass substrates. *J. Neurosci. Methods* 50:385–397.

Lom, B. and Hockberger, P. E. (1995) Laminin-1 is non-preferred and anti-adhesive for developing cerebellar cells. *Society of Neuroscience Abstracts*, Vol. 21:1037.

Lopez, G. P., Albers, M. W., Schreiber, S. L., Carroll, R., Peralta, E., and Whitesides, G. M. (1993) Convenient methods for patterning the adhesion of mammalian cells to surfaces using self-assembled monolayers of alkanethiolates on gold. *J. Am. Chem. Soc.* 115:5877–78.

Massia, S. P. and Hubbell, J. A. (1990) Covalent surface immobilizations of Arg-Gly-Asp- and Tyr-Ile-Gly-Ser-Arg-containing peptides to obtain well-defined cell-adhesive substrates. *Anal. Biochem.* 187:292–301.

Massia, S. P. and Hubbell, J. A. (1992) Immobilized amines and basic amino acids as mimetic heparin-binding domains for cell surface proteoglycan-mediated adhesion. *J. Biol. Chem.* 267:10133–41.

Massia, S. P., Rao, S. S., and Hubbell, J. A. (1993) Covalently immobilized laminin peptided Tyr-Ile-Gly-Ser-Agr (YIGSR) supports cell spreading and co-localization of the 67-kilodalton laminin receptor with a-actinin and vinculin. *J. Biol. Chem.* 268:8053–59.

Matsuzawa, M., Potember, R. S., Stenger, D. A., and Krauthamer, V. (1993) Containment and growth of neuroblastoma cells on chemically patterned substrates. *J. Neurosci. Methods* 50:253–260.

McKeehan, W. J. and Ham, R. G. (1976) Stimulation of clonal growth of normal fibroblasts with substrata coated with basic polymers. *J. Cell Biol.* 71:727–34.

Mikulski, B. A. (1994) Science in the national interest. *Science* 264:221–222.

Neumann, A. W. and Good, R. J. (1979) Techniques of measuring contact angles. In *Surface and Colloid Science,* ed. R. J. Good and R. A. Stromberg, pp. 31–91. New York: Plenum.

Pawlenko, S. (1986) *Organosilicon Chemistry.* New York: Walter de Gruyter.

Peppas, N. A. and Langer, R. (1994) New challenges in biomaterials. *Science* 263:1715–1720.

Plueddemann (1982) *Silane Coupling Agents.* New York: Plenum.

Ranieri, J. P., Bellamkonda, R., Jacob, J., Vargo, T. G., Gardella, J. A., and Aebischer, P. (1993) Selective neuronal cell attachment to a covalently patterned monoamine fluorinated ethylene propylene films. *J. Biomed. Mater. Res.* 27:917–925.

Rao, Y., Wu, X.-F., Gariepy, J., Rutishauser, U., and Siu, C.-H. (1992) Identification of a peptide sequence involved in homophilic binding in the neural cell adhesion molecule NCAM. *J. Cell Biol.* 118:937–49.

Ruoslahti, E. (1991) Integrins. *J. Clin. Invest.* 87:1–5.

Ruoslahti, E. and Pierschbacher, M. D. (1987) New perspectives in cell adhesion: RGD and integrins. *Science* 238:491–497.

Singhvi, R., Kumar, A., Lopez, G. P., Stephanopoulos, G. N., Wang, D. I. C., Whitesides, G. M., and Ingber, D. E. (1994) Engineering cell shape and function. *Science* 264:696–698.

Singhvi, R., Stephanopoulos, G., and Wang, D. I. C. (1994) Review: effects of substratum morphology on cell physiology. *Biotech Bioengin.* 43:764–771.

Soekarno, A. and Hockberger, P. E. (1994) Image analysis of neuronal pathfinding on microfabricated substrates. *SPIE Proceedings*, Vol. 2137:29–38.

Soekarno, A., Lom, B., and Hockberger, P. E. (1993) Pathfinding by neurblastoma cells in culture is directed by preferential adhesion to positively charged surfaces. *NeuroImage* 1: 129–44.

Soekarno, A., Lom, B., and Hockberger, P. E. (1995) Growth cones and fibroblasts exhibit substrate preference in the absence of differential adhesion. *Society of Neuroscience Abstracts*, Vol. 21:1021.

Takeuchi, M. (1991) Cadherin cell adhesion receptors as a morphogenetic regulator. *Science* 251: 1451–1455.

Verschueren, H. (1985) Interference reflection microscopy in cell biology: methodology and applications. *J. Cell Sci.* 75: 279–301.

Wasserman, S. R., Tao, Y.-T., and Whitesides, G. M. (1989) Structure and reactivity of alkylsiloxane monolayers formed by reaction of alkyltrichlorosilanes on silicon substrates. *Langmuir 5*: 1074–1087.

Weiss, R. (1991) Breast implant fears put focus on biomaterials. *Science* 252: 1059–1060.

Williams, D. F. (1987) Tissue–biomaterial interactions. *J. Material Science* 22:3421–3445.

Yamada, K. M., Aota, S. Akiyama, S. K., and LaFlamme, S. E. (1992) Mechanisms of fibronectin and integrin function during cell adhesion and migration. *Cold Spring Harbor Symp. Quant. Biol.* 57:203–212.

Yavin, E. and Yavin, Z. (1974) Attachment and culture of dissociated cells from rat embryo cerebral hemispheres on polylysine-coated surfaces. *J. Cell Biol.* 62:540–46.

17

Microcontrol of neuronal outgrowth

HELEN M. BUETTNER

Department of Chemical and Biochemical Engineering
Rutgers – The State University of New Jersey
Piscataway, NJ 08855

17.1 Introduction

The ability to control outgrowth by nerve cells has important implications for improving nerve regeneration, investigating nerve development, understanding neural behavior, and incorporating aspects of neural structure and function into advancing medical and computer technologies (see, for example, Curtis et al., 1992). Ultimately, it is desirable to control growth at the single-cell level and at micron dimensions, that is, dimensions on the order of the cell features involved in the process. Achieving this goal requires a quantitative understanding of the fundamental mechanisms by which neurons develop and function, and how these mechanisms govern the response of the cell to specific environments.

Nerves develop at the cellular level by the outgrowth of long, slender processes from the cell body (Alberts et al., 1989). These processes, the axons and dendrites (collectively referred to as neurites), eventually extend throughout the body, connecting neurons and innervation targets in a vast neural network that defines the basic structure of the nervous system. Neurite outgrowth is guided by the growth cone – a sensory motile apparatus at the growing neurite tip. A spread and flattened structure, the growth cone radiates fine spikes, or filopodia, around its lamellipodial periphery (see Figure 17.1). The interaction of filopodia with elements of the extracellular environment is believed to play an important role in guiding neurite outgrowth and, thus, the development of neural architecture.

One key means by which filopodia may be involved in neurite outgrowth is through the remote sensing of discrete cues in their environment. Both in vitro and in vivo studies have shown that filopodial contact with remote cues can initiate rapid neurite advance to the point of contact (Hammarback and Letourneau, 1986; Myers and Bastiani, 1993; O'Connor et al., 1990). Proper placement of such cues in series faithfully reproduces stereotyped pathways of extension during development (Caudy and Bentley, 1986). However, when cues are small and sparsely located, as is typically the case, contact is a probabilistic event, prompting questions as to how successful development is ensured and whether other more subtle guidance mechanisms are also required in concert.

These questions can be addressed most effectively with the aid of a quantitative model for growth cone behavior in micropatterned environments. The framework for such a model has recently been proposed (Buettner et al., 1994) based on parameters

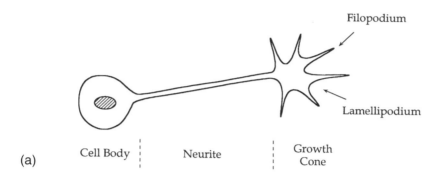

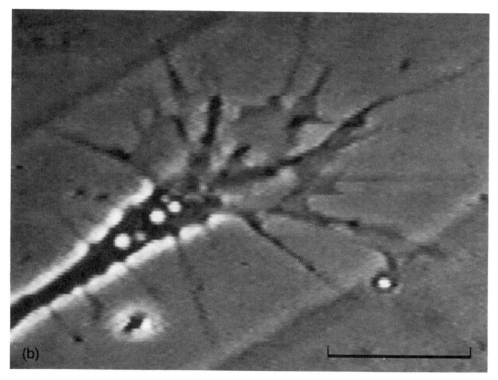

Figure 17.1. (a) Sketch of relevant neuronal feaures during neurite outgrowth. Outgrowth occurs at the growth cone, the tip of a growing neurite. The growth cone consists of thin filopodia that extend and retract from a central cytoplasmic region and lamellipodial veils that move back and forth, usually between extended filopodia. (b) Micrograph of a chick dorsal root ganglion (DRG) growth cone obtained with a 100x objective, 1.6x projection lens. Bar = 10 μm.

of growth cone movement and filopodial dynamics that can be measured experimentally. The model can be used to simulate neurite outgrowth in a variety of systems, and it thus serves as a conceptual framework for investigating the outgrowth response to microheterogeneities. In addition, it provides a basis for designing appropriate experimental systems, particularly microfabricated environments, for the further study and application of neurite outgrowth. The combination of precisely engineered, well-defined microfeatured environments and quantitative approaches to analyzing cell behavior in these environments offers tremendous potential for furthering our insight into the development and function of neurons as well as of a wide variety of other cell types whose behavior may depend on similar mechanisms.

17.1.1 Mechanisms of growth cone advance

Growth cones can advance through an environment by several key mechanisms (see, for example, O'Connor et al., 1990). In the first of these, which will be referred to here as lamellipodial advance, the forward movement of the growth cone appears to occur primarily through the net forward flow of lamellipodial veils. Tracing the growth cone centroid over time during this type of advance reveals a random walk trajectory analogous to that observed for particle diffusion. Although filopodial activity is present, its relationship to the overall growth cone movement is not readily discernible. In the second type of advance, called guideposting or filopodial dilation, filopodia play a more clearly defined role in directing growth cone movements. In this phenomenon, filopodia can detect an isolated region of favorable substrate (that is, the guidepost) and cause the growth cone to move toward it in a rather direct fashion following filopodial contact. This can result in the growth cone's bridging less-favorable substrate between two favorable regions, or it can serve to redirect the movements to an extremely favorable region within a good substrate. Filopodia probably also contribute to growth cone tracking of narrow lanes of substrate, although their contribution has not been demonstrated as clearly as in guideposting.

17.1.2 Microstructural paradigms

Growth cone migration on two-dimensional patterned substrates has been studied using two main experimental paradigms, which are distinguished from one another by the characteristic dimensions of the substrate microfeatures. Both situations involve using pairs of substrates that favor neurite outgrowth to different degrees. Many pairs incorporate the extracellular matrix protein laminin as the more permissive substrate, laminin being one of the most permissive substrates available for neuronal culture in vitro. Examples include laminin–glass (Clark et al., 1993; see also Chapters 16 and 20), laminin–denatured laminin (Hammarback et al., 1985, 1988), laminin–albumin (Hammarback and Letourneau, 1986), and laminin–collagen (Gundersen, 1987). Other work has paired molecules such as fibronectin, collagen, and albumin with glass as the less-permissive substrate (see Chapter 16). The effectiveness of certain nonphysiological patterns has also been demonstrated (Kleinfeld et al., 1988; Letourneau, 1975; see also Chapter 16).

In the first main experimental paradigm, the more permissive substrate is placed in large, central regions with narrow spaces of nonpermissive or less-permissive substrate between them (see Figure 17.2). This arrangement is useful for looking at

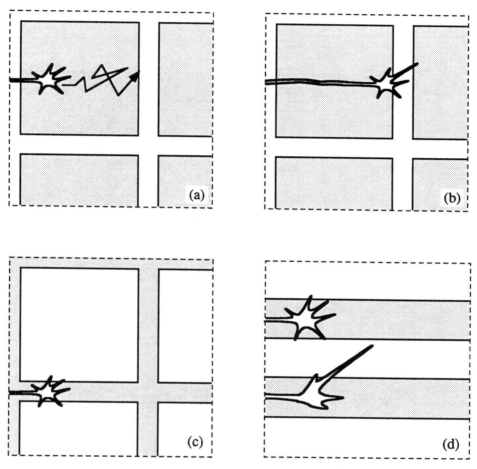

Figure 17.2. Mechanisms of growth cone advance and microstructural paradigms. (a) On large islands of permissive substrate (gray regions), the growth cone exhibits a random walk advance like that seen on homogenous surfaces of the same substrate. (b) Upon reaching the permissive–nonpermissive border, the growth cone may advance across the nonpermissive gap via filopodial dilation. The combination of (a) and (b) represents a guideposting paradigm. (c) If the permissive substrate is confined to narrow regions on the order of the characteristic growth cone dimension, then the growth cone will track these lanes. Although growth cone tracking typically involves some random walk behavior, it is restricted by the boundaries of the track, possibly as a result of filopodial interactions with the nonpermissive edges. (d) Both tracking and guideposting can be observed in a simpler geometry involving alternating lanes of permissive and nonpermissive substrate. Similar responses to physical patterns formed by etching the solid culture support are possible (see Chapter 20).

guideposting if the width of the narrow strip is on the order of the average filopodium length. In the second paradigm, the less-favorable substrate is placed in the
islands and the permissive substrate in the narrow strips between them; this provides
a model environment for neurite tracking. The same phenomena can be observed in
a simpler geometrical arrangement of parallel lanes, depending on the dimensions of
the lanes. Several basic results have been obtained with these systems. Hammarback
and Letourneau (1986) observed that when laminin was placed in islands between
lanes of albumin, growth cone advance across the refractory albumin increased as the
width of the albumin decreased or as laminin concentration increased. Clark et al.
(1993) showed that growth cones track widely spaced laminin strips, but not narrowly spaced ones. Gundersen (1987) reported a stabilization of filopodia on laminin
relative to collagen for growth cones present at a laminin–collagen border. All of
these observations support the notion of filopodial involvement, and the dependence
on laminin concentration suggests a receptor-mediated process. In particular, the
behavior reported in these types of systems is consistent with a probabilistic process
in which chance filopodial contact with certain substrate features is a deciding factor
in growth cone movement. Similar mechanisms may function in the case of three-
dimensional microstructured substrates formed by etching grooves in a solid support
(Clark et al., 1987, 1990). More specific details of growth cone advance in a
micropatterned environment are suggested by casting these concepts in quantitative
terms designed to accommodate the randomness of growth cone activity and to permit closer examination of filopodial dynamics and lamellipodial advance.

17.2 Quantitative framework

The simplest quantitative description of growth cone advance in a microstructured
environment can be constructed for the guideposting scenario represented by
Figures 17.2a and 17.2b, where large islands of permissive substrate are separated by narrow strips of nonpermissive substrate (Buettner et al., 1994). In this case,
the growth cone can undergo considerable migration on the permissive substrate
without any awareness that the second substrate exists. Near the substrate border,
filopodial contact with a neighboring region of permissive substrate may induce
filopodial dilation. Thus, we can conceptually divide the migration into two key
periods of advance: (1) movement across a homogeneous region according to the
random walk characteristics of lamellipodial advance, and (2) filopodial dilation
across the nonpermissive region once filopodial contact has been made. Modeling
lamellipodial advance requires an appropriate description of the random walk trajectory; determining when and where filopodial dilation occurs requires depiction
of the filopodial structure of the growth cone as a function time. In addition, criteria must be set to translate filopodial contact into guidepost detection.

The growth cone trajectory has been modeled previously (Buettner et al., 1994)
using an equation suggested by experimental observation of fibroblast motion (Dunn

and Brown, 1987), to which growth cone motility has often been likened. Subsequent experimental measurements of the trajectory suggest a similar, but slightly different, model of lamellipodial advance (Buettner, 1994; also see Section 17.3.2). The time evolution of filopodial structure has also been considered from an experimental perspective, as described later. The criteria for target detection have yet to be characterized experimentally, however, and are therefore argued from qualitative observations.

Based on the observation that growth cones cross nonpermissive substrate more frequently for higher concentrations of permissive substrate, we assume a receptor-mediated interaction between filopodium and substrate in which a critical number of substrate receptors on the filopodium surface must be bound for growth cone recognition of the neighboring region to occur (Buettner et al., 1994). If the number of receptors per filopodium is relatively constant and receptors are concentrated at the filopodium tip, then this criterion is equivalent to a critical number of filopodia contacting the neighboring permissive substrate and can be expressed as

$$
\begin{aligned}
p(G_c) \quad &= 1 \quad \text{if } f \geq f^*, \\
&= 0 \quad \text{if } f < f^*,
\end{aligned}
\tag{17.1}
$$

where $p(G_c)$ is the probability that a growth cone at the border will detect and respond to a neighboring permissive region; f is the number of contacting filopodia; and f^* is the critical threshold. This condition can be evaluated using a mathematical description of filopodial dynamics that provides the spatial coordinates of each filopodium on the growth cone over time.

17.3 Experimental characterization of growth cone dynamics

17.3.1 Methods

The experimental data for characterizing lamellipodial advance and filopodial dynamics are obtained from time-lapse video sequences of growth cone advance. As previously described, video sequences are typically obtained using a phase contrast microscope equipped with a 100x objective and a videocamera, and recorded at about 1 frame/sec on a time-lapse videocassette recorder (Buettner, 1994; Buettner et al., 1994; Buettner and Pittman, 1991). Individual frames are then digitized as 8-bit computer images (256 gray levels) at a fixed time interval on the order of seconds. Alternatively, images can be digitized directly to disk. Each digital growth cone image is then traced, and the outline is stored as a binary image for subsequent analysis. To date, outlining has been a time-consuming, manual task; however, newly developed image analysis software can complete this step and carry out subsequent analysis semi-automatically (Gwydir et al., 1994).

Lamellipodial advance is characterized by tracking the geometrical center of the lamellipodial region, that is, the growth cone outline excluding filopodial projections, and using time series analysis to fit an equation to the motion observed.

Defining the x and y components of motion as parallel and perpendicular to the current axonal orientation, respectively, one can use autocorrelation analysis (Wei, 1990) to identify a specific form of the general equation

$$z_i = \phi_1 z_{i-1} + \phi_2 z_{i-2} + \cdots + a_i + \theta_1 a_{i-1} + \theta_2 a_{i-2} + \cdots + b_0 \qquad (17.2)$$

for each component, where z_i, the x or y coordinate of the centroid during the ith time interval, is related to the value of the same variable during the $i-1$st, $i-2$nd, . . . intervals through the constants $\phi_1, \phi_2, \ldots$ plus some random variation in the $i-1$st, $i-2$nd, . . . intervals. This random variation is represented by the white noise terms $a_i, a_{i-1}, a_{i-2}, \ldots$, weighted by the constants $\theta_1, \theta_2, \ldots$. The term b_0 represents a constant bias in the movement. By fitting a model of this form to experimental data, specific values of the parameters $\phi_1, \phi_2, \ldots$ and $\theta_1, \theta_2, \ldots$ can be determined and terms for which the values are close to zero can be eliminated to simplify Eq. (17.2) to an appropriate specific form.

Filopodial dynamics are determined by tracking each filopodium throughout its lifetime from the point of initiation. The probability of filopodial initiation is then determined from the distribution of initiation times, and the probability of initiation at a given site is obtained from the spatial distribution of initiation sites. The kinetics of filopodial extension and retraction are measured by plotting the trajectories of individual filopodium tips on a given growth cone during the observation sequence. Filopodium length is calculated as the straight-line distance between the position of the filopodium tip at any time and its initial position,

$$l(t_i) = [(x(t_i) - x_0)^2 + (y(t_i) - y_0)^2]^{1/2}, \qquad (17.3)$$

where (x_0, y_0) represents the x–y coordinate of the tip at time t_0 (initiation), and $(x(t_i), y(t_i))$ represents the x–y coordinate of the tip at a later time t_i. The length at maximum extension is given by the maximum value of $l(t_i)$ measured.

17.3.2 Results

Lamellipodial advance. Analysis of growth cone trajectories for several different situations listed in Table 17.1 yields the following form of Eq. (17.2) for the x and y components of movement:

$$z_i = z_{i-1} + a_i + b_0, \qquad (17.4)$$

where the random term, a_i, is normally distributed with a mean of zero and a variance of σ. Thus, a_i can be rewritten as $a_i = \sigma \Delta t^{1/2} \mathcal{N}(0,1)$, where $\mathcal{N}(0,1)$ represents a random variable taken from a normal distribution with a mean of zero and a variance of $\sigma^2 \Delta t$. Recognizing that the step size in a specific interval

Table 17.1. *Experimental parameters of filopodial dynamics and lamellipodial advance.*

Parameter	Chick DRG	Rat SCG[1]	Range[2]
Lamellipodial Advance:[3]			
Drift Velocity, μ_x (μm/sec)	0.021	0.037	—
R.W. Variance, σ_x (μm/sec$^{1/2}$)	0.10	0.12	—
R.W. Variance, σ_y (μm/sec$^{1/2}$)	0.15	0.06	—
Filopodial Dynamics:[4]			
Rate of initiation, ν (sec^{-1})	0.015	0.014	0.01–0.05
Rate of extension, r_e (μm/sec)	0.17	0.05	0.05–0.2
Rate of retraction, r_r (μm/sec)	0.05	0.03	0.05–0.2
Max time, Tmax $(= r/\theta)$ (sec)	140	73	—
Shape parameter, r	2–3	0.8	—

[1] Postnatal Day 1, 5–7 μg/ml laminin on glass; experimental methods as described in Buettner et al. (1994).
[2] Published data, various sources.
[3] Chick DRG data: Embryonic Day 8, laminin, laminin on collagen, collagen, laminin–collagen
border. Laminin was applied at 50 μg/ml and allowed to dry either on a plain glass coverslip or on a glass coverslip previously coated with a collagen film and air dried. Laminin–collagen borders were created by placing dots or stripes of 50 μg/ml laminin on a dried collagen film as previously described (Gundersen, 1987).
[4] Chick DRG data: Embryonic Day 7, 40 μg/ml laminin on glass; experimental methods as described in Buettner et al. (1994).

is given by components $\Delta x = x_i - x_{i-1}$ and $\Delta y = y_i - y_{i-1}$, and that b_0 is typically negligible in the y direction, Eq. (17.4) can be rewritten in component form as

$$\Delta x = \mu_x + \sigma_x \, \Delta t^{1/2} \, \mathcal{N}(0,1), \tag{17.5a}$$

$$\Delta y = \sigma_y \Delta t^{1/2} \mathcal{N}(0,1). \tag{17.5b}$$

Equation (17.5b) describes unbiased, random step sizes in the direction perpendicular to the axon and reflects a purely diffusive motion. In the x direction, parallel to the axon, the steps are random but with a positive mean, μ_x, indicating that the growth cone tends to make a net positive advance. This type of motion is termed diffusion with drift (Berg, 1983).

Table 17.1 shows values of the experimental parameters for lamellipodial advance appearing in Eqs. (17.5a) and (17.5b) for two systems: chick dorsal root ganglion (DRG) and rat superior cervical ganglion (SCG). These numbers are preliminary in nature, with a small sample number and with standard deviations of ≥50%. Thus it is difficult to state at this point whether subtle differences will emerge in values measured for different neuron types and on different substrates, although this can be anticipated on the basis of previous observations (see, for example, Buettner and Pittman, 1991).

Filopodial dynamics. Experimental observation of chick DRG and rat SCG growth cones reveals a typical filopodium lifetime that is characterized by the parameters indicated in Figure 17.3 (Buettner et al., 1994). Filopodia initiate with a mean rate ν, extend at a rate r_e to a maximum length L_{max}, and then retract at a rate r_r. Initiation can be described as a Poisson event, such that filopodia are equally likely to initiate at any time. Once initiated, they grow out and shrink

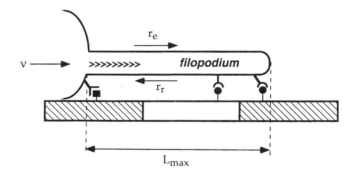

Parameters of Filopodium Lifecycle

- rate of initiation, ν
- rates of extension and retraction, r_e and r_r
- maximum length, L_{max}

Possible Underlying Processes

- cytoskeletal dynamics
- receptor mediation
- cell adhesion

Figure 17.3. Filopodial dynamics of the migrating growth cone can be described in terms of several key parameters characterizing the filopodium lifecycle, including the rates of filopodial initiation, extension and retraction, and the maximum filopodial length. Although precise relationships have not yet been established, these parameters are likely to correspond to underlying biophysical mechanisms, such as the polymerization and depolymerization of actin filaments in the filopodium core as well as the receptor-mediated interactions between cell-surface receptors and substrate molecules.

back at fairly constant rates, but they extend to random lengths. The extension time (which is proportional to the maximum filopodial length) fits a gamma distribution characterized by two parameters, r and θ; the mean time to maximum extension is given by r/θ.

Experimental parameters measured for chick DRG and rat SCG filopodia are listed in Table 17.1. The filopodia on chick DRGs are generally longer than those on rat SCGs, which is reflected in a higher rate of extension and a longer extension time for chick DRGs. The third column of data shows the ranges of these parameters that have been reported for various systems. Commonly, only one or two of these parameters is reported in a given source; the only complete sets at present are those represented in data columns 1 and 2.

17.4 Testable predictions

17.4.1 Visual descriptions of growth cone motility

A number of testable predictions can be made within the basic framework of the model described here. One of the simplest, but in many ways extremely useful, things we can do is to provide a visual representation of the model results. Figure 17.4a shows one realization of the trajectory traced by a model growth cone for typical parameter values, with the position of the growth cone plotted at 10-sec

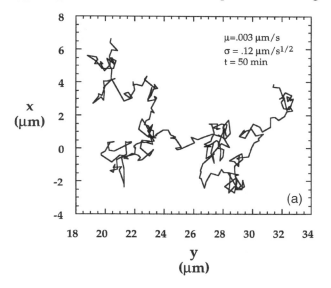

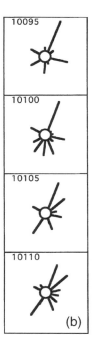

Figure 17.4. (a) Sample trajectory for a model growth cone during 50 min of migration on a permissive substrate. (b) Sequence of the filopodial configuration at 5-sec intervals for a model growth cone. The circular center of the growth cone represents a 5-μm diameter.

intervals. Figure 17.4b illustrates a realization of the filopodial configuration generated for a model growth cone over several 5-sec intervals. Comparing this type of visual information for model growth cones with the information for experimental growth cones enables evaluation of model characteristics that are not as readily apparent by other means. For example, seemingly minor degrees of correlation in the location of filopodia around the growth cone can lead to model configurations that look significantly different from one another. In addition to depicting the visual history of the growth cone movement in fixed images such as Figure 17.4, the model can also easily be used to generate animated motion that can be compared directly to the video sequences of growth cone motility obtained experimentally

17.5 *Quantitative growth cone response to micropatterned environments*

Time to reach a border. In the guideposting paradigm involving alternating regions of a substrate pair, considerable neuronal outgrowth may occur on a homogeneous, permissive region of substrate with dimensions much larger than the growth cone itself. In the absence of growth cone contact with the second substrate, progress across the first substrate can be modeled by the random walk described in Eqs. (17.5a) and (17.5b) with parameters appropriate for the specific growth surface. One straightforward prediction we can make in this situation is the length of time that it should take for the neurite to approach the border of a region of dimension L. Figure 17.5a shows the average time for a neurite to reach the border from a random starting point in the interior of a square substrate region, assuming parameter values that are typical of the systems we have measured. If the distribution of step sizes is narrowly distributed around zero (that is, if σ is small), or if the drift velocity in the x direction is small, then the neurite requires a very long time to encounter the border. Interestingly, the greatest sensitivity to variations in these parameters occurs in the range of experimental values that we have measured. This may be advantageous developmentally, enabling finely tuned responses to slight variations in the growth environment. In addition, parametric sensitivity yields an ideal situation for hypothesis testing with true variations in behavior less likely to be obscured by experimental error.

Mean time for a neurite to track a distance L. Although it is likely that filopodial contact with the second substrate is important in growth cone tracking of narrow strips of a more favorable substrate (Clark et al., 1993), a phenomenological description of tracking can be obtained by modifying the preceding calculation slightly. Rather than asking how long the neurite takes to reach any of the four borders of the substrate region, the question of interest becomes how long it takes to extend between two parallel borders a distance L apart. A preliminary answer to this

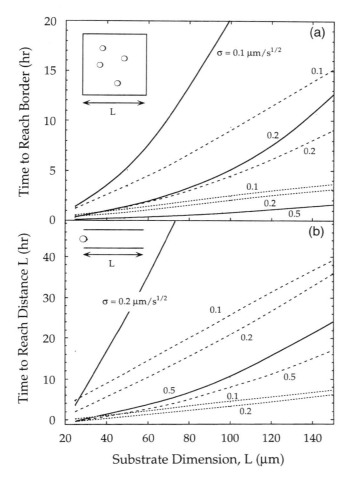

Figure 17.5. (a) Average time for a model growth cone to reach the border of a square region of dimension L from a random starting point within the region. $\sigma_x = \sigma_y = \sigma$ in this calculation. The different sets of curves represent drift velocities of $\mu_x = 0$ μm/sec (——), $\mu_x = 0.001$ μm/sec (- - - -), and $\mu_x = 0.005$ μm/sec (. . . . .) for the value of σ indicated. The curve for $\sigma = 0.5$ μm/sec$^{1/2}$ is the same for all values of μ_x. (b) Average time for a model growth cone to track a distance L along a narrow strip of substrate. Legends for the curves are the same as in (a).

question is provided by assuming that a growth cone cannot cross either edge of the track between the two ends (see Figure 17.5b). Modeling the growth cone trajectory with Eqs. (17.5a) and (17.5b) results in very little dependence on the width of the track, because the axon remains fairly aligned with the long axis of the track, and the distance traveled parallel to axonal orientation is independent of perpendicular motion.

Mean time for filopodial detection. The probability of filopodial contact with a

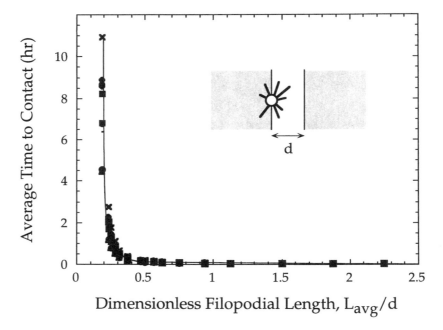

Figure 17.6. Mean time for filopodial contact to occur with a remote target across a dimensionless distance d/L_{avg}.

second substrate region, which appears to be important to both guideposting and tracking, can be characterized in terms of the mean time required for filopodial contact to occur. As shown in Figure 17.6, the average time for one filopodium to contact a region located a distance d away depends primarily on the ratio of the average filopodial length per growth cone, L_{avg} to d. When plotted as a function of this dimensionless filopodial length, filopodial contact time for the entire parameter space indicated in column 3 of Table 17.1 falls nearly along a single curve that asymptotes to zero at high values of L_{avg}/d and rises sharply for values less than 0.5.

17.6 Summary and Conclusions

The control of neuronal outgrowth is a complex problem with tremendous biological, medical, and technological implications. Although neurobiologists have provided numerous insights into factors that participate in neural development and serve to guide the growing neurite, our ability to manipulate outgrowth at will remains a goal rather than a reality. The work discussed here draws on two powerful techniques, microfabrication and mathematical modeling, to propose a basic approach for investigating neuronal outgrowth and its regulation in microfeatured environments. Applications range from the repair of nerve injuries to the construction of next-generation bioartificial organs and biologically based computers.

The primary features of the proposed model are (1) the probabilistic encounter between growth cone filopodia and microstructural features due to the randomness of filopodial dynamics, and (2) the random walk advance of the growth cone as a whole. The description of these processes provides realizations of the growth cone trajectory in a given environment. By examining a large number of realizations, it becomes possible to predict the average migration behavior of the growth cone under specific conditions in quantitative terms. On the other hand, because the model is constructed around a single growth cone, it also provides an ideal framework for extending the investigation to biophysical mechanisms of growth cone activity, including receptor mediation, signaling events, and cytoskeletal dynamics. To obtain quantitative predictions with the model, it is necessary to measure the model parameters experimentally. Preliminary data exist as described here, but more are required before parameter values can be established with a high level of confidence.

The next important step in this work is to employ micropatterned substrates to begin to test the model predictions. This will inevitably lead to some refinement and more model testing. However, the end result of the approach described here should be a much more directed iteration on our understanding of neuronal outgrowth than has previously been possible.

Acknowledgments

Supported by the Whitaker Foundation and by a Hoechst Celanese Young Investigator Award.

References

Alberts, B., Bray, D., Lewis, J., Raff, M., Roberts, K., and Watson, J. D. (1989) *Molecular Biology of the Cell*. New York: Garland Publishing, Inc.

Berg, H. C. (1983) *Random Walks in Biology*. Princeton, NJ: Princeton University Press.

Buettner, H. M. (1995) Nerve growth dynamics: quantitative models for nerve growth and regeneration. *Ann. NY Acad. Sci.* 745:210–221.

Buettner, H. M. and Pittman, R. N. (1991) Quantitative effects of laminin concentration on neurite outgrowth *in vitro*. *Dev. Biol.* 145:266–276.

Buettner, H. M. Pittman, R. N., and Ivins, J. K. (1994) A model of neurite extension across regions of non-permissive substrate: simulations based on experimental measurement of growth cone motility and filopodia dynamics. *Dev. Biol.* 163:407–422.

Caudy, M. and Bentley, D. (1986) Pioneer growth cone steering along a series of neuronal and non-neuronal cues of different affinities. *J. Neurosci.* 6:1781–1795.

Clark, P., Britland, S., and Connolly, P. (1993) Growth cone guidance and neuron morphology on micropatterned laminin surfaces. *J. Cell Sci* 105:203–212.

Clark, P., Connolly, P., Curtis, A. S. G., Dow, J. A. T., and Wilkinson, C. D. W. (1987)

Topographical control of cell behaviour: I. Simple step cues. *Development* 99:439–448.

Clark, P., Connolly, P., Curtis, A. S. G., Dow, J. A. T., and Wilkinson, C. D. W. (1990) Topographical control of cell behaviour: II. Multiple grooved substrata. *Development* 108:635–644.

Curtis, A. S. G., Breckenridge, L., Connolly, P., Dow, J. A. T., Wildinson, C. D. W., and Wilson, R. (1992) Making real neural nets: design criteria. *Med. & Biol. Eng. & Comput.* 30:CE33–CE36.

Dunn, G. A. and Brown, A. F. (1987) A unified approach to analysing cell motility. *J. Cell Sci. Suppl.* 8:81–102.

Gundersen, R. W. (1987) Response of sensory neurites and growth cones to patterned substrata of laminin and fibronectin in vitro. *Dev. Biol.* 121:423–431.

Gwydir, S. H., Buettner, H. M., and Dunn, S. M. (1994) Non-rigid motion analysis of the growth cone using continuity splines. *I.T.B.M.* 15:308–321.

Hammarback, J. A. and Letourneau, P. C. (1986) Neurite extension across regions of low cell-substratum adhesivity: implications for the guidepost hypothesis of axonal pathfinding. *Dev. Biol.* 117:655–662.

Hammarback, J. A., McCarthy, J. B., Palm, S. L., Furcht, L. T., and Letourneau, P. C. (1988) growth cone guidance by substrate-bound laminin pathways is correlated with neuron-to-pathway adhesivity. *Dev. Biol.* 126:29–39.

Hammarback, J. A., Palm, S. L., Furcht, L. T., and Letourneau, P. C. (1985) Guidance of neurite outgrowth by pathways of substratum-adsorbed laminin. *J. Neurosc. Res.* 13:213–220.

Kleinfeld, D., Kahler, K. H., and Hockberger, P. E. (1988) Controlled outgrowth of dissociated neurons on patterned substrates. *J. Neurosci.* 8:4098–4120.

Letourneau, P. C. (1975) Cell-to-substratum adhesion and guidance of axonal elongation. *Dev. Biol.* 44:92–101.

Myers, P. Z. and Bastiani, M. J. (1993) Growth cone dynamics during the migration of an identified commissural growth cone. *J. Neurosci.* 13:127–143.

O'Connor, T. P., Duerr, J. S., and Bentley, D. (1990) Pioneer growth cone steering decisions mediated by single filopodial contacts in situ. *J. Neurosci.* 10:3935–3946.

Wei, W. W. S. (1990) Time Series Analysis: Univariate and Multivariate Methods. Reading, Mass.: Addison-Wesley.

18

Microfabricated surfaces in signaling for cell growth and differentiation in fungi

H. C. HOCH

Department of Plant Pathology, Cornell University
New York State Agricultural Experiment Station, Geneva, NY 14456

R. J. BOJKO, G. L. COMEAU, and D. A. LILIENFELD

National Nanofabrication Facility, Knight Laboratory
Cornell University, Ithaca, NY 14853

18.1 Introduction

Microfabricated surfaces and devices have been used to study important problems in medicine and in cell and molecular biology that otherwise could not have been effectively addressed with more conventional tools. Among the first applications of microfabricated devices created through microlithography techniques was the development of substrata bearing specific topographical features used to help understand how fungal pathogens of agronomically important plants grow and develop specialized structures needed to invade their hosts (Hoch et al., 1987). Since that initial study, microfabricated substrata have been used in many endeavors aimed at elucidating the parameters surrounding fungal growth and growth orientation, as well as the mechanisms involved in signaling for infection structure development (Hoch et al., 1993; Kwon and Hoch, 1991; Kwon et al., 1991a, 1991b; Terhune and Hoch, 1993; Terhune et al., 1993). Although applications of submicrometer devices to plant and fungal cell biology thus far have been limited in scope, their use in mammalian cell research and in medicine has been much more extensive. For example, they have been used to investigate the development and egress of red blood cells from bone marrow, where they are formed, into the blood stream (Lichtman and Waugh, 1988; Waugh et al., 1984; Waugh and Sassi, 1986), the attachment and biocompatibility of cells and tissues to orthopedic and dental implants (Chehroudi et al., 1990, 1991; Chesmel, 1991; Inoue et al., 1987; also see Chapter 19), and the biomechanics of muscle cell contraction as well as the dynamic forces of cytoskeletal action (Krueger and Denton, 1992; also see Chapters 21 and 22). In this chapter, we will review applications of microfabricated substrata to fungal cell biology. We will also discuss the need for such defined substrata and the problems that can be addressed, fabrication techniques, examples of fabricated devices and substrata, and the kinds of devices that would be desirable in future investigations.

18.1.1 Fungal cell systems: rust fungi

Microfabricated substrata have been used most extensively to investigate the biology of rust fungi and, in particular, how these fungi prepare to invade their host. Rust fungi constitute a group of fungi that cause serious plant diseases referred to as "rust" because the spore masses that appear on the surface of the diseased plant usually occur as rust-colored powdery areas. The spores are the asexual reproductive units that are dispersed by wind and rain to healthy plant surfaces. Under appropriate environmental conditions (namely, free surface water, and temperatures near 17 °C), the spores germinate by forming a germ tube or short threadlike hypha that grows on the leaf surface (Figure 18.1a). Rust fungi have a fastidious requirement for the site at which they can infect the host – namely, stomata, the pores in the leaf epidermis through which water vapor can escape the interior of the leaf but, more important, through which O_2 and CO_2 are exchanged as a part of the normal physiological function of the plant. The stomatal aperture is surrounded by a pair of specialized cells, called guard cells, that have a role in regulating the opening of the stomatal aperture via changes in cell turgor (Figure 18.1b). The rust cell grows, directed by the physical features of the leaf surface, toward the stomate where it ceases growth and forms a specialized infection structure termed an appressorium (Figure 18.1a). It is from the appressorium that the fungus eventually gains ingress into the leaf interior through the stomatal aperture.

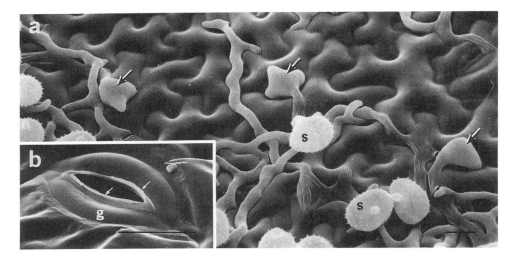

Figure 18.1. (a) The rust fungus, *Uromyces appendiculatus*, germinates and grows from spores (**s**) as short tubelike cells on the surface of a "fake" plant leaf surface. The surface, made of a polystyrene replica of the real leaf surface, retains all the physical features of the leaf to such an extent that the fungal cells sense and respond by forming infection structures called appressoria (*arrows*) over stomates. (b) A stomatal complex, over which the fungus develops an appressorium. A stomate of *Phaseolus vulgaris* consists of two guard cells (g), each of which bears a cuticular "lip" (*arrows*) and surrounds the stomatal pore or aperture (from Allen et al., 1991c). Bar scales = 40 μm and 10 μm, respectively.

The entire process, from spore germination to appressorium formation, can occur within three hours. We have known for many years that the rust fungus senses the physical features of the substratum to initiate formation of the appressorium, because surfaces made of specially treated collodion (nitrocellulose), bearing poorly defined topographies, effectively induce appressoria (Dickinson, 1949; Wynn, 1976). Furthermore, polystyrene or other plastic polymer replicas of the leaf surface made from a silastic "negative" template made of the leaf surface also induce appressoria over imprints of stomata, as exemplified in Figure 18.1a. Our interest has been in ascertaining (1) the topographical "signal" that induces appressorium development in rust fungi, that is, the nature and location of the inductive signal in relation to the stomatal apparatus; (2) the physical parameters (height, width, shape, etc.) of the inductive signal; (3) how a fungal cell, encased in a wall, senses a topographical signal; (4) how fungal cell growth is directed by topography; and (5) how long the signal needs to be present in order for signal reception to be complete. These and other concerns led us to microfabricate substrata bearing specific topographies that would help elucidate the cell biology of rust fungi. The rust fungal system we have emphasized is that of *Uromyces appendiculatus*, the causal agent of bean rust; as discussed later, other rust fungi are studied with these approaches as well. Rust fungi are particularily adaptable to studies involving growth on artificial substrata because the entire growth and differentiation process occurs within 2–6 h; in addition, the fungus grows well in distilled deionized water, because all necessary nutrients for 12–16 h of growth are contained within the spore (= urediospore).

18.2 Microfabricated substrata

In general, microfabrication of substrata bearing topographical patterns consists of a two-step process: micropatterning and creating a template in silicon or other suitable material, followed by replication of the template pattern in a plastic polymer, such as polystyrene or polycarbonate. Both electron beam and light lithography have been employed to create the silicon-based templates; the choice depends on the application and the resolution of the final product.

18.2.1 Photolithography

Microfabrication of silicon-based topographical patterns was performed using photolithography when the dimensions of the patterns were greater than 1.0 μm; with care, dimensions could effectively be made as small as 0.4 μm (Figure 18.2). Standard doped 75-mm-diameter single crystal <100> silicon wafers were used for the substrates. These were coated with 200 nm of silicon dioxide using plasma-enhanced chemical vapor deposition (PECVD) with a 13-MHz RF discharge employing silane and nitrous oxide. This SiO_2 layer was then patterned by photolith-

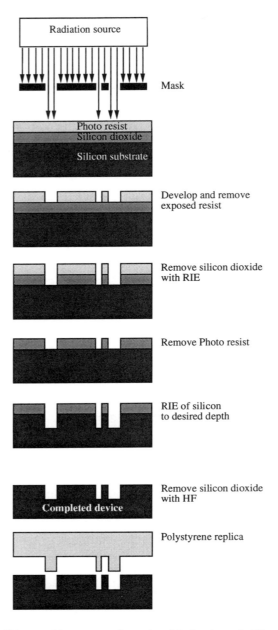

Figure 18.2. Photolithographic process for microfabrication of silicon-based templates and molded polystyrene replicas.

ography and etching. A layer of photosensitive polymeric resist was spin-coated onto the wafer; after baking to remove residual solvents, the layer was exposed using ultraviolet contact photolithography, in which an ultraviolet light source is projected onto the wafer surface. The pattern is defined by a mask (a glass plate with the pattern defined on it in opaque chrome) placed in contact with the wafer surface so that only the desired pattern is exposed. During development, the photoresist material that had been uv exposed is dissolved in an alkaline developer. The exposed silicon dioxide layer not covered by the remaining photoresist material is then exposed to a fluorinated (CHF_3) reactive-ion etch (RIE) plasma. Through both physical bombardment and chemical reaction at the surface by fluorine ions, this RIE step will selectively remove the silicon dioxide layer with respect to the photoresist, thereby exposing the underlying silicon surface. The photoresist etch mask is then removed using organic solvents or an oxygen plasma, and the wafer is exposed to a chlorinated RIE (Cl_2) plasma. The silicon dioxide now acts as an etch mask; the chlorine ions attack the open silicon areas and a vertical-walled trench of desired dimensions is obtained. A final step of wet etching in hydrofluoric acid removes the silicon dioxide etch mask and provides a hydrophobic silicon surface with the appropriate topography.

Many patterns have been created on which the rust fungus *Uromyces appendiculatus* has been studied to yield informative and useful information regarding cell growth and differentiation. Most frequently, patterns of lines or grids of varying spacing and depths have been created (Figures 18.3a and 18.3b). For microscope studies where it was important to return to a specific site, alphanumeric "indexed" patterns have proven to be very useful for re-location (Figure 18.3c). Other useful patterns

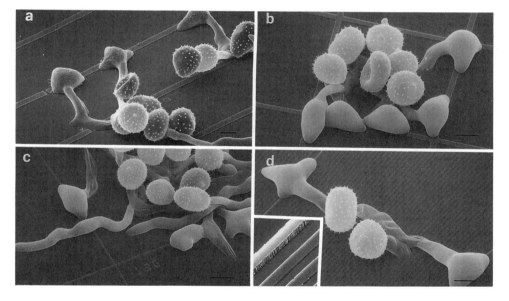

Figure 18.3. Polystyrene replicas bearing various topographies and patterns on which germlings of *Uromyces* were grown. (a) Parallel ridges, 0.5 μm high x 4.0 μm wide, induced appressorium formation (from Hoch et al., 1987). (b) Ridges, 0.45 μm high, arranged in a grid pattern spaced 60 μm apart serve as the most efficient topographical pattern for appressorium formation. (c) Alphanumeric-indexed topographical patterns are used to return to the same site. (d) Parallel ridges, 0.1 μm high and spaced 1.0 μm apart, do not induce appressorium development in the fungus, but rather orient cell growth toward the higher (0.5 μm) inductive ridges. (d inset) Detailed view of the topographical pattern of 18.3d. Bar scales = 10 μm.

have included double-height topographies created by processing the silicon wafer through the lithographic process two times, etching to a greater depth the second time. Such processes have been used to create topographies that orient cell growth via closely spaced parallel ridges toward higher topographies inductive for cell differentiation (Figure 18.3d, inset). The usefulness of these guiding topographies becomes apparent when orientation of the cells is important for such microscopic purposes as microinjection studies. The physical approach of the micropipette for injection is often limited; thus control of cell orientation becomes essential (Corrêa and Hoch, 1994).

18.2.2 Electron beam lithography

During the course of our investigations, we have frequently needed more complex patterns or patterns with dimensions below the resolution of that obtainable with conventional photolithographic methods. Electron beam lithography has been the method of choice for these patterns. The process is similar to the photolitographic protocol described in Section 18.2.1. The primary difference is that, instead of being effected by a broad-beam ultraviolet light source projected through a patterned chrome mask, the exposure is effected by a finely focused beam of electrons, deflected under computer control to trace out the desired pattern into the resist material on the wafer surface. The minimum feature size of photolithography is limited by diffraction of the exposing light, whereas the limit in electron beam lithography is beam size, which can be focused to less than 10 nm. Features with dimensions less than 100 nm can be routinely produced using electron beam lithography, and features as small as 20 nm are possible under certain conditions.

Microtopographies created using electron beam lithography include 0.25-μm-wide x 0.5-μm-deep features. These features were used as a negative from which replicated, 0.5-μm-high polystyrene ridges of corresponding width have been produced to emulate the guard cell "lips" (Figure 18.1b) that protrude above the stomatal apparatus. Such thin ridges were used to ascertain deformation phenomena that occur *in planta* in which the ridges (= guard-cell "lips" of the stomata) collapsed as appressoria formed over them (Figure 18.4a–j). Other important topographical patterns include those of closely spaced ridges similar to those created by light lithography (Figure 18.3d) used to study cell growth orientation. Using electron beam lithographic methods, the defining pattern can be made with far greater resolution – for example, parallel line patterns 0.5 μm apart (Figure 18.5a; also see Figure 6a in Hoch et al., 1987), as well as the creation of bull's-eye targets that are similarly used to direct cell growth to a specific site (Figure 18.5b). Fungal cells such as the rust fungi normally exhibit a nonoriented random growth habit when grown on smooth, featureless substrata, as depicted in Figure 18.7c (also see Figure 6a in Allen et al., 1991a); however, closely spaced parallel ridges orient cell growth in a highly directed linear pattern (Figures 18.3d and 18.5a).

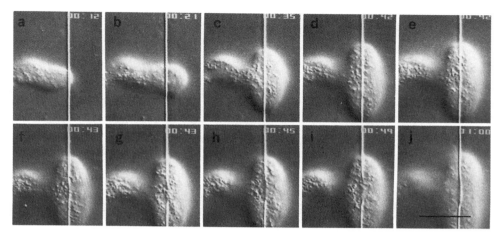

Figure 18.4. A temporal sequence of light micrographs depicting appressorium development by *Uromyces* over an inductive polystyrene ridge (0.5 µm high and 0.25 µm wide). Time depicted in the upper right corner is in minutes that began from an arbitrary time prior to cell contact with the ridge. The polystyrene ridge remained unchanged during the early stages of appressorium formation (a–d), but became clearly distorted and collapsed (f–j) a few moments later. (From Terhune et al., 1993.) Bar scale = 15 µm.

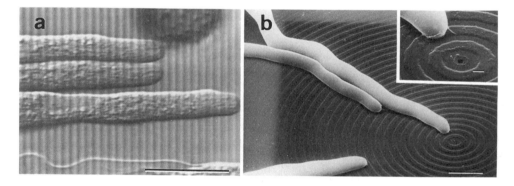

Figure 18.5. (a) Oriented growth of *Uromyces appendiculatus* over 2.0-µm-wide by 0.5-µm-high polystyrene ridges spaced 0.5 µm apart. (b, b inset) Bull's-eye target used to direct cell growth to a centered 0.5-µm-diameter hole. Darker square area is the "window" of silicon nitride with overlaid PMMA (polymethyl methacrylate) ridges. Lighter area to the left and above is backed by silicon (see Figure 18.6) (from Hoch et al., 1993). Bar scales = 30 µm and 10 µm (1.0 µm for inset), respectively.

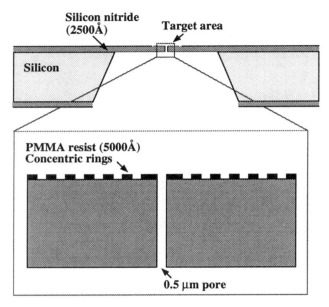

Figure 18.6. Schematic depiction of the bull's-eye target of Figure 18.5b. To fabricate the device, a double-polished silicon wafer was coated on both sides with 2500 Å of low-stress silicon nitride using a low-pressure chemical vapor deposition (LPCVD) process employing ammonia and dichlorosilane at 850 °C. Photolithographic processing was used to create 100-μm-square areas in the back-side silicon nitride, etched with a fluorinated RIE (CF4). Anisotropic etching of the exposed silicon with potassium hydroxide solution left a 100-μm-square silicon nitride membrane on the front side. Electron beam lithography and RIE were used to etch a small pore (0.5 μm in diameter) through the center of the membrane. A second electron beam lithography step was used to define concentric rings around the etched pore in a 5000-Å-thick PMMA layer.

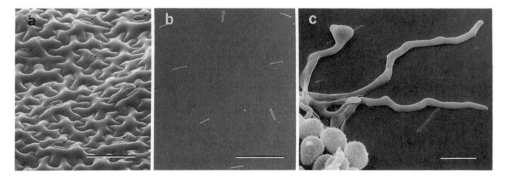

Figure 18.7. (a) Scanning-electron micrograph of a leaf surface depicting the spatial distribution of stomata (*arrows*) from which their *x*–*y* coordinates were determined and used to microfabricate "artificial" stomata (b) with similar distributions. Such artificial stomata distributed on an otherwise smooth surface can be used to study the influence of leaf surface topography on orienting cell growth toward stomata (c). Bar scales = 50 μm, 50 μm, and 30 μm, respectively.

18.2.3 *Replication of the topography*

Fungal cells can be grown either directly on the topographical patterns created in the silicon medium or on plastic replicas made from the topographical features inherent on the silicon templates. The kind of fungal cells that are best adapted to these "solid" substrata are those that grow from spores having endogenous nutrient reserves such as those found in the rust fungi, although nutrients supplied from an exogenous medium also can be used (Clay et al., 1994; Gow et al., 1994). Many other phylloplane (leaf surface) fungal pathogens are similarly suitable for growth in the absence of exogenous nutrients. In addition, most of these fungi that grow on leaf surfaces covered with a cuticle and wax layer exhibit a preference for a surface with low wettability characteristics, that is, a surface that is hydrophobic. Silicon is generally suitable for these kinds of studies, because it is highly hydrophobic until it becomes oxidized. Most plastic polymers are also hydrophobic, although like silicon, their surfaces can be made hydrophilic through appropriate chemical or physical treatments.

A disadvantage of using silicon as the substratum for cell growth and differentiation studies is its opacity to light for routine light microscope examination; we have overcome this problem, however, by treating the fungal cells with fluorophores, such as Calcofluor, that bind to wall components of the fungus and thus allow for observations using epi-fluorescence microscopy. This approach is suitable only for studies in which the cells are not living. Alternatively, polystyrene and polycarbonate replicas are well suited to either transmission-light or epi-fluorescence-light microscopy. These materials have very little inherent autofluorescence and are thus very amenable to fluorescence microscopy. Phase-contrast optics are well suited for transmission-light microscopy of plastic replicas; however, Normarski differential-interference contrast (DIC) presents a few problems that can be overcome with proper adaptation. If polymeric plastics are too thick (for example, >150 μm), the crystalline nature of the plastic interferes with the orientation of the polarized light as required for proper DIC; hence the image is greatly degraded. We have overcome this by (1) using very thin solvent-cast plastic films (<25 μm thick) that are ultimately mounted on a microscope slide and covered with a cover glass, or (2) using heat-pressed polystyrene membranes of similar thickness that serve both as the topography-bearing substratum and the cover slip.

Solvent-cast replicas. Polystyrene membranes with well-defined topographies were produced by casting a thin film of dissolved polystyrene, such as Styron 685D (Dow Chemical, Midland, Mich.) (20% w/v in ethyl acetate or other appropriate solvent) onto 7.5-cm-diameter silicon wafer templates bearing specific patterns microfabricated into them by either optical or electron beam lithography. Once the solvent had completely evaporated from the polystyrene, the resulting membranes were floated off the templates in a 45 °C water bath and then collected, with the topographied side up, on glass microscope slides. Polycarbonate dissolved in chloroform can be similarly cast on the templates.

Heat-pressed replicas. Similar polystyrene membranes with well-defined topographies were prepared by melting polystyrene pieces (from petri plates or other sources) between a silicone-treated glass plate (7.5 cm in diameter) and a silicon wafer bearing the desired microfabricated topography. For sturdiness, the wafer was backed with another glass plate. The entire "sandwich" was clamped with binder clips and heated to 200 °C for about 40 min. Thickness of the polystyrene membranes was controlled by placing three to four thickness gauges – for example, small pieces of a No. 1 glass coverglass (= 150 μm thick) or aluminum foil (= 25 μm thick) – between the glass plate and wafer. Once cooled, the glass plate and wafer were separated, and the polystyrene replica was removed from the wafer by immersion in warm water. Nontopographied (control) polystyrene membranes could be similarly prepared by heat pressing polystyrene between two smooth glass plates.

Replication of wafer topography in polystyrene by either method was highly precise, and features <50 nm in width could be readily duplicated, as illustrated in Figure 18.8. In this example, the microfabrication process was flawed, which resulted in a silicon template with the outline of the desired pattern but with the midsection poorly produced (Figure 18.8a). Pores 50–200 nm in diameter and 0.83 μm deep resulted and were replicated as a positive topography in polystyrene (Figure 18.8b); however, most of the polystyrene pillars did not remain erect or as distinct pillars because of the surface tension created between the water interface and the

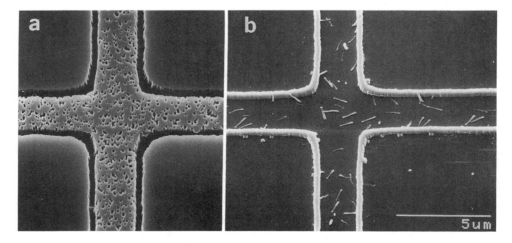

Figure 18.8. Depiction of the detail that can be replicated in plastic polymers from silicon templates. (a) A flawed lithographic process yielded the outline of what was to have been the intersection of two 0.83-μm-deep grooves in silicon. (b) Along with the outline of the ridges, pores 50–200 nm in diameter and 0.83 μm deep resulted, and were replicated as pillars in polystyrene. Bar scale = 5 μm.

hydrophobic polystyrene as the membrane was lifted from the surface. Nevertheless, the results illustrate that polystyrene can be used to create accurate three-dimensional replicas of topographical features.

18.3 Preparation of rust fungi for growth on inert microfabricated substrata

Growth of fungi on hydrophobic surfaces requires slightly different procedures than those normally used for mammalian cell culture. Spores of many fungi, including the rust fungi, are extremely hydrophobic. When they are deposited onto the various substrata and submerged in aqueous growth media, the spores usually clump together or float off the substratum. Fortunately, many fungal spores secrete an extracellular matrix (ECM) upon hydration. Thus, by hydrating rust spores deposited on substrata by misting with distilled water and leaving the spores in a humidified environment for 10–30 min, the ECM that is formed will serve as a "glue" when they are dried. For most studies, such spore-laden membranes were floated spore-side down on the surface of distilled water (or nutrient media for other fungi) for the duration of the cell growth period (Allen et al., 1991a, 1991b). For some fungi, such as *Candida albicans*, the patterned polystyrene membranes were coated with a nutrient medium by gently touching them to the surface of nutrient agar. Sufficient nutrient carry-over was obtained from the agar to sustain cell growth on the nonporous polystyrene membranes (Gow et al., 1994).

18.4 Microfabricated substrata for elucidation of the signal for appressorium formation in rust fungi

The plant surface is by no means a smooth and featureless terrain. It is replete with a wide array of physical features that the rust fungus must discriminate to decipher the signal that triggers infection structure formation (Terhune et al., 1991; Wynn, 1976; Wynn and Staples, 1981). As noted in Section 18.1.1, the fungal cell recognizes unique features inherent in the architecture of the stomatal apparatus through which it ultimately gains ingress into the plant. With the use of artificial plastic replicas of the plant surface we and, previously, Wynn (1976) have learned that the rust fungus recognizes a physical feature in the vicinity of the stoma. With the use of microfabricated substrata in which topographical features were controlled to a high degree, we were able to elucidate the parameters of the inductive topography. In brief, *Uromyces appendiculatus* recognizes 0.5-μm-high features for appressorium formation (Figure 18.9). More specifically, the fungus perceives abrupt changes in a substratum elevation between 0.4 and 0.8 μm in height (Allen et al., 1991a, 1991b; Hoch et al., 1987). Topographical features significantly less than 0.25 μm in height or greater than 1.2 μm in height were generally not inductive. The width of the topographical feature, for example, a ridge, was not important. For example, fea-

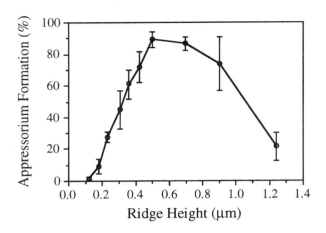

Figure 18.9. Appressorium formation by *Uromyces appendiculatus* (race D85 C1-5) on 10 different ridge heights (modified from Allen et al., 1991b).

tures 100 μm or more in width, such as 100 x 100-μm plateaus arranged in a checkerboard pattern, were equally inductive as 0.25-μm-wide ridges, as long as the height was within the 0.4- to 0.8-μm range (Hoch et al., 1987). Germlings growing either onto or off the plateaus developed appressoria only at the edge of the plateaus. Taken together, these data indicated to us that the signal for cell differentiation in *U. appendiculatus* is an elevation change of approximately 0.5 μm and that two closely spaced acute angles must be present. Clearly, one angle, such as that associated with the top or bottom of plateaus or ridges (for example, 5.0 μm in height), is not sufficient to signal appressorium formation.

The formation of appressoria in response to microfabricated topographies has been determined not only for *U. appendiculatus*, but also for other rust fungi whose plant hosts include such plants as maize, strawberry, mint, wheat, and peanut (Adendorff and Rijkenberg, personal communication; Allen et al., 1991a; Stark-Urnau and Mendgen, 1993). Of 27 species examined by Allen and co-workers (Allen et al., 1991b), 8 species, *U. vignae, Melampsora medusae, Puccinia antirrhini, P. calcitrapae, P. carduorum, P. melanocephala, P. substriata,* and one isolate of *P. recondita,* developed appressoria on ridges within an optimal height range of approximately 0.4–0.8 μm, similar to that observed for *U. appendiculatus*. A broader range of optimum ridge heights (up to 2.25 μm) was observed to be inductive for appressorium formation with *P. polysora, P. menthae, P. hieracii, P. sorghi, P. arachidis,* and *Physopella zeae.* Seven species, including *Coleosporium asterum, C. tussilaginis, Phragmidium potentillae, Puccinia coronata, P. graminis* f. sp. *tritici, P. graminis* f. sp. *avenae,* and *Tranzchelia discolor,* did not form significant numbers of appressoria on microfabricated topographies of defined heights. All these observations were made on substrata with microfabricated ridges (2.0 µm wide) spaced 60 µm apart. In similar studies by Read and colleagues (Collins and Read, personal communication; Read et al., personal communication), it was noted that when the topographical pattern was condensed so that the ridges are very close together, for example, 1.5–2.5 µm apart and 2.0 µm high, appressoria readily formed (>70%) in several rust fungi (*P. hordei*, *Puccinia coronata*, *P. graminis* f. sp. *tritici*, and *P. graminis* f. sp. *avenae*) that usually were not thigmoresponsive to single ridges. Apparently, these fungi require multiple or repeat signals to trigger appressorium formation.

The use of a defined signal (0.5-µm-high ridge) for infection structure formation has allowed us to study in detail the temporal and spatial aspects of the developmental process. For example, it was determined that *Uromyces* germling apices sensed inductive topographies within 4 min after initial contact (Kwon and Hoch, 1991), a time frame far sooner, and more precise, than had been previously reported. These conclusions were based on observed changes in the rate of germling growth and the timing of apical tip swelling that began 4 min after contact with a ridge. It was surmised that the area along the germling that senses such topographies was located on the substrate side of the cell, within 6.0 µm of the apex. Recently we have confirmed that supposition by other methods (Corrêa and Hoch, 1994). Also, by using substrata with a highly defined microfabricated topography, the temporal aspects of DNA synthesis during appressorium development was elucidated and found to occur only after the first nuclear (mitotic) division that was in preparation for the second mitosis that normally occurs in the substomatal vesicle. These and other results indicate that the nuclei in fungal germlings were already in the G2 phase when spore germination began, with DNA synthesis for the first round of mitosis having occurred during sporogenesis (Kwon and Hoch, 1991).

18.4.1 Influence of topographical pattern on cell growth habit by Uromyces

Not only must the rust fungus recognize the appropriate site (stomate) at which it ceases growth and develops an appressorium, it must find this site without growing aimlessly over the leaf surface and thus expending unnecessary energy. This raises the question, is growth oriented (toward the stomata) by topography or is growth random? Clearly, growth can be oriented by topography. Using polystyrene replicas of silicon wafers containing uniformly but variously spaced parallel topographical ridges, we found that rust germlings do indeed respond with oriented growth. As already noted, the germlings grew very straight and perpendicular from 0.2-µm-high by 1.4-µm-wide ridges spaced 0.5–15.5 µm apart (Figures 18.3d and 18.5a); however, the straightness of the germlings diminished as the spacing between the orienting ridges was increased (Hoch et al., 1987). Spacing the ridges more than 30 μm apart generally did not serve as efficient signals for the fungus to continue growing perpendicular from the ridges, although the germlings did grow perpendicular *from* these ridges for a distance, sometimes up to 50 μm, at which time they began to exhibit nonoriented growth patterns. Many of the valleys on the bean leaf surface that are formed by the merging of the epidermal leaf cells, namely, regions over the anticlinal walls, are often arranged in semiconcentric patterns around the stomata. These valleys are appropriately spaced to serve as signals to guide the rust germlings to the stomata. On graminaceous (grass) leaves, the epidermal cells are arranged in linear arrays, and growth of rust fungi on these surfaces is considerably more oriented into straight growth patterns, much as it is on closely spaced ridges (see Figure 5.5 in Allen et al., 1991c; Staples and Hoch, 1996). It is interesting to note that fungi grow oriented perpendicular from ridges, whereas mammalian cells, such as fibroblasts, "grow" oriented along ridges (see Chapter 20). The two cell types move by different mechanisms: fibroblasts migrate rather than actually grow; filamentous fungi grow by apical extension. Growth of fungi is by deposition of wall and plasma-membrane materials at the cell apex. Such polarized cell growth is directed by poorly defined features of the cell apex; however, it is the positioning of the cluster of apical vesicles, which contribute materials involved in cell extension, that dictates cell growth direction. It is not well understood which cell components control the position of the apical vesicle cluster, but such components may include the orientation of previously polymerized wall filaments, the orientation and conformation of cytoskeletal components such as microtubules and F-actin, and contributions of the physical constraints of cell shape at the apex.

"Artificial" stomata have been microfabricated to help address the question of whether growth is random or directed toward stomata. Stomatal number, orientation, and size were assigned *x–y* coordinates as determined from scanning-electron micrographs of host leaves, and corresponding 0.5-μm-high topographical features were prepared in an otherwise smooth, featureless silicon wafer by light lithography (Figures 18.7a and 18.7b). Growth of the fungus on these substrata resulted in appressorium formation only when the germlings encountered an artificial stomata (Figure 18.7c). Comparison of the number of appressoria formed on these microfabricated surfaces with the number formed either on a polystyrene replica of a leaf (from which the coordinates for artificial stomata were derived) or on a real leaf of similar phenological characteristics, indicated that growth was oriented toward stomata. For example, the number of appressoria formed by 16 h on surfaces bearing microfabricated artificial stomata was approximately 20%, compared to >90% on replicas or on the real leaf. Not only have closely spaced parallel ridges been useful in resolving orientation characteristics of rust fungi, they also have been the basis of orienting cell growth to a particular site for subsequent study, as with bull's-eye patterns.

18.4.2 Use of microfabricated topographies in other aspects of Uromyces rust cell biology

Aside from serving as signals for appressorium formation, microfabricated ridges have been used to elucidate other aspects of rust fungal cell biology. For example, they have been used to help ascertain why similar structures associated with the stomatal complex, such as guard-cell cuticular lips, are not normally observed protruding into the appressorium that they presumably triggered to form. The normally erect stomatal guard-cell lips were usually observed prostrate at most stages of appressorium development; there were no persistent or significant indentations into the fungal cell that might have been caused by the topographical features. One would expect to see such a protrusion into the fungal cell especially if the cell wall is a static structure once it is polymerized; thus this lack of morphological evidence has been a concern for some time. To address this problem, 0.25-μm-wide polystyrene or polycarbonate ridges (0.5 μm high) were replicated from silicon templates with corresponding topographical features created by electron beam lithography. These thin ridges thus mimicked the stomatal guard-cell lips of *Phaseolus vulgaris*, the host plant of *U. appendiculatus*. These artificial lips induced appressoria and became deformed (flattened) approximately 30 min after initial contact by the germ tube apex, as recorded and observed with time-lapse-video light microscopy (Terhune et al., 1993). The collapsed nature of the ridges was further evaluated by both transmission and scanning-electron microscopy. These results suggest that mechanical forces imposed by a combination of cell turgor pressure and adhesion of the appressorium to the substratum were responsible for deformation of the inductive topography both in vitro as well as *in planta*.

As with most phytopathogenic fungi, adhesion of rust fungi to plant surfaces is essential for the establishment of a successful host–parasite relationship. *Uromyces appendiculatus* must have intimate contact with the substratum in order for the cell to sense the inductive signal, and we had surmised that the degree of hydrophobicity of the substratum was important in this respect. Topographical signals microfabricated into either silicon wafers or quartz plates were used to assess the role of cell adhesion in signal perception for appressorium formation, particularily when the wettability of the substratum could be controlled. To accomplish this, we used both silicon, which invariably becomes oxidized, and quartz (SiO_2) substrata bearing 0.5-μm-high/deep ridges/grooves that were treated with a range of silanes known to influence hydrophobicity (Terhune and Hoch, 1993). We observed a direct correlation between substratum wettability and the degree of appressorium development. The choice of substrate material (silicon or quartz) depended on how the cells were to be observed microscopically. Because the quartz is transparent, transmission modes of optical microscopy could be employed, whereas silicon was used only for epi-reflective (such as fluorescence) microscopy. A distinct advantage of either microfabricated substrate is that the surface was uniform among silane-treated samples, a feature that had not been possible previously.

Much of our research efforts have been directed toward addressing the question of how such cells as the rust fungus sense a minute change in surface topography, let alone distinguish between similar topographies that are only tenths of microns different in height. There are several possible mechanisms by which a cell could sense topography; one of these is that the cell's plasma membrane stretches as the cell grows over, and conforms to, a ridge. During the stretching process, molecular-size channels or pores in the membrane may be stretched open, thereby allowing an efflux or influx of specific ions out of or into the cell. The new balance of ions within the cell might activate a cascade of biochemical events that, in turn, cause a physiological response culminating in the formation of an appressorium. Such "stretch-activated" ion channels have been shown to exist in the plasma membrane of many cell types, including that of the rust fungus (Zhou et al., 1991), but a means of studying them during the activation process in vivo has been lacking. To address this problem, we microfabricated a device that could help in correlating such channel openings with appressorium formation. The device, a bull's-eye target arrangement consisting of a series of 1.0-μm-wide concentric rings spaced 1.0 μm apart, is used to direct fungal growth to the exact center (Figure 18.5b). As already noted and from previous studies (Hoch et al., 1987), we determined that ridges spaced very close together could be used to direct fungal cell growth. Thus, using a concentric ring pattern, it should be possible to direct growth to the center of the pattern where a hole of, for example, 0.5-μm diameter, or some other subdevice such as an electrode, could be placed to study various aspects of the cell's biology. In the instance of a bull's-eye target with a centered hole, the application of a slight vacuum exerted though the hole has stimulated appressorium formation. In effect, the cell's plasma membrane likely was stretched as the cell grew over the targeted hole. Changes in electrical resistance due to different states of membrane channel openings can thus be studied. Microfabrication of these devices was a multistep process; in the first step, 100-μm-square free-standing membranes of 200-nm-thick silicon nitride were created on a silicon wafer by wet-etching the underlying silicon away completely. By direct-write electron beam lithography, a hole (0.25–3.0 μm in diameter), positioned with alignment marks, was etched through the silicon nitride. The wafer was then recoated with PMMA (polymethyl methacrylate) resist, and direct-write electron beam lithography was then used again to expose a series of concentric rings about the centered hole.

18.5 Other fungal cell systems

The use of microfabricated topographies has been directed primarily toward studies of those fungi, namely, rust fungi, that have been clearly documented as sensing physical features of the substratum. However, several researchers have recently explored the thigmotropic sensing capabilities of other fungi, including *Magnaporthe grisea, Cochliobolus sativus,* and *Candida albicans. Magnaporthe grisea*, a pathogen of rice and other grasses, was found to be not responsive to topographical patterns similar to those used for rust fungi (Lee and Dean, 1994). Hyphal cells of this fungus grew over polystyrene ridges 0.18–3.16 μm high, without exhibiting any notable growth or differentiation pattern that could be attributed to the topographical features. In contrast, Gow and colleagues noted that the animal pathogen *Candida albicans* was frequently responsive to substratum topography (Gow, 1993; Sherwood et al., 1992). In studies where polystyrene ridges were placed in the growth environment of *C. albicans*, the hyphae of the fungus grew in close contact with the substratum, particularily at the junction formed at the substratum base and the wall of the ridge (Gow et al., 1994). The highest ridges had the most influential effects on the fungal cells. The only other nonrust fungus to be studied with regard to substratum topography is *Cochliobolus sativus*, an important pathogen of wheat and barley. Clay et al. (1994) determined that the fungus is induced to form appressoria (42% at 18 h) on polystyrene surfaces with 1.6-μm-high ridges, compared to 16% appressoria on similar surfaces but with 0.5-μm-high ridges. This latter rate was not significantly different than that noted on smooth non-topographied surfaces. Clay and associates concluded that the 1.6-μm-high topographical features were similar to the cell junctures of the host plants, the sites where most appressoria are formed *in planta*.

18.6 Future uses and needs of microfabricated substrata and devices in fungal cell biology

Microfabrication of tools to answer questions directed toward understanding the biology of fungal cells can be a very valuable asset for the research scientist. There are many questions about how fungal cells grow, differentiate, and function that to date have not been addressed because the appropriate instrument, device, or surface has not been available. Nano- and microfabrication methodologies represent possible approaches.

Fungal-cell biologists have long sought a means of determining turgor pressure within fungal cells by nondestructive means. Current methods involve insertion of a micropipette into the cell. Cell pressure is determined by a pressure transducer connected to the micropipette. The method is generally destructive to the cell. Alternatively, cells are placed in a concentration series of osmotically active solutes, and the highest concentration at which the cell does not collapse is considered osmotically equal to that of the cell's cytoplasm. Both methods have serious shortcomings that might be overcome with innovative nanofabrication approaches. Some fungi, such as *Magnaporthe grisea*, have been reported to exert pressures as great as 8.0 MPa (1160 psi) to gain entry into their host leaf (Howard et al., 1991). Such pressures have been indirectly determined by incipient plasmolysis (with a concentration series of solutes), and it would be particularly useful if pressure transducers could be fabricated to determine such pressures on very small cells (5–8 μm in diameter) by either internal or external means. Equally useful would be surface pressure devices that could be used to monitor penetration pressures exerted by the fungal cell as it penetrates the cuticle of its host.

Monitoring the extracellular environment for specific ions, metabolic products, and so forth, provides valuable clues to cell growth, function, and health. Generally, there are good methods to monitor many molecular species around cells; for example, the vibrating ion-specific probe (Smith, 1994) is very good for determining ionic currents generated by specific ions around functioning cells. However, for fungal cells as well as mammalian cells, surface signaling is an important parameter to study. There is no good method for monitoring the environment at the cell–substratum interface, a place where the cell is perceiving both chemical and topographical signals. Microfabricated sensors over which the cells might grow would be valuable tools to study the physiology of the cell in these normally inaccessible locations. Similarly, microfabricated tools to control and/or monitor surface charges and densities would be invaluable. Experimental cell biologists could also make great use of devices used to control ionic and chemical domains around the living cell.

Integration of the cellular sciences with nano- and microfabrication resources represents an important alliance in research, an approach that provides the opportunity to decipher many cell functions that could not otherwise be elucidated.

Acknowledgments

We thank J. S. Lamboy for reviewing and making useful suggestions regarding the preparation of this manuscript. We further acknowledge the USDA Competitive Grants Program and the National Science Foundation, through which much of the research summarized herein was supported.

References

Allen, E. A., Hazen, B. E., Hoch, H. C., Kwon, Y., Leinhos, G. M. E., Staples, R. C., Stumpf, M. A., and Terhune, B. T. (1991a) Appressorium formation in response to topographical signals by 27 rust species. *Phytopathology* 81:323–331.

Allen, E. A., Hoch, H. C., Stavely, J. R., and Steadman, J. R. (1991b) Uniformity among races of Uromyces appendiculatus in response to topographic signaling for appressorium formation. *Phytopathology* 81:883–887.

Allen, E., Hoch, H. C., Steadman, J. R., and Stavely, J. R. (1991c) Leaf surface features and the epiphytic growth of fungi. In *Microbial Ecology of Leaves*, ed. J. H. Andrews and S. S. Hirano, pp. 87–110. New York: Springer-Verlag.

Chehroudi, B., Gould, T. R. L., and Brunette, D. M. (1990) Titanium-coated micromachined grooves of different dimensions affect epithelial and connective-tissue cells differently *in vitro*. *J. Biomed. Mater. Res.* 24:1203–1220.

Chehroudi, B., Gould, T. R. L., and Brunette, D. M. (1991) A light and electron microscopic study of the effect of surface topography on the behavior of cells attached to titanium-coated percutaneous implants. *J. Biomed. Mater. Res.* 25:387–406.

Chesmel, K. D. (1991) Mediation of the local host response by the physical aspects of a biomaterial's surface. PhD dissertation, University of Pennsylvania, Philadelphia, PA.

Clay, R. P., Enkerli, J., and Fuller, M. S. (1994) Induction and formation of *Cochliobolus sativus* appressoria. *Protoplasma* 178:34–47.

Corrêa Jr., A. and Hoch, H. C. (1993) Microinjection of urediospore germlings of *Uromyces appendiculatus. Exp. Mycol.* 17:241–252.

Dickinson, S. (1949) Studies in the physiology of obligate parasitism. II. The behavior of the germ-tubes of certain rusts in contact with various membranes. *Annals of Botany* 13:219–236.

Gow, N. A. R. (1993) Nonchemical signals used for host location and invasion by fungal pathogens. *Trends Microbiol.* 1:45–50.

Gow, N. A. R., Perera, T. H. S., Sherwood-Higham, J., Gooday, G. W., Gregory, D. W., and Marshall, D. (1994) Investigation of touch-sensitive responses by hyphae of the human pathogenic fungus *Candida albicans. Scanning Microscopy* 3:705–710.

Hoch, H. C., Bojko, R. J., Comeau, G. L., and Allen, E. A. (1993) Integrating microfabrication and biology. *Circuits and Devices* 9:16–22.

Hoch, H. C., Staples, R. C., Whitehead, B., Comeau, J., and Wolf, E. D. (1987) Signaling for growth orientation and cell differentiation by surface topography in *Uromyces. Science* 235:1659–1662.

Howard, R. J., Ferrari, M. A, Roach, D. H., and Money, N. P. (1991) Penetration of hard substrates by a fungus employing enormous turgor pressures. *Proc. Natl. Acad. Sci.* 88:11281–11284.

Inoué, T., Cox, J. E., Pilliar, R. M., Melcher, A. H. (1987) Effect of the surface geometry of smooth and porous-coated titanium alloy on the orientation of fibroblasts *in vitro. J. Biomed. Mater. Res.* 21:107–126.

Krueger, J. W. and Denton, A. (1992) High resolution measurement of striation patterns and sarcomere motions in cardiac muscle cells. *Biophys.* J. 61:129–144.

Kwon, Y. and Hoch, H. C. (1991) Temporal and spatial dynamics of appressorium development in *Uromyces appendiculatus. Exper. Mycol.* 15:116–131.

Kwon, Y. H., Hoch, H. C., and Aist, J. R. (1991a) Initiation of appressorium formation in *Uromyces appendiculatus*: Organization of the apex, and the responses involving microtubules and apical vesicles. *Can. J. Botany* 69:2560–2573

Kwon, Y. H., Hoch, H. C., and Staples, R. C. (1991b) Cytoskeletal organization in *Uromyces* urediospore germling apices during appressorium formation. *Protoplasma* 165:37–50.

Lee, Y. H. and Dean, R. A. (1994) Hydrophobicity of contact surface induces appressorium formation in *Magnaporthe grisea. FEMS Microbiology Letters* 115:71–75.

Lichtman, M. A. and Waugh, R. E. (1988) Red cell egress from bone marrow: Ultrastructural and biophysical aspects. In *Regulation of Erythropoiess*, ed. E. D. Zanjani, M. Tavassoli, and J. L. Ascensao, pp. 15–35. New York: PMA Publishing Corp.

Sherwood, J., Gow, N. A. R., Gooday, G. W., Gregory, D. W., and Marshall, D. (1992) Contact sensing in *Candida albicans*: a possible aid to epithelial penetration. *J. Med. Vet. Mycol.* 30:461–469.

Smith, P. J. S., Sanger, R. H., and Jaffe, L. F. (1994) The vibrating Ca^{2+} electrode: A new technique for detecting plasma membrane regions of Ca^{2+} influx and efflux. *Methods in Cell Biology* 40:115–134.

Staples, R. C. and Hoch, H. C. (1996) Physical and chemical clues for spore germination and appressorial formation. In *The Mycota: Plant Relationships*, vol. VI, ed. G. Carroll, and Tudzynski. Berlin: Springer-Verlag (in press).

Stark-Urnau, M. and Mendgen, K. (1993) Differentiation of aecidiospore- and ure-dospore-derived infection structures on cowpea leaves and on artificial surfaces by *Uromyces vignae*. *Canad. J. Bot.* 71:1236–1242.

Terhune, B., Allen, E., Hoch, H. C., Wergin, W., and Erbe, E. (1991) Morphology and ontogeny of stomata in *Phaseolus vulgaris*. *Canad. J. Bot.* 69:477–484.

Terhune, B. T., Bojko, R. J., and Hoch, H. C. (1993) Deformation of stomatal guard cell lips and microfabricated artificial topographies during appressorium formation by *Uromyces*. *Exp. Mycol.* 17:70–78.

Terhune, B. T. and Hoch, H. C. (1993) Substrate hydrophobicity and adhesion of Uromyces urediospores and germlings. *Exp. Mycol.* 17:253–273.

Waugh, R. E., Hsu, L. L., Clark, P., and Clark, A. J. (1984) Analysis of cell egress in bone marrow. In *White Cell Mechanics: Basic Science and Clinical Aspects*, ed. H. J. Meiseiman, M. A. Lichtman, and P. L. LaCelle, pp. 221–236. New York: Alan R. Liss Inc.

Waugh, R. E. and Sassi, M. (1986) An *in vitro* model of erythroid egress in bone marrow. *Blood.* 36:250–257.

Wynn, W. K. (1976) Appressorium formation over stomates by the bean rust fungus: Response to a surface contact stimulus. *Phytopathology* 66:136–146.

Wynn, W. K. and Staples, R. C. (1981) Tropisms of fungi in host recognition. In *Plant Disease Control: Resistance and Susceptibility*, ed. R. C. Staples and G. H. Toenniessen, pp. 45–69. New York: Wiley Interscience.

Zhou, X.-L., Stumpf, M. A., Hoch, H. C., and Kung, C. (1991) A mechano-sensitive cation channel in the plasma membrane of the topography sensing fungus, *Uromyces*. *Science* 253:1415–1417.

19

Effects of surface topography of implant materials on cell behavior in vitro and in vivo

D. M. BRUNETTE

Department of Oral Biology, Faculty of Dentistry
The University of British Columbia
Vancouver, B.C., Canada

19.1 Introduction

There have been two major strategies in the selection of the design of surfaces for dental and orthopedic implants; the first is based on the desirability of obtaining tissue integration through mechanical retention and the second on the principle that art should imitate nature. In the first, tissue integration has often been obtained through the macroscopic as well as the microscopic structure of the implant. An example of the role of macroscopic design is the highly successful implant design introduced by Brånemark (Brånemark et al., 1977), in which the implant is screw-shaped and contains a hole through which bone can grow. Both elements of the design would be expected to lead to mechanical stabilization of the implant in the surrounding bone tissue. At the microscopic level, the Brånemark implant is machined and contains a rich variety of grooves and pits that might also aid implant retention. The development of the microscopic features of the surface has been driven by the availability of materials-fabrication technology that has resulted in a wide variety of porous or textured surfaces being tested, including void metal composites (Karagianes et al., 1976), fibers (Jasty et al., 1993), sintered spheres (Deporter et al., 1990; Pilliar, 1988), and plasma-sprayed (Schroeder et al., 1981) or machined surfaces (Ratner, 1993). Each of these surfaces contains irregularities that would be expected to aid in tissue retention, but the problem is that each is associated with a specific topography and it is difficult to vary shape factors systematically to arrive at an optimal result. Ratner (1993) has characterized this approach to biomaterials as trial-and-error optimization that results from acting on the statement, "let's implant it and see what happens." As has been repeatedly emphasized by Kasemo and co-workers (Kasemo and Lausmaa, 1988), the surface status of a particular implant material may vary widely depending on its preparation and handling history. There is little detailed understanding of the interaction between surfaces and biological tissues to make it possible to predict with accuracy how a particular change in surface properties will affect the in vivo function of the implant; therefore, any new surface-treatment procedure should in principle require new clinical documentation (Kasemo, 1983). But it is hard to acquire such understanding about implant surface topography if the materials used to produce the surfaces are not flexible enough so that factors such as the depth and geometry of features can be varied systematically. The result is that the design of the surface is frozen at the point at which acceptable results are obtained and procedures are applied ritualistically. Thus, at a recent conference of the Academy of Osseointegration, Brånemark – though being able to point with justifiable pride at the success of the implant system that his team had developed – conceded that we still do not know how these implants work; that is, we do not know the biologic mechanisms whereby they produce their effects.

The most direct example of the application of the principle that fabrication art should yield implant surfaces that imitate nature is the porous hydroxyapatite implants that replicate the topography of a marine invertebrate (Porites coral) (Schoenaers et al., 1986). Another example would be the Tübingen implant (Schulte, 1984), which incorporates topographic features modeled on bone lacunae in its Al_2O_3 surface in the hope that cells of osteogenic lineage would find such an environment hospitable for the production of bone. Indeed, despite the widespread success of dental implants that directly attach to bone (that is, osseointegrated), some practitioners developed other systems, such as the blade-vent implant. In this example, the dental implant is attached to alveolar bone through a pseudoperiodontal ligament, on the grounds that the best replacement for a natural tooth is that which approximates nature's principles as closely as possible (Linkow, 1987). More recently, in his presidential address to the Biomaterials Society, Ratner (1993) advocated an engineering approach to achieving biointegration, and he suggested that it is worthwhile to examine how nature handles specificity and rapid-reaction kinetics, noting that three design themes – order, recognition, and mobility – are iterated and exploited throughout biology. However, using principles of design to achieve the optimal structure differs from attempting to reproduce the original structure. The attempts to replicate the original structure, that is, art imitating nature, however, suffer from the dual handicaps of being approximate and pessimistic. Imitation is by necessity approximate, because the complexity and variability of natural systems make them impossible to replicate with perfect fidelity. For example, the pseudoperiodontal ligaments formed on bladevent implants would not be expected to include the epithelial cell rests of Malassez even though these cells have been implicated as being important in maintaining periodontal ligament width. The approach is pessimistic because it assumes that implants designed by man cannot surpass in performance the original structures produced by nature. But biological structures are fabricated under many constraints, and they have some intrinsic limitations; mammals, for example, are limited in their ability to regenerate new organs. It is, therefore, in no way certain that the best replacement is that which most closely resembles the original structure.

Thus, these two approaches, optimization of mechanical retention and imitation of nature, to designing surfaces are not ideal, and they have led to less-than-ideal solutions. In doing so, they illustrate the general principle, enunciated by Ratner, that "existing biomaterials, although demonstrating generally acceptable clinical success, look like dinosaurs poised for extinction in light of the winds of change blowing through the biomedical, biotechnological, and physical sciences" (Ratner, 1993). Ratner has advocated the view that biomaterials should be engineered to produce the desired result rather than simply being based on a trial-and-error optimization approach. The tools Ratner envisioned as playing a role in the production of engineered biomaterials include nano- and micropatterning (Ratner, 1993). In 1983, my laboratory introduced the application of photolithographic techniques to produce surfaces for the study of cell behavior – in particular, contact guidance. This approach and refinements to it have been taken up by a number of laboratories, but it is still safe to say that it has not been exploited to its full potential. In this review, I will illustrate some of the uses of micromachined surfaces to produce surface topographies that affect cell behavior. Substratum surface topography can influence cell behavior on implants in at least seven ways, including cell adhesion, cell selection, mechanical interlocking, orientation and topographic (also known as contact) guidance, tissue organization, cell shape, and production of local microenvironments. These aspects will be considered separately, although they often interact.

19.2 Cell adhesion

Most experiments investigating the relationship between substratum surface topography and cell adhesion have used traditional materials-processing methods, such as electropolishing, polishing with various size grits, grit blasting, or machining, to produce the surfaces. An important parameter in determining cell adhesion is the total area available for cell attachment. Establishing the area for many of the surfaces already mentioned is problematic, because the area is both difficult to calculate (because the surface is irregular) and difficult to measure directly (Bowers et al., 1992). Thus the most appropriate data, the number of cells attached per unit area, cannot be readily obtained. Nevertheless, it is clear that the effects of substratum surface topography on cell adhesion depend on cell type. Osteoblast-like cells demonstrated significantly higher levels of cell attachment on rough sandblasted surfaces with irregular morphologies than they did on smooth surfaces (Bowers et al., 1992). In contrast, more human gingival fibroblasts attached to electropolished surfaces than did to etched or sandblasted surfaces (Konnonen et al., 1992). Micromachined surfaces have also been compared to smooth surfaces for their ability to support cell attachment, and the regular geometry of micromachined surfaces enables the surface areas of the grooved surfaces to be calculated. More oral epithelial cells attached to a surface with V-shaped grooves produced by micromachining than to a smooth control surface. The relationship was true whether the surface attaching the cells was titanium or epoxy. Moreover, it was found that the increase in epithelial-cell attachment on grooved surfaces was not simply the result of the grooved surfaces' having a greater surface area (Chehroudi et al., 1988, 1989, 1990). The flexibility of the micromachining process could be further exploited in studies of cell adhesion by examining systematically the effects of groove size, shape, and spacing, or by investigating pits or dolines and their spacing.

19.3 Cell selection

Differences in cell adhesion could be used for the selection of specific cell populations. For example, if osteoblasts prefer roughened surfaces to smooth ones and fibroblasts prefer smooth surfaces to roughened ones, then a roughened surface on an implant surface might be used to select preferentially for osteoblasts. Thus, the roughened surface might be placed on locations on the implant where osseointegration was desired. Conversely, a smooth surface might be applied in other locations to enhance fibroblast adhesion. There is a need, however, to consider all the cell types that might be in contact with the device. A potential problem for dental implants and other devices that penetrate a stratified squamous epithelium is the downgrowth of the epithelium leading to extrusion of the device. Thus, the surface intended to contact fibroblasts should not be so attractive to epithelium that the epithelium displaces fibroblasts and migrates down the implant. One possible problem with roughened surfaces in the region in which the cells contact bone might be that such surfaces have the potential to attract osteoclasts or their monocyte precursors that would resorb bone. Evidence for such a possibility is the pioneering study of Rich and Harris (1981), who found that fibroblasts shunned rough surfaces and preferred smooth ones. Macrophages, however, preferred rough surfaces to smooth ones, a behavior described as "rugophilia" in contrast to the fibroblasts' "rugophobia." Moreover, Murray et al. (1989) demonstrated that rough surfaces on a variety of polymers, glass, and stainless steel encouraged macrophages to release mediators that stimulate bone resorption. Although some studies do not support the concept of any detrimental effect of rough or textured surfaces placed in bone (Buser et al., 1990; Meffert et al., 1987; Thomas and Cook, 1985), one recent report (Unwin and Stiles, 1993) concluded that the majority of roughened titanium-alloy femoral implants produced endosteal lysis. Thus, there is at least the possibility that the use of a roughened surface to select for osteoblasts might promote a gain in osseointegration in the short term, but produce a long-term loss if the surface subsequently favored the selection of bone-resorptive cells. There is also the possibility, indeed perhaps the likelihood, that not all roughened surfaces have the same effects on osteoblasts and osteoclasts, and that a surface could be designed that would not attract both types of cell. Micromachining would be an obvious approach to obtaining such a surface, because surface topography can be precisely controlled and systematically varied until optimal results are obtained.

Another means whereby cells can be specifically selected is through the simple mechanism of having the surface contain pores that exclude structures above a certain dimension. Chvapil et al. (1984), for example, found that the decisive factor for tissue permeation of a collagen-based sponge was that the size of the pores needed to exceed 100 μm. More compact preparations were simply encapsulated (Chvapil et al., 1984). Micromachining, which can produce a variety of sizes of pits or pores, could also be used to select cells or groups of cells by size exclusion.

Finally, Dow et al. (1987) have shown that micropatterning can be used to produce cell traps, that is, surfaces to which cells attach but from which they cannot move. An example of a cell trap is an etched spiral; cells near the center of the spiral are unable to spread completely and they are unable to move. Because different cell populations might differ in the spreading required to locomote, cell traps might be designed to collect specific cell populations.

In summary, micromachining might be used to select specific cell populations by varying surface roughness, spreading area, and porosity.

19.4 Mechanical interlocking

Tissue attachment is a prerequisite for the success of some types of implants; therefore, many surfaces have been developed for implants that are intended to produce mechanical interlocking. For example, Pilliar and colleagues (Deporter et al., 1990; Pilliar, 1985, 1988) have obtained porous (50–200-μm pore size) surfaces on implants by sintering Ti-alloy spheres onto a solid core. These surfaces have the major advantage that they can resist forces that might remove the implant regardless of the direction in which the force is applied. Such mechanical stability is important because even small, so-called micromotions can have profound biological effects. For example, micromotion at a dental implant–tissue interface can adversely affect osseointegration, resulting in a fibrous tissue interface (Pilliar et al., 1986). The size of the irregularities that might be important in the mechanical stabilization of osseointegrated implants has been estimated to be as small as 1 μm (Kasemo, 1983). Similarly, mechanical interlocking may be of importance in determining the fate of implants attached to soft tissue, because it appears that interfacial friction between the implant and adjacent tissues is a determining factor in a continued inflammatory response to such implants. In soft tissue, however, it might be expected that mechanical properties of the cell–implant interaction may be of more importance in stabilizing the device than is the case with mineralized tissue, where the extracellular components probably dominate. Campbell and Von Recum (1989) have systematically examined microtopography and soft-tissue response using a filter material with average pore diameters between 0.42 and 14 μm. They concluded that a defined surface topography of 1–2 μm appears to allow direct fibroblast attachment to the surface, independent of its chemical or electrochemical nature.

It is thus of interest to examine how cell processes interact with implant surfaces at a detailed level. Scanning electron microscopy of fibroblasts (Brunette, 1986b; Meyle et al., 1993) and epithelial cells (Brunette, 1986c) growing on micromachined surfaces has clearly demonstrated that cell processes extend into the grooves, leading to extensive contact, and could possibly produce significant mechanical interlocking. Meyle et al. (1993) have speculated that the intimate interdigitation of the cell body with surface contours could increase the shear stress required to detach cells from a surface up to the point where cell rupture occurs. Such observations may explain the higher number of cells attached to roughened surfaces noted in some in vitro studies. In most assays of cell attachment, an attached cell is arbitrarily and operationally defined according to the shear force it must resist to avoid being dislodged (Grinnel, 1978). Mechanical interlocking of cells and surface would be expected to help the cells resist detachment. Micromachining has the potential to produce surfaces that utilize this effect.

19.5 Topographic (or contact) guidance

The most extensive application of micromachined surfaces to biologic problems at present has been their use in studies of topographic guidance (Brunette, 1986a, 1986b; Clark et al., 1987, 1992). Topographic (also called contact) guidance refers to the tendency of cells to be guided in their direction of locomotion by the shape of the substratum. In general, moving fibroblasts in vitro are fan-shaped with sheet-like protrusions termed lamellae that extend from the cell surface and also attach to the substratum. Time-lapse observations of locomoting cells indicate that the cells move in the direction of the largest lamella, which is termed the leading lamella. The consistency of this phenomenon, at least on two-dimensional surfaces, enables one to infer the direction of locomotion from cell orientation. In practice, contact guidance is inferred as being operative if a greater proportion of cells aligns with a topographic feature than would be predicted by chance.

Topographic guidance was one of the first phenomena observed in cell culture (Harrison, 1914); it was systematically and extensively studied by Weiss, who coined the name "contact guidance" and used substrata as diverse as plasma clots, fish scales, and engraved glass (Weiss, 1959; Weiss and Garber, 1952; Weiss and Taylor, 1956). Detailed studies on the relationship between surface topography and cell behavior were made possible by the introduction of the use of microfabrication and, more recently, nanofabrication methods that enable precise control over topographic features to be obtained (Brunette 1986a, 1986b; Brunette et al., 1983; Clark et al., 1990, 1991; Dunn and Brown, 1986; Meyle et al., 1993; Wood, 1988). Topographic guidance is a very robust phenomenon that can be exhibited by diverse cell types on diverse surfaces, including glass and silica (Dunn, 1982; Dunn and Heath, 1976), polymers such as polyvinyl chloride (Rovensky and Slavnaya, 1974), epoxy (Chehroudi et al., 1988), araldite (Meyle et al., 1993), polystyrene (Ohara and Buck, 1979), polycarbonate (Clark et al., 1990, 1991), metals such as titanium and titanium alloys (Brunette, 1988; Chehroudi et al., 1988, 1990; Inoue et al., 1987), extracellular matrices such as oriented collagen fibers (Dunn and Ebendal, 1978), extracellular-matrix fibrils from amphibian gastrulae (Nakatsuji and Johnson, 1984), fractured enamel (Jones and Boyde, 1979), and fish scales (Weiss and Taylor, 1956).

The ability of micromachined surfaces to produce cell orientation and contact guidance in vitro is illustrated in Figure 19.1, in which truncated V-shaped grooves clearly produced alignment of fibroblasts and epithelial cells. Other shapes effective in producing guidance include rectangular profiles (Brunette, 1986b; Clark et al., 1990, 1991; Meyle et al., 1993), cylindrical fibers (Curtis and Varde, 1964), and hemispherical grooves and ridges (Rovensky and Slavnaya, 1974). Even very fine

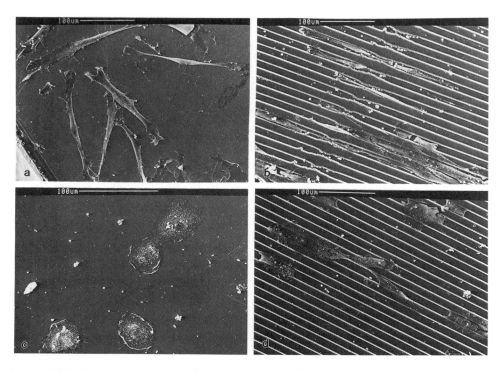

Figure 19.1. Scanning electron micrographs of (a and b) human gingival fibroblasts and (c and d) porcine epithelium derived from the cell of Malassez on smooth (a and c) and grooved (b and d) surfaces. Both types of cell elongate and orient in the direction of the grooves.

structures down to the dimensions of extracellular-matrix fibrils (≈130 nm) can produce alignment; surprisingly, however, the proportion of cells aligned, as well as the length of the cells, depends markedly on groove depth, with deeper grooves producing larger effects. Groove depth and spacing interact so that the more close-ly spaced the topographic cues, the more the cells will react to features of a given depth (Clark et al., 1991). A similar conclusion was reached by Weiss (1945) one-half century ago on the basis of his studies of cell alignment on scratched mica plates, where he noted that the deeper the scratches, the more marked the align-ment. Microfabrication techniques enable the limits and characterization of topo-graphic effects to be determined precisely.

Micromachined surfaces have also been used to investigate the mechanism of contact guidance, because they offer the possibility of producing uniform surfaces with precisely defined topographies. My laboratory has been interested in the intracellular distribution of cytoskeletal elements such as microtubules, microfil-aments, intermediate filaments, and adhesive structures of cells cultured on grooved surfaces, because the cytoskeletal and adhesive structures have been implicated in theories suggesting mechanisms for topographic guidance (Dunn and Heath, 1976; Ohara and Buck, 1979). Figure 19.2 shows the distribution of actin and microtubules in trypsinized cells spreading on a grooved titanium-coat-ed surface. Interestingly, it is found that the various filaments segregate preferen-tially with time into subcellular locations defined by the substratum. Microtubules are the first cytoskeletal element to reform and align, and they are first observed in the deepest portion of the groove. In contrast, actin-containing microfilaments form later and are typically first located close to the groove-wall/ridge edge (Oakley and Brunette, 1993). The ability of grooved surfaces to control cytoskele-ton distribution probably accounts for their marked effects on cell shape.

The flexibility of microfabrication processes enables investigators to set chal-lenges to cells by exposing them to topographic cues that point the cells in different directions. For example, sets of grooves can be placed at right angles to each other. Culture of human gingival explants on such surfaces demonstrated that the direction of actively migrating cell outgrowths was changed when the cells encountered the grooves running in a transverse direction. Figure 19.3 illustrates the behavior of cells close to the intersection of two sets of grooves oriented at right angles to each other. Some of the cells develop two leading lamellae migrating at right angles so that the cell body becomes oriented at a 45° angle to both patterns. Such observations indi-cate that the control of cell migration by topographic cues appears to be localized to the leading lamellae and need not involve microfilament bundles that traverse the cell from the lamellae to a perinuclear location, as has been suggested in an early theory of contact guidance. Another means of exposing the cells to opposing directional cues is to etch grooves in silicon wafers with a {1, 1, 0} orientation, which can pro-duce surfaces with major grooves defined by the photolithographic technique and minor grooves at a 55° angle to the major grooves. When cells are cultured on such

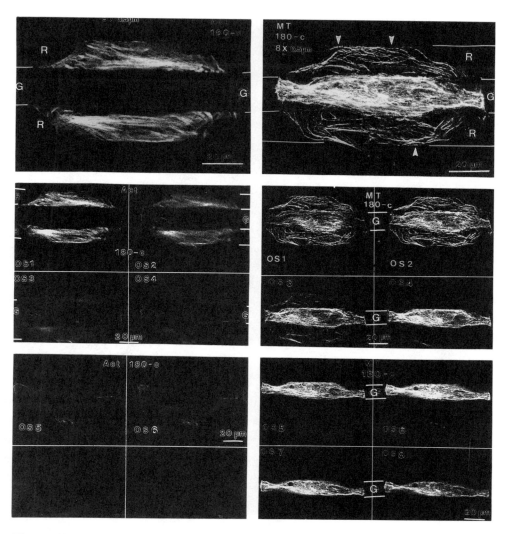

Figure 19.2. Confocal microscopy of human gingival fibroblasts that had spread for 180 min on a grooved (G) substrata. The two images at the top of the figure are composites comprising all the optical slices. The lower images show optical slices (OS) at various levels. Actin and microtubules show strikingly different spatial distributions, with microtubules being found in the deepest portion of the grooves (OS7 and S1) and actin being prominent at the ridge level.

surfaces, it is found that they are guided by the larger grooves, following the minor grooves only when the larger grooves are not present (Brunette, 1986a).

The studies cited were done in vitro where the cell's primary option is to interact with the substratum on its ventral surface. In vivo, cells have the opportunity to make contact with surfaces in all three dimensions, and in such circumstances cell behavior can differ. For example, Noble (1987) devised a method for studying the locomotion of cells in a three-dimensional hydrated collagen lattice. In the

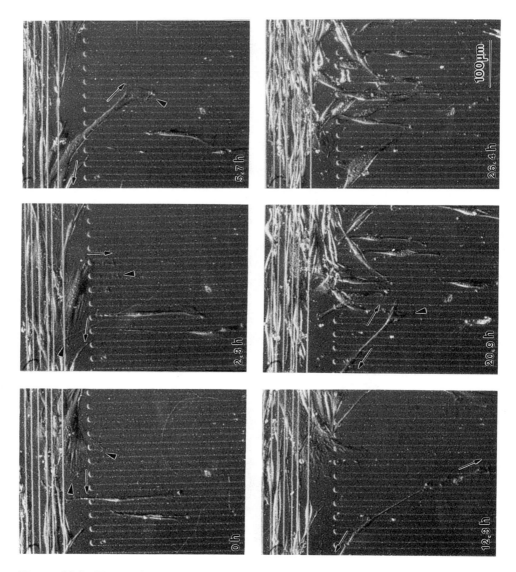

Figure 19.3. Frames from a time-lapse cinefilm illustrating the behavior of human gingival fibroblasts moving on substrata with cues directing the cells to move in different directions. At time 0, fibroblast migration is in the direction indicated by the arrow that is that specified by the top set of grooves. At 2.3 h this fibroblast has established another lamella on the perpendicularly oriented bottom set of grooves. The two lamellae migrate at a 90° to each other, stretching the cell body between them at a roughly 45° angle.

collagen gel, the broad leading lamella that is prominent on two-dimensional surfaces was rarely, if at all, seen. It is thus by no means a given that topographic guidance would also occur in vivo, but the migration of mesenchyme cells in the transparent, developing fin paddle of the killifish *Aphyosemion scheeli* provides the opportunity of examining the possibility. Thorogood and Wood (1987) observed the migrating mesenchymal cells using Nomarski interference contrast microscopy and found that they did migrate along the collagen fibrils within the fin. However, the cells did not produce a broad lamellipodium but rather were led by a number of aligned filopodia. Although these observations are compatible with contact guidance's being operative in vivo during development, it is difficult to be certain because the environment contains a multitude of possible directional cues, such as differential adhesiveness, chemotaxis, or simple exclusion of the cells by the physical properties of the extracellular matrix.

Another approach to determining whether topographic guidance can occur in vivo is to implant artificial surfaces with varying surface topographies percutaneously. The normal response of the epithelium (Winter, 1974) would be to attempt to exclude such an implant by migrating downward. If topographic guidance were operative, it would be expected that placing horizontal grooves on the implant would impede this vertical downgrowth relative to a smooth surface. Conversely, placing vertical grooves on the implant would be expected to speed vertical downgrowth. Both these predictions have been confirmed experimentally (Chehroudi et al., 1990). Figure 19.4 shows epithelial cells interdigitating into the horizontal grooves of a percutaneous device that was implanted in a rat. Epithelial downgrowth was inhibited on such a surface. Some data comparing vertical and horizontal grooves are given for various parameters in Figure 19.5. Recession (a direct measure of downgrowth) was inhibited, and epithelial attachment was found to vary in the expected order of vertical grooves > smooth > horizontal grooves. The length of epithelial attachment, also

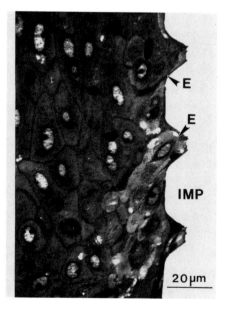

Figure 19.4. Epithelial cells interdigitating into truncated V-shaped grooves on a titanium-coated percutaneous implant. (E = epithelium; Imp = implant)

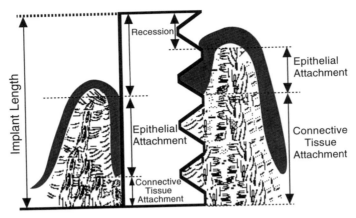

Figure 19.5a. Illustration of definition of measurements of epithelial- and connective-tissue response to a percutaneous implant. Figure also illustrates the inhibition of epithelial downgrowth caused by horizontally oriented grooves.

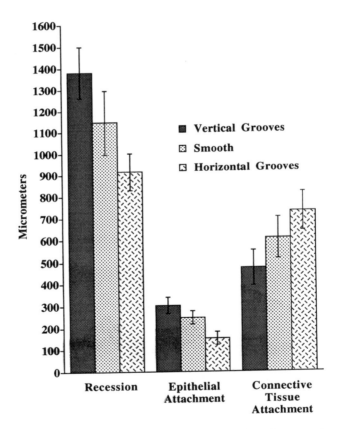

Figure 19.5b. Measurements of recession and epithelial- and connective-tissue attachment on percutaneous implants in place for 7 days. Data indicate that epithelial downgrowth is inhibited on horizontally oriented grooved surfaces and increased on longitudinally oriented grooved surfaces. Data from Cheroudi et al. (1990).

as expected, followed the same order. The length of connective-tissue attachment, however, was in the reverse order, as expected, because the inhibition of downgrowth allowed for the preservation of greater connective-tissue attachment. It thus appears highly likely that topographic guidance can occur on micromachined artificial substrata in vivo. Such micromachined grooved surfaces may have a clinical application in improving the performance of percutaneous devices, because they have the ability to inhibit directly (by means of topographic guidance) the downgrowth of epithelium, which has been a major problem with these devices.

19.6 Tissue organization

Micromachined grooved surfaces also appear to have the ability to effect the organization of tissues adjacent to implants. For example, the connective tissue in contact with V-shaped grooves on a titanium-coated surface becomes oriented so that the long axis of the fibroblasts is at an oblique angle to the long axis of a percutaneous implant (Figure 19.6). Such behavior stands in contradistinction to the more usual phenomenon, in which fibroblasts are aligned parallel to the surface and form a connective-tissue capsule whose function is generally regarded as being to wall off the implant from surrounding tissue. Thus, the micromachined surface organizes and integrates with adjacent tissue rather than being separated from it. The probable basis for the behavior is a phenomenon, also extensively studied by Weiss (1959, which he called the two-center effect. Weiss believed that cells exerted tension on the fibrin fibers found in his explant cultures, bringing the fibers into alignment

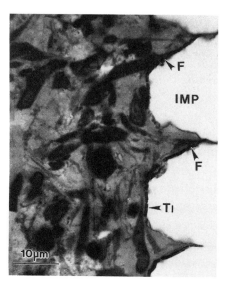

Figure 19.6. Fibroblasts (F) inserting into the grooves of a titanium-coated epoxy percutaneous implant in place for 7 days. (Imp = implant; Ti = titanium)

between the explants. Subsequently, cells migrated along the fibers, eventually forming a densely populated cellular bridge connecting the two explants. The ability of cells to exert force against their substratum has since been studied extensively in an elegant series of experiments by Harris and co-workers (Harris, 1982; Harris et al., 1981). They have speculated that the contraction of collagen networks could be involved in morphogenic functions and may be a cause of the apparent contraction of collagen networks around wounds, burns, and surgically implanted prostheses. It thus appears that cell traction can result in tissue organization, in particular into ligament-like collagen-containing structures. The virtue of placing micromachined grooved surfaces on implants is that they can orient the cells so that the implant interacts with the tissue rather than being excluded from it. Once again, the surfaces that have been used so far to produce tissue organization have been simple ones such as grooves or pits; even within this limited class of surface topographies, only a restricted range of dimensions has been employed.

19.7 Microenvironment

Stoker et al. (1990) have introduced the term "designer microenvironments" to describe the purposeful alteration of conditions so that desired cell functions can be maintained and studied. A typical in vitro application of the concept would be the use of extracellular-matrix components, such as basement-membrane matrices, to promote epithelial-cell differentiation (Stoker et al., 1990). However, the topography of cell interactions can also markedly alter cell function. For example, the subcapsule of the thymus contains microenvironments comprised of "baskets" of epithelial cells filled with thymocytes (van Ewijk, 1991). It appears that thymocytes, when differentiating, sequentially enter different microenvironments. A specialized microenvironment may also explain the results obtained by Selye and co-workers (Seyle, 1962; Selye et al., 1960), who implanted glass cylinders of various shapes into rats. Bone and cartilage formation occurred in these implants, but only in some sizes and shapes of cylinders. A microenvironment favoring differentiation could be the result of cells conditioning their environment by the release of cytokines or extracellular-matrix molecules and the products being more concentrated in the local area, or simply the specific geometry's facilitating cell–cell interactions, or other mechanisms. Whatever the molecular mechanisms, microfabrication methods enable the construction of topographies of precisely defined dimensions that might produce specific microenvironments. We have observed that mineralized bonelike tissue nodules are formed on some micromachined surfaces when these surfaces are implanted subcutaneously in rats. Figure 19.7 shows an example of such nodules. The interesting aspect of the mineralized tissue formation is that, on a micromachined pit, for example, the mineralization appears to start toward the center of the pit well away from the walls or bottom. Thus it appears that immediate surface interactions are less likely to be involved than is the creation of a microenvironment. Bone-like tissue production can

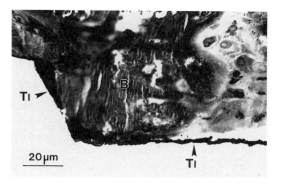

Figure 19.7. Mineralized bonelike tissue nodules (B) formed on titanium-coated (Ti) epoxy implants placed subcutaneously for 63 days.

also occur in vitro. Cell populations of rat calvarial cells also form nodules of bonelike tissue in vitro. The initial stages of the process are shown in Figure 19.8, in which small calcified nodules of mineralized material are formed on the micromachined surface. These observations are similar to those of Davies and Lowenberg and colleagues, who have studied similar cell populations on machined Ti surfaces (Lowenberg et al., 1991). The frequency of formation of bonelike nodules appears to be higher on grooved and pitted surfaces than on smooth surfaces.

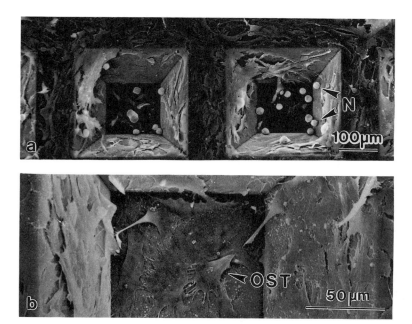

Figure 19.8. Osteoblast-like (OST) cells derived from rat newborn calvaria cultured on a titanium-coated surface with large pits produced by micromachining. Small globules or nodules (N) of calcified material are deposited on the surface.

19.8 Cell shape

A current theme in cell biology is the relationship between cell shape and cell function. Cell shape can regulate cell growth (Folkman and Moscona, 1978), cytoskeleton gene expression (Ben-Ze'ev, 1987), collagenase and stromelysin gene expression (Werb et al., 1986), radiation-induced DNA unwinding (Olive and MacPhail, 1992), extracellular-matrix metabolism (McDonald, 1989), and differentiation (Watt et al., 1988). Surface topography can markedly alter cell shape in vitro, an effect that was noted for grooved surfaces in the early observations of Weiss and analyzed at a sophisticated mathematical level for cells on micromachined surfaces by Dunn and Brown (1986). However, anchorage-dependent cells growing in vitro have little option but to attach to their substratum, and the question arises whether surface topography of implants affects cell shape in vivo, where cells can attach to other cells as well as to the extracellular matrix. One approach to examining the shape of cells attached to implants in vivo is to obtain serial sections, digitize the cell outlines, and use computer techniques to reconstruct the shape of the cells. This has been done for cells attached to percutaneous implants with smooth or micromachined grooved surfaces (Figure 19.9). The technique is time consuming and tedious, so very few cells have been examined at present; preliminary results indicate, however, that the shape and orienta-

Figure 19.9. Computer-assisted three-dimensional reconstructions of some epithelial cells adjacent to a percutaneous implant.

tion of epithelial cells adjacent to grooved implant surfaces differ significantly from those of cells attached to smooth surfaces. In particular, the cells adjacent to grooved surface bulge away from the surface more than do those on smooth surfaces, and they tend to be oriented at an angle to the grooved surface rather than flattened against it. The further question arises whether such shape changes are sufficient to alter cell function.

The answer to this question for cells in vivo is not known, but preliminary experiments of cell behavior in vitro indicate that cell metabolism can be affected at the very basic level of mRNA levels and metabolism. Figure 19.10 shows the results of an experiment in which the mRNA levels for the major cell-adhesive protein, fibronectin, as determined by Northern hybridization, is compared for cells growing on smooth or micromachined grooved surfaces. A clear difference in fibronectin mRNA levels in the two cultures is evident. Differences can also be demonstrated for proteinase secretion; an oral-epithelial cell line secreted more tissue-plasminogen activator when cultured on a grooved surface than when cultured on a smooth control surface (Hong and Brunette, 1987). From the point of view of designing surface topographies for implant function, this result may be considered as good news/bad news. The good news is that it seems possible to select specific topographies that produce a cell shape that enhances the production of a specific protein. The bad news is that the secretion of other proteins might

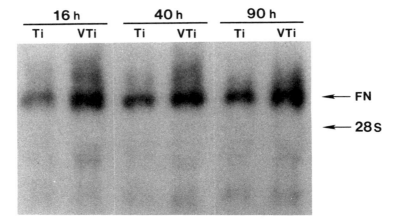

Figure 19.10. Northern-hybridization analysis of fibronectin gene expression of the cells cultured on grooved and smooth surfaces. Fibronectin (FN) mRNA levels of human gingival fibroblasts cultured on grooved titanium-coated surfaces (VTi) and smooth titanium-coated surfaces (Ti) for 16, 40, and 90 h were determined by Northern hybridization (n = 3). Total cellular RNA was fractionated on 1.2% (w/v) agarose, 2.2 M formaldehyde gels, transferred onto a nylon membrane, and hybridized with [^{32}P]dCTP-labeled cDNA fragments of human fibronectin. Fibronectin mRNA was identified at the bands of 7.9 kb in size. Equal sample loading was confirmed by the density of 28S ribosomal bands stained by ethidium bromide in gels before transfer. The duration of film exposure was 6 h.

also be enhanced, and some of these (such as the proteinases, which degrade connective tissue) might have deleterious effects on implant integration. Thus, considered at the molecular level, the regulation of cell function by altering surface topography could be a very complex affair.

19.9 Conclusion

This brief review indicates only a few of the possible uses of microfabrication methods in the study of the effects of topography on cell behavior in vitro and in vivo. A major advantage of microfabrication techniques is their versatility in producing surfaces with varying shapes and their close control over dimensions. Yet this advantage has not been exploited to any great degree; as these studies illustrate, only relatively simple topographies such as grooves and pits have been fabricated, and often these have been constructed over only a limited range of dimensions. Moreover, only a few biological responses have been examined, even though dramatic effects have often been obtained. Thus there is a good possibility that topographic control of cell responses will be an effective tool in the bioengineering of implants.

References

Ben-Ze'ev, A. (1987) The role of changes in cell shape and contacts in the regulation of cytoskeleton expression during differentiation. *J. Cell Sci. Suppl.* 8:293–312.

Bowers, K. T., Keller, J. C., Randolph, B. A., Wick, D. C., and Michaels, C. M. (1992) Optimization of surface micromorphology for enhanced osteoblast responses in vitro. *Int. J. Oral Maxillofac. Imp.* 7(3):302–310.

Brånemark, P.-I., Hansson, B.O., Adell, R., Breine, U., Lindström, J., Hallen and Ohman, A. (1977) *Osseointegrated Implants in the Treatment of the Edentulous Jaw: Experience from a 10-year Period.* Stockholm: Almqvist Wiksell International.

Brunette, D. M. (1986a) Fibroblasts on micromachined substrata orient hierarchically to grooves of different dimensions. *Exp. Cell Res.* 164:11–26.

Brunette, D. M. (1986b) Spreading and orientation of epithelial cells on grooved substrata, *Exp. Cell Res.* 167:203–217.

Brunette, D. M. (1988) The effects of implant surface topography on the behavior of cells. *Int. J. Oral Maxillofac. Imp.* 3:231–246.

Brunette, D. M., Kenner, G. S., and Gould, T. R. L. (1983) Grooved titanium surfaces orient growth and migration of cells from human gingival explants. *J. Dent. Res.* 62(10):1045–1048.

Buser, D., Weber, H.P., and Lang, N.P. (1990).Tissue integration of non-submerged implants. *Clin. Oral Implants Res.* 1:33–40.

Campbell, C. E. and Von Recum, A. F. (1989) Microtopography and soft tissue response. *J. Invest. Surg.* 2:51–74.

Chehroudi, B., Gould, T. R. L., and Brunette, D. M. (1988) Effects of grooved epoxy substratum on epithelial cell behavior *in vitro* and *in vivo. Biomed. Mater. Res* 2:459–473.

Chehroudi, B., Gould, T. R. L., and Brunette, D. M. (1989) Effects of grooved titanium-coated implant surface on epithelial cell behavior in vitro and in vivo. *J. Biomed. Mater. Res.* 23:1067–1085.

Chehroudi, B., Gould, T. R. L., and Brunette, D. M. (1990) Titanium-coated micromachined grooves of different dimensions affect epithelial and connective-tissue cells differently in vivo. *J. Biomed. Mater. Res.* 24(9):1203–1219.

Chvapil, M., Holusa, R., Kilment, K., and Sttoll, M. (1984) Some chemical and biological characteristics of a new collagen-polymer compound material. *J. Biomed. Mater. Res.* 18:323–336.

Clark, P., Connolly, P., Curtis, A. S. G., Dow, J. A. T., and Wilkinson, C. D. W. (1987) Topographical control of cell behavior: I. Simple step cases. *Development* 99:439–448.

Clark, P., Connolly, P., Curtis, A. S. G., Dow, J. A. T., and Wilkinson, C. D. W. (1990) Topographical control of cell behavior: II. Multiple grooved substrata. *Development* 108:635–644.

Clark, P., Connolly, P., Curtis, A. S. G., Dow, J. A. T., and Wilkinson, C. D. W. (1991) Cell guidance by ultrafine topography in vitro. *J. Cell Sci.* 99:73–77.

Clark, P., Connolly, P., and Moores, G. R. (1992) Cell guidance by micropatterned adhesiveness in vitro. *J. Cell Sci.* 103:287–292.

Curtis, A. S. G. and Varde, M. (1964) Control of cell behavior: Topological factors. *J. Natl. Cancer Inst.* 33:15–26.

Deporter, D. A., Watson, P. A., Pilliar, R. M., Chipman, M. L., and Valiquette, N. (1990) A histological comparison in the dog of porous-coated vs. threaded dental implants. *J. Dent. Res.* 69(5):1138–1145.

Dow, J. A. T., Clark, P., Connolly, P., Curtis, A. S. G., and Wilkinson, C. D. W. (1987) Novel methods for the guidance and monitoring of single cells and simple networks in culture. *J. Cell Sci. Suppl.* 8:55–79.

Dunn, G. A. (1982). Contact guidance of cultured tissue cells: a survey of potentially relevant properties of the substratum. In *Cell Behaviour*, ed. R. Bellairs, A. S. G. Curtis, and G. Dunn, pp. 247–280. London: Cambridge University Press.

Dunn, G. A. and Brown, A. F. (1986) Alignment of fibroblasts on grooved surfaces described by a simple geometric transformation. *J. Cell Sci.* 83:313–340.

Dunn, G. A. and Ebendal, T. (1978) Contact guidance on oriented collagen gels. *Exp. Cell Res.* 111:475–479.

Dunn, G.A. and Heath, J. P. (1976) A new hypothesis of contact guidance in tissue cells. *Exp. Cell Res.* 101:1–1014.

Folkman, J. and Moscona, A. (1978) Role of cell shape in growth control. *Nature* 273:345–349.

Grinnel, F. (1978) Cellular adhesiveness and extracellular substrata. *Int. Rev. Cytol.*:65–174.

Harris, A. K. (1982) Traction, and its relation to contractions in tissue cell locomotion. In *Cell Behaviour*, ed. R. Bellairs, A. S. G. Curtis, and G. Dunn, pp. 109–134. London: Cambridge University Press.

Harris, A. K., Stopak, D., and Wild, P. (1981) Fibroblast traction as a mechanism for collagen morphogenesis. *Nature* 290(5803):249–251.

Harrison, R. G. (1914) The reaction of embryonic cells to solid structures. *J. Exp. Zool.* 14:521–544.

Hong, H. L. and Brunette, D. M. (1987) Effect of cell shape on proteinase secretion. *J. Cell Sci.* 87:259–267.

Inoue, T., Cox, J. E., Pilliar, R. M., and Melcher, A. H. (1987) Effects of the surface geometry of smooth and porous-coated titanium alloy on the orientation of fibroblasts in vitro. *J. Biomed. Mater. Res.* 21:107–126.

Jasty, M., Bragdon, C. R., Haire, T., Mulroy, R. D., and Harris, W. H. (1993) Comparison of bone ingrowth into cobalt chrome sphere and titanium fiber mesh porous coated cementless canine acetabular components. *J. Biomed. Mater. Res.* 27:639–644.

Jones, S. J. and Boyde (1979) Colonization of various natural substrates by osteoblasts in vitro. *Scann. Electron Microscopy* 11:529–538.

Karagianes, M. T., Westerman, R. E., Rasmussen, J. J., and Lodmell, A. M. (1976) Development and evaluation of porous dental implants in miniature swine. *J. Dent. Res.* 55:85–93.

Kasemo, B. (1983) Biocompatibility of titanium implants: surface science aspects. *J. Prosthet. Dent.* 49(6):832–837.

Kasemo, B. and Lausmaa, J. (1988) Biomaterial and implant surfaces: on the role of cleanliness, contamination, and preparation procedures. *J. Biomed. Mater. Res.* 22: 145–158.

Konnonen, M., Hormia, M., Kivilahti, J., Hautaniemi, J., and Thesleff, I. (1992) Effect of surface processing on the attachment, orientation, and proliferation of human gingival fibroblasts on titanium. *J. Biomed. Mater. Res.* 26.

Linkow, L. (1987) *Without Dentures: The Miracle of Dental Implants.* Montreal: Book Centre.

Lowenberg, B., Chernecky, R., Shiga, A., and Davies, J. E. (1991) Mineralized matrix production by osteoblasts on solid titanium *in vitro*. *Cells Mat.* 1(2):177–187.

McDonald, J. A. (1989) Matrix regulation of cell shape and gene expression. *Cur. Opin. Cell Biol.* 1:995–999.

Meffert, R. M., Block, M. S., and Kent, N. J. (1987) What is osseointegration? *Int. J. Periodont. Res. Dent.* 4:9–21.

Meyle, J., Gultig, K., Wolburg, H., and Von Recum, A. F. (1993) Fibroblast anchorage to microtextured surfaces. *J. Biomed. Mater. Res.* 27:1553–1557.

Murray, D. W., Rae, T., and Rushton, N. (1989) The influence of the surface energy and roughness of implants on bone resorption. *J. Bone Joint Surg.* 71-B:632–637.

Nakatsuji, N. and Johnson, K. E. (1984) Experimental manipulation of a contact guidance system in amphibian gastrulation by mechanical tension. *Nature* 307:453–455.

Noble, P. B. (1987) Extracellular matrix and cell migration: Locomotion characteristics of MOS-11 cells within a three-dimensional hydrated collagen lattice. *J. Cell Sci.* 87:241–248.

Oakley, C. and Brunette, D.M. (1993) The sequence of alignment of microtubules, focal contacts, and actin filaments in fibroblasts spreading on smooth and grooved titanium substrata. *J. Cell Sci.* 106:343–354.

Ohara, P. T. and Buck, R. C. (1979) Contact guidance in vitro. *Exp. Cell Res.* 121:235–249.

Olive, P. L. and MacPhail, S. H. (1992) Radiation-induced DNA unwinding is influenced by cell shape and trypsin. *Rad. Res.* 130:241–248.

Pilliar, R. M. (1985) Implant stabilization by tissue ingrowth. In *Tissue Integration in Oral and Maxillofacial Reconstruction*, ed. D. van Steenberghe, pp. 60–76. Proceedings of an International Congress, Brussels.

Pilliar, R. M. (1988) Tissue integration in oral and maxillo-facial construction. In *Experta Med.*, ed. D. van Steenberghe, pp. 247–280.

Pilliar, R. M., Lee, J. M., and Maniatopoulos (1986) Observations of the effects of movement on bone ingrowth into porous-surfaced implants. *Clin. Orthop.* 208:108–113.

Ratner, B. D. (1993) New ideas in biomaterials science—a path to engineered biomaterials. *J. Biomed. Mat. Res.* 27:837–850.

Rich, A. M. and Harris, A. K. (1981) Anomalous preferences of cultured macrophages for hydrophobic and roughened substrata. *J. Cell Sci.* 50:1–7.

Rovensky, Y. A. and Slavnaya, I. L. (1974) Spreading of fibroblast-like cells on grooved surfaces. *Exp. Cell Res.* 84:199–206.

Schoenaers, J. H., Holmes, R. E., Finn, R. A., and Bell, W. H. (1986) Healing in interconnected porous hydroxyapatite blocks: a long-term histological and histomorphometric analysis. In *Tissue Integration in Oral and Maxillo-facial Reconstruction* (Excerpta Medica), pp. 96–100. Amsterdam: Correct Clinical Practice no. 29.

Schroeder, A., van der Zypen, E., Stich, H., and Sutter, R. (1981). The reactions of bone, connective tissue, and epithelium to endosteal implants with titanium-sprayed surfaces. *J. Maxillofacial Surg.* 9:15–25.

Schulte, W. (1984) The intra-osseous AI2O3 (Frialit) Tübingen implant. Developmental status after eight years (I). *Oral Surg.* 1:9–26.

Selye, H. (1962) The dermatologic implications of stress and calciphylaxis. *J. Inves. Dermatol.* 39:259–275.

Selye, H., Lemire, Y., and Bajusz, E. (1960) Induction of bone, cartilage and hemopoietic tissue by subcutaneously-implanted tissue diaphragms. *Roux' Archiv für Entwicklungsmechanik* 151:572–585.

Stoker, A. W., Streuli, C. H., Martins-Green, M., and Bissell, M. J. (1990) Designer microenvironments for the analysis of cell and tissue function. *Curr. Opin. Cell Biol.* 2:864–874.

Thomas, K. A. and Cook, S. D. (1985) An evaluation of variables influencing implant fixation by direct bone apposition. *J. Biomed. Mater. Res.* 19:875–901.

Thorogood, P. and Wood, A. (1987) Analysis of in vitro cell movement using transparent tissue systems. *J. Cell Suppl.* 8:395–413.

Unwin, A. J. and Stiles, P. J. (1993) Early failure of titanium alloy femoral components: a quantitative radiological analysis of osteolytic and granulomatous change. *J. Royal Soc. Med.* 86:460–463.

van Ewijk, W. (1991) T-cell differentiation is influenced by thymic microenvironments. *Ann. Rev. Immunol.* 9:591–615.

Watt, F. M., Jordan, P. W., and O'Neill, C. H. (1988) Cell shape controls terminal differentiation of human epidermal keratinocytes. *Proc. Natl. Acad. Sci.* 85:5576–5580.

Weiss, P. (1959) Interactions between cells. *Rev. Mod. Phys.* 31(2):449–454.

Weiss, P. (1959) Cellular dynamics. *Rev. Mod. Phys.* 31:11–20.

Weiss, P. (1945) Experiments on cell and axon orientation in vitro: the role of colloidal exudates in tissue organization. *J. Exp. Zool.* 100:353–386.

Weiss, P. and Garber, B. (1952) Shape and movement of mesenchyme cells as functions of the physical structure of the medium. Contributions to a quantitative morphology. *Proc. Natl. Acad. Sci.* 264–280.

Weiss, P. and Taylor, A. (1956) Fish scale as substratum for uniform orientation of cells in vitro. *Anat. Rec.* 124: 381.

Werb, Z., Hembry, R. M., Murohy, G., and Aggeler, J. (1986) Commitment to expression of the metalloendopeptidases, collagenase and stromelysin: Relationship of inducing events to changes in cytoskeletal architecture. *J. Cell Biol.* 102:697–702.

Winter, G. D. (1974) Transcutaneous implants: reactions of the skin implant interface. *J. Biomed. Mater. Res. Symp.* 5:99–113.

Wood, A. (1988) Contact guidance on microfabricated substrata: the response of teleost fin mesenchyme cells to repeating topographical patterns. *J. Cell Sci.* 90:667–681.

Cell and neuron growth cone behavior on micropatterned surfaces

PETER CLARK

Department of Anatomy and Cell Biology, St. Mary's Hospital Medical School
Imperial College of Science, Technology and Medicine
Norfolk Place, London W2 1PG, United Kingdom

20.1 Introduction

In the early to mid-1980s, a number of biomedical research groups began to use the microfabrication techniques of the electronics industry to make micropatterned devices with which to examine the motile behavior of cultured cells (Brunnette, 1986a,1986b; Clark et al., 1987, 1990 1991; Dunn and Brown, 1986; Hirono et al., 1988; Kleinfeld et al., 1988; O'Neill et al., 1986, 1990; Wood, 1988; see also Chapters 15, 16, 17, and 19). These techniques provided definition and precision of microscopic cues previously not available to biologists who, for many years, had been concerned with how the local microenvironment influenced the shape and directionality of locomotion of animal cells (see Curtis and Clark, 1990, for review). Micropatterning is proving to be extremely useful in modeling the kinds of microtopographic and/or adhesive cues that cells and growth cones may encounter in vivo, and also in patterning artificial networks of cells. This chapter reviews the work of the author and his colleagues in examining the effects on cell behavior of a variety of topographic and adhesive cues.

20.2 Topography

20.2.1 Single step cues

The first micropatterned features we examined were single step cues (Clark et al., 1987). Oxygen-reactive ion etching was employed to produce step features of various heights in Perspex. Baby hamster kidney (BHK) fibroblasts were cultured on these patterns, and their behavior on encountering steps was determined by time-lapse video microscopy. We found that cells approaching steps from the upper or lower parts of the structure were able to cross the step by climbing or descending, although the proportion of cells that did so was dependent on step height; crossing was reduced as step height increased. Cells were seen to align at these steps at both the upper and lower edges. Again, alignment was dependent on step height, although opposite to the crossing response, because the proportion of cells aligning increased with increasing step height. Whereas no difference showed in the degree of crossing from upper to lower or from lower to upper surfaces, on patterns where a difference in adhesiveness for cells was apparent between the upper step surface (more adhesive) and the wall and lower surface (less adhesive), a bias toward crossing from upper to lower surfaces became apparent.

Neuron growth cone crossing of steps was also seen to be dependent on step height (Figure 20.1E). On a suitably adhesive surface, the neurites provide a convenient record of the path taken by growth cones. The crossing of neurites of chick embryo cortical neurons over steps was found to decrease with increasing height. Comparing the reactions of a number of different cell types to a 5-μm step, it was apparent that the response was dependent on cell type. Neutrophil leukocytes were relatively unperturbed by this feature and freely climbed and descended.

20.2.2 Grooved surfaces

Multiple parallel grooved surfaces could be considered to be a closely packed array of single steps. We examined, in two studies, the effects of the density and magnitude of such cues by culturing cells on multiple parallel grooved substrata with varying spacing and depth. In the first study (Clark et al., 1990), grooved surfaces in Perspex with period spacings of 4–24 μm (that is, feature sizes of 2–12 μm) and depths of 0.2–2.0 μm were fabricated using conventional photolithography. The alignment of BHK fibroblasts (Figures 20.1A and 20.1B) and Madin Darby canine kidney (MDCK) epithelial cells (Figures 20.1C and 20.1D), and the orientation of neurite outgrowth from chick embryo neurons, were determined on grooved surfaces (Figure 20.1F). The alignment of fibroblasts to these patterned substrata was dependent on both groove spacing and groove depth; the alignment increased with decreasing spacing and increasing depth, with the latter having a stronger effect. On shallow grooved surfaces, lamellipodia appeared to conform to the underlying substratum.

Single MDCK epithelial cells were more sensitive to grooved surfaces, because they were found to align on all but the shallowest of grooves and were less dependent on groove spacing. When these epithelial cells were part of a colony, their responses to substratum topography were markedly altered, and the cells were less aligned, if at all. Cells at colony margins spread over a number of grooves and ridges. Chick embryo neurons' neurite outgrowth was aligned by 2-μm-deep grooves, but less so by 1-μm-deep structures.

In the second study (Clark et al., 1991), laser interferometry was used to initially define a pattern for fabrication of submicron-period grooved surfaces. These ultrafine structures, defined in fused quartz, were 260-nm-period multiple parallel gratings (130-nm grooves, 130-nm spaces) of various depths (100, 210, and 400 nm). BHK fibroblasts were aligned and elongated by these surfaces, although these responses increased with depth (Figure 20.2A). Most single epithelial cells

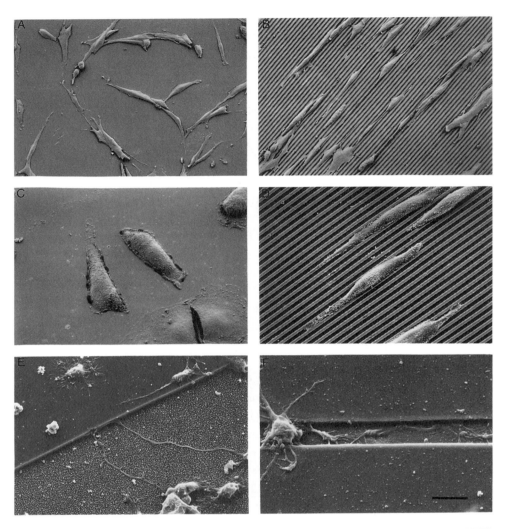

Figure 20.1. Scanning electron micrographs of cells' responses to topography. (A) BHK fibroblasts on an unpatterned surface. (B) BHK fibroblasts aligned on 6 μm-period, 2 μm-deep grooves. (C) MDCK epithelial cells on an unpatterned surface. (D) MDCK epithelial cells on a 6 μm-period, 2 μm-deep grooved surface. (E) Chick embryo cortical neuron being reflected by a 4 μm step. (F) Chick embryo cortical neuron at a 2 μm-deep, 7 μm-wide groove. Bar represents 60 μm in A, B, and E; 30 μm in C and D; and 12 μm in F. Parts A, B, and D are from Clark et al. (1990); E and F are from Clark et al. (1987).

were aligned and elongated by these surfaces, with the degree of elongation increasing with increased etch depth. On the other hand, when the cells were part of a colony they were typically unaligned, with the colonies appearing no different from those on planar substrata (Figure 20.2B). Chick embryo neurons were not affected by substrata of any of the depths tested. Neurite outgrowth was not oriented by these ultrafine substrata.

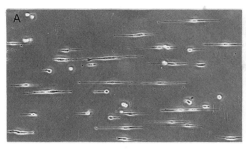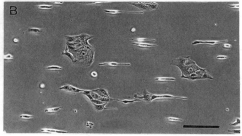

Figure 20.2. Phase-contrast micrographs of cells on 260 nm-period, 200 nm-deep grooved surfaces. (A) BHK fibroblasts aligned to the groove directions. (B) MDCK epithelial cells are aligned to the groove direction as single cells but less so, if at all, as cells in colonies. Bar represents 200 μm.

20.2.3 Factors affecting cells' responses to topography

When taken together, the observations made in these studies provide interesting information on how cells detect, and respond to, the morphology of their environment. On encountering a topographic feature, the response of a cell is probabilistic. The reaction of a cell is dependent on feature magnitude, feature density, cell–cell interactions, and cell type.

Increasing feature magnitude (for example, step height or groove depth) will increase the responsiveness of a population of cells to topography (Clark et al., 1987). Increasing the density of features (such as decreasing the spacing of multiple parallel grooves) also increases the responsiveness to features of given magnitude: single step cues of, for example, approximately 30% of fibroblasts will align (if only transiently) at a 1-μm step (Clark et al., 1987), whereas 6-μm-period, 1-μm-deep grooves align 80% of the same cell type (Clark et al., 1990). Seventy percent of fibroblasts and 100% of MDCK epithelial cells were aligned by ultrafine grooves 210 nm deep (Clark et al., 1991). Earlier data suggest that a single 210-nm step would have little, if any, effect on the cells. It would seem that parallel cues reinforce one another; that is, when a cell encounters more than one edge at a time, its behavior is modified more than if only one cue of the same magnitude were involved.

Epithelial cells are extremely sensitive to topography as single cells, but much less so when they form part of a colony (Clark et al., 1990, 1991). The formation of cell–cell contacts appears to allow these cells to override the guiding influence of the substratum. At high density, however, fibroblasts' responses to guiding topography may be amplified (Curtis and Clark, 1990). On a surface that only partially guides a population of fibroblasts at lower densities, the guidance response is more apparent when the cells are cultured to confluence. It was suggested that contact inhibition of locomotion, which can lead to local alignment of cells, enhances the degree of alignment already present.

Neutrophil leukocytes were found to be much less responsive to single cues and multiple parallel cues than are fibroblasts or neurons (Clark et al., 1987, 1990), and, whereas fibroblasts and epithelial cells were sensitive to ultrafine-grooved surfaces, neurons were not (Clark et al., 1991). Earlier studies have shown that transformed (cancerous) cells are less susceptible to topography than the equivalent nontransformed cells (Fisher and Tickle, 1981; McCartney and Buck, 1981; Rovensky et al., 1971).

Proposed mechanisms by which cells detect topography generally involve the cytoskeleton. Dunn (1982) suggested that the relatively inflexible actin stress fibers at a fibroblast leading lamella must shorten to accommodate a curved or angled surface; this shortening reduces the ability of that protrusion to exert traction, so that relatively small angles (only 17° greater than the horizontal plane) acted as a barrier to cells. O'Hara and Buck (1979) proposed that cells bridged grooved surfaces, which aligned focal adhesions at the cell–substratum interface; this in turn leads to the alignment of actin stress fibers and ultimately the cells themselves. Brunette (1986a, 1986b) and Clark et al. (1987, 1990) provided strong evidence that cells can accommodate angles of greater than 17° (for example, 90°), and that topographies do not act as absolute barriers for cells. Clark et al. (1987) suggested that accommodating topography could lead to regions of a cell's cytoskeleton being isolated from the remainder of the cell, because this isolated region was unable to exert significant "pull." The response of cells to the ultrafine grooved surfaces (Clark et al., 1991) provided evidence that contradicted O'Hara and Buck's (1979) proposed mechanism. The degree of alignment and elongation of cells increased with increasing groove depth; therefore, simple bridging of cells could not be occurring, because there should be no difference in bridging 200-nm- or 400-nm-deep grooves (Clark et al., 1991).

After examining the effects of multiple parallel grooved substrata, Clark et al. (1990, 1991) suggested that the response to topography could be accounted for by a model stating that a cell's direction of locomotion (and shape) is determined by the net action of protrusions. Protrusions in any direction on an unpatterned surface will have an equal probability of successfully establishing the axial polarity of a cell; therefore, cells will be randomly oriented. On a patterned surface, a topographical feature (or features) will reduce the success of protrusions in a given direction (or directions), thus leading to nonrandom orientation; that is, the more successful protrusions are likely to be the ones whose cytoskeleton is not distorted or isolated by topography. Elongation of cells parallel to multiple grooved surfaces may result from a cell's attempts to minimize its cytoskeletal distortion. Similarly, cells at the margins of epithelial cell colonies were seen to bridge over grooved surfaces without cell alignment (Clark et al., 1990). Support for this model is also provided by the fact that less susceptible cells (such as neutrophil leucocytes and transformed cells) generally have less-well-organized cytoskeletal structures, often lacking actin stress fibers.

20.3 Patterned adhesiveness

20.3.1 Organosilane patterning and cell adhesion

Differential adhesiveness is a cue that is known to be capable of guiding cells and growth cones (see Clark et al., 1992, 1993, for references). We felt that patterned adhesiveness could prove to be more useful than topography for patterning cells and neurons. To this end, we adapted a method for patterning organosilanes to silicon wafers published by Kleinfeld et al. (1988) for patterning adhesiveness for neurons. We simplified the procedure, used conventional glass, and were able to pattern a hydrophobic silane to give patterns of differential adhesiveness (Britland et al., 1992). The hydrophobic silanated surface provided a less adhesive surface than adjacent untreated glass, and also could be used as a mask for the patterning of other organosilanes, such as aminosilanes, to provide greater adhesiveness (Britland et al., 1992; Kleinfeld et al., 1988). Britland et al. (1992) quantified the attachment and spreading of fibroblasts to the variously treated surfaces and showed that after pro-longed exposure to serum-containing medium, hydrophobic surfaces acquired adhesiveness for cells, although this adhesiveness was less than found on untreated glass or aminosilane-derivatized glass. This is presumably as a result of eventual adsorption of serum attachment factors such as vitronectin and fibronectin.

20.3.2 Fibroblast and epithelial cell behavior on adhesive patterns

BHK fibroblasts, when seeded onto patterns of adjacent large areas of hydropho-bic and untreated glass and allowed to settle for a short period before unattached cells were rinsed away, adhered and spread on the untreated glass but rarely on the hydrophobic glass (Britland et al., 1992). When these patterned cells were cul-tured further to increase cell density, they reached confluence on the untreated glass and, although crowded at the boundary between the adhesive and nonadhe-sive areas, did not cross to the hydrophobic area (Figures 20.3A and 20.3B).

Patterns of alternating strips of hydrophobic glass and untreated glass, of iden-tical dimensions to the grooves used in the topographic studies (see Section 20.2.2), were used to examine the behavior of BHK fibroblasts and MDCK epithelial cells cultured on multiple parallel adhesive cues (Clark et al., 1992). The BHK cells became aligned to these patterns, although alignment was dependent on the pattern spacing. On 4-μm-period patterns (2-μm adhesive tracks separated by 2-μm nonad-hesive tracks) BHK cells were significantly oriented to a small degree, but there was a gradual increase in alignment with increasing pattern period; patterns of 24- and 50-μm periods aligned nearly the whole cell population (Figures 20.3C and 20.3D).

MDCK epithelial cell populations were completely aligned by the whole range of pattern spacing, but their elongation, which increased with increasing pattern period, was dependent on period (Clark et al., 1992). In many instances, when cells

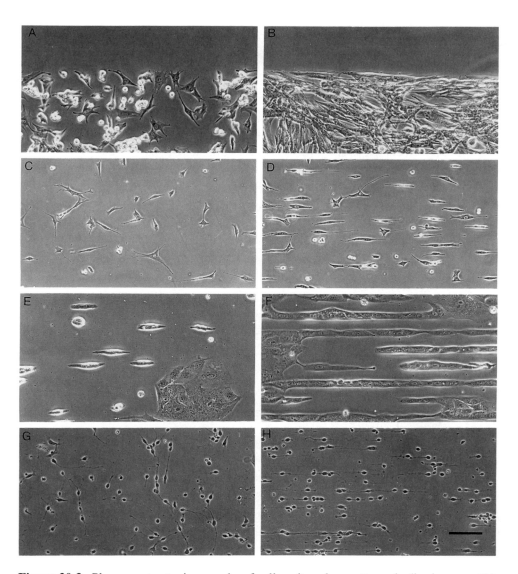

Figure 20.3. Phase-contrast micrographs of cells cultured on patterned adhesiveness. (A) BHK fibroblasts at an area of adjacent adhesive/nonadhesive boundary approx. 1 hour after plating. (B) As in (A), but 24 h after plating. (C) BHK fibroblasts on a 4 μm-period adhesive pattern: little alignment is evident. (D) BHK fibroblasts aligned on a 24 μm-period adhesive pattern. (E) MDCK epithelial cells align as single cells, but not as colonies, on a 6 μm-period adhesive pattern. (F) MDCK epithelial cells forming ordered chains on a 50 μm-period pattern of adhesiveness; this order breaks down at certain points and the cells are less aligned. (G) Chick embryo cortical neurons on an unpatterned laminin-coated surface. (H) Chick embryo cortical neurons aligned to a 24 μm-period pattern of adhesiveness. Bar represents 100 μm. E is from Clark et al. (1992). G and H are from Clark et al. (1993).

reached higher densities they formed colonies, often as a row of cells only one cell wide. At higher densities, some colonies formed that spanned the nonadhesive tracks; the cells at the center of these colonies were often not aligned, although the edges of such colonies were often "squared off" (Figures 20.3E and 20.3F).

20.3.3 Neuron guidance on laminin micropatterns

When neurons were seeded onto patterns of hydrophobic and untreated glass onto which laminin had been passively adsorbed, the pattern of attachment was unexpected (Clark et al., 1993). Neurons adhered to, and exhibited neuritic outgrowth on, what were previously the hydrophobic regions. Subsequent immunolocalization confirmed that laminin preferentially adsorbed to the hydrophobic regions of these patterns.

Because most neurons generally have more stringent adhesive requirements, and because culture surfaces are normally required to be precoated with adhesive materials, we were able to exploit the differential adsorption of laminin to make patterns of differential adhesiveness for neurons (Clark et al., 1993). Chick embryo cortical neurons, as well as mouse spinal cord and dorsal root ganglion neurons (DRGs), were cultured on multiple parallel repeat tracks of laminin of various periods of the same dimensions as those used in the earlier studies. The orientation of neurite outgrowth was found to be dependent on pattern periodicity; orientation increased with increasing periodicity. Patterns of 4 and 6 µm did not significantly orient neurite outgrowth, although 24- and 50-µm-period patterns aligned the growth of the entire neurite population (Figures 20.3G and 20.3H). Whereas 4-µm patterns (2-µm tracks of laminin separated by 2 µm of nonadhesive track) did not orient neurite outgrowth, single isolated 2-µm tracks did entirely constrain the extension of neurites.

One interesting finding was that DRGs cultured on an unpatterned adhesive surface developed a multipolar, highly branched morphology, but when the same cells are cultured on a pattern that guides their neurite outgrowth, the cells are bipolar with little or no branching being evident. During their development in vivo, DRGs adopt a bipolar morphology in an intermediate phase of their differentiation to their typical pseudo-unipolar morphology (see Clark et al., 1993, for discussion).

Tracks of different sizes altered the morphology of chick embryo cortical neuron growth cones (Clark et al., 1993). On unpatterned adhesive surfaces, growth cones were flattened lamellar structures bearing a number of filopodia. As tracks became narrower, the growth cones became simpler, so that on single 2-µm tracks many growth cones appeared to consist of a single tapering filopodium. On narrow-period multiple parallel patterns that did not guide growth cones, individual growth cones were seen to span nonadhesive tracks. At the boundaries between large areas of adjacent adhesive and nonadhesive regions (and also at the edges of the widest tracks), it could be seen that filopodia extended to the boundary, but the lamellopodial veils never did, even though laminin was available to them. This resulted in neurites' maintaining a distance from such boundaries; that is, the neurites themselves did not reach the edges.

20.3.4 *Mechanisms of adhesive guidance*

It would appear that cells and growth cones respond to adhesive patterns more absolutely than they do to topography. Even though adjacent hydrophobic areas eventually acquire adhesiveness for BHK fibroblasts, and even though population pressure due to contact inhibition of locomotion and crowding at adhesive pattern boundaries will favor movement to less crowded sites, the cells do not cross to the less adhesive region. The adhesive difference that exists is enough to prevent such migration (Britland et al., 1992). Similarly, growth cones do not cross these boundaries (Clark et al., 1993).

Adhesive patterns will therefore be more useful for making defined patterns of cells and neurites, but geometrical limitations are a factor in the design of such patterns. Narrowly spaced multiple tracks did not guide cells as strongly as on wider multiple tracks and single narrow tracks where alignment was absolute. This was interpreted as indicating that cells and growth cones are capable of bridging across narrow nonadhesive regions and are thus able to override some cues. Any track expected to guide a cell or growth cone must be sufficiently distant from neighboring adhesive regions to prevent bridging.

The responses of DRGs suggest that patterned adhesiveness could be involved in controlling the morphology of neurons by constraining the initiation of neurites and determining their degree of branching.

20.4 Topography vs. patterned adhesion

Both microtopographic cues and micropatterned adhesiveness are capable of profoundly influencing the behavior of cells and growth cones, but some important differences in their responses are apparent:

(1) Cells and growth cones respond to topography in a probabilistic manner, but patterned adhesiveness can elicit an absolute response such that cells can conform their shape to, and can be confined or guided by, more adhesive surfaces. This suggests that differential adhesiveness will provide a more accurate mechanism for the guidance of cells and neurites.

(2) The differences in the responses of cells or growth cones to either topographic or adhesive multiple parallel cues have a number of consequences. Cells and growth cones are increasingly sensitive to topographic cues as the density of the cues increases; that is, as the repeat period of multiple cues decreases, the guidance response

increases. On the other hand, as the period of multiple parallel adhesive cues decreases, so too does the guidance response. In other words, the two cues could be mutually antagonistic. This indicates that O'Hara and Buck's (1979) suggestion – that grooves act by aligning adhesions – could not operate, because the more closely packed the adhesions, the weaker the response. These data also indicate that the responses of cells to ultrafine grooves and aligned extracellular matrix (or indeed an axon fascicle), in vitro and in vivo, are not responses to aligned adhesion but to the topographic cues presented (see Clark et al., 1993, for discussion).

Although these studies have highlighted a number of important limitations to the effectiveness of the micropatterned cues described here, these cues are clearly capable of profoundly modulating the behavior of cells; similar cues are likely to operate in vivo during development and regeneration (Clark et al., 1993; Curtis and Clark, 1990).

References

Britland, S. T., Clark, P., Connolly, P., and Moores, G. R. (1992) Micropatterned substratum adhesiveness: a model for morphogenetic cues controlling cell behaviour. *Exp. Cell Res.* 198:124–129.

Brunette, D. M. (1986a) Fibroblasts on micromachined substrata orient hierarchically to grooves of different dimensions. *Exp. Cell Res.* 164:11–26.

Brunette, D. M. (1986b). Spreading and orientation of epithelial cells on grooved substrata. *Exp. Cell Res.* 167:203–217.

Clark, P., Britland, S., and Connolly, P. (1993) Growth cone guidance and neuron morphology on micropatterned laminin surfaces. *J. Cell Sci.* 105:203–212.

Clark, P., Connolly, P., Curtis, A. S. G., Dow, J. A. T., and Wilkinson, C. D. W. (1987) Topographical control of cell behaviour: I. Simple step cues. *Development* 99:439–448.

Clark, P., Connolly, P., Curtis, A. S. G., Dow, J. A. T. and Wilkinson, C. D. W. (1990) Topographical control of cell behaviour: II. multiple grooved substrata. *Development* 108:635–644.

Clark, P., Connolly, P., Curtis, A. S. G., Dow, J. A. T., and Wilkinson, C. D. W. (1991) Cell guidance by ultrafine topography in vitro. *J. Cell Sci.* 99:73–77.

Clark, P., Connolly, P., and Moores, G. R. (1992) Cell guidance by micropatterned adhesiveness *in vitro. J. Cell Sci.* 103:287–292.

Curtis, A. S. G. and Clark, P. (1990) The effects of topographic and mechanical properties of materials on cell behaviour. *Crit. Rev. Biocompat.* 5:343–362.

Dunn, G. A. (1982) Contact guidance of cultured tissue cells: A survey of potentially relevant properties of the substrutum. In *Cell Behaviour,* eds. R. Bellairs, A. Curtis and G. Dunn, pp. 247–280. Cambridge University Press.

Dunn, G. A. and Brown, A. F. (1986) Alignment of fibroblasts on grooved surfaces described by a simple geometric transformation. *J. Cell Sci.* 83:313–340.

Fisher, P. E. and Tickle, C. (1981) Differences in the alignment of normal and transformed cells on glass fibres. *Exp. Cell Res.* 131:407–409.

Hirono, T., Torimitsu, K., Kawana, A., and Fukada, J. (1988) Recognition of artificial microstructures by sensory nerve fibres in culture. *Brain Res.* 446:189–194.

Kleinfeld, D., Kahler, K. H., and Hockberger, P. E. (1988) Controlled outgrowth of dissociated neurones on patterned substrates. *J. Neurosci.* 8:4098–4120.

McCartney, M. D. and Buck, R. C. (1981) Comparison of the degree of contact guidance between tumor cells and normal cells in vitro. *Cancer Res.* 41:3064–3051.

O'Hara, P. T. and Buck, R. C. (1979) Contact guidance in vitro. A light, transmission, and scanning microscopic study. *Exp. Cell Res.* 121:235–249.

O'Neill, C., Jordan, P., and Ireland, G. (1986) Evidence for two distinct mechanisms of anchorage stimulation in freshly explanted 3T3 Swiss mouse fibroblasts. *Cell* 44:489–496.

O'Neill, C., Jordan, P., Riddle, P., and Ireland, G. (1990) Narrow linear strips of adhesive substratum are powerful inducers of both growth and total focal contact area. *J. Cell Sci.* 95:577–586.

Rovensky, Y. A., Slavnaja, I. L., and Vasiliev, J. M. (1971) Behaviour of fibroblast-like cells on grooved surfaces. *Expl. Cell Res.* 65:193–201.

Wood, A. T. (1988) Contact guidance on microfabricated substrata: the response of teleost fin mesenchyme cells to repeating topographical patterns. *J. Cell Sci.* 90:667–681.

21

Force generation by the microtubule-based motor protein kinesin

F. GITTES,

Department of Physics, Biophysics Research Division, University of Michigan
Ann Arbor, MI 48109-1055

E. MEYHÖFER, S. BAEK, D. COY,
B. MICKEY, and J. HOWARD

Department of Physiology and Biophysics, Center for Bioengineering
University of Washington, Seattle, WA 98195

21.1 Introduction

21.1.1 Mechanical processes in cells

The physiology of eukaryotic cells depends on various means of transport; this is clearly necessary because of the localization of different functions, for example, in the nucleus, the mitochondria, and other organelles. A great many molecular processes (for example, the common biochemical ones) are intrinsically nonmechanical, because their spatial movements take place through the action of diffusion. However, some of the most dramatic movements are obviously "mechanical" rather than diffusive, because they involve a coordinated motion of relatively large objects over relatively large distances. Such movement is not left to diffusion but is carried out by the movement of certain molecules along the cytoskeleton, a system of filamentous proteins that runs throughout cells. The most dramatic manifestation of macroscopic mechanics by a single cell is the contraction of a muscle fiber; this too is actually a specialization of the cytoskeletal mechanical system. An even more fundamental example is the movement of the chromosomes in metaphase and anaphase of cell division. Another example is the transport and active arrangement of organelles and vesicles within the cell. Such large-scale motion may be described in terms of forces and work, which are often inappropriate concepts for biochemical processes.

21.1.2 Protein motors and the cytoskeleton

Nearly all large-scale cellular motions seem to be based on a certain general mechanical arrangement, in which a "motor protein" cyclically hydrolyzes ambient fuel molecules such as ATP (adenosine triphosphate) as the protein moves along a specific filament of the cytoskeleton (that is, along either an actin filament or a microtubule). In muscle, for example, myosin motors move along actin filaments, which are long, twisted chains of globular actin subunits. Another class of motor proteins, including kinesin and dynein, moves along microtubules, which are long, hollow cylinders helically polymerized from protein subunits of tubulin. Microtubules form the mitotic spindle along which the chromosomes separate (see Goldstein, 1993), and axonemes of sperm tails are built out of microtubules and the motor protein dynein. The microtubule protein is virtually identical in all these functions, and both actin and tubulin are very highly conserved across species. However, the motor proteins seem to be specific for different tasks.

All motor proteins have one or several globular head regions, which contain the sites that interact with the filament and the sites that hydrolyze ATP. Presumably, the protein undergoes some conformational change during its chemical cycle that couples the ATP hydrolysis to a displacement of the binding site along the filament. There are a growing number of distinct motor proteins that may be fairly specific to their tasks. The longest-studied motor protein is two-headed myosin (myosin II), which moves along actin filaments in contracting muscle; the spatial structure of its head region has recently been published (Rayment et al., 1993a, 1993b) and is leading to an increased understanding of motor molecules.

The motor molecule we are studying is kinesin, which moves along microtubules (Brady, 1985; Vale et al., 1985). Kinesin was originally detected in squid axoplasm, and it is thought to be responsible for the transport of vesicles down axons (which are full of microtubules), and also from the Golgi stacks to the cell membrane. In appearance, it is a Y-shaped molecule, with two small (10-nm) heads connected to a long and compliant tail region that presumably binds to its cargo. Kinesin is a remarkably compact machine; in fact, a truncated head region with only about 400 amino acids is by itself able to generate force and move microtubules (Yang et al., 1990). The family of kinesin-related proteins has grown quickly (for a table of eight kinesin-related proteins and a discussion of others, see Bloom, 1992).

21.2 Mechanics of kinesin and microtubules in vitro

21.2.1 In vitro motility experiments

Because of the difficulty of extrapolating down from muscle mechanics, and because of the discovery of many nonmuscle motor molecules, there has been a considerable body of research in recent years seeking to determine the forces and displacements that can be generated by single motor molecules in motion experiments in vitro, using purified motor molecules and filament proteins.

In a standard in vitro motility experiment, one purifies both the motor molecules and the filament proteins (either actin monomers or tubulin dimers), repolymerizes the filaments, and observes the gliding motion of the filaments across a glass surface to which the motor molecules have adhered. The first in vitro motility assay, using myosin-coated fluorescent beads that moved along actin cables from the alga

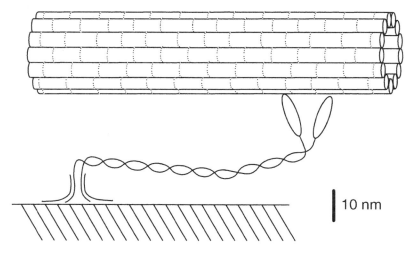

Figure 21.1. Schematic drawing of a kinesin molecule interacting with a microtubule. Only a short portion of the microtubule is shown. The dotted lines indicate boundaries between oriented dimers of tubulin, each containing one α and one β subunit. The tail of the kinesin molecule is shown attached to a solid substrate, as for in vitro experiments where the microtubule is moved. In vivo, the tail might be attached to a vesicle that the kinesin is transporting.

Nitella, was performed by Sheetz and Spudich (1983); the first gliding-filament experiments were done with actin and myosin by Kron and Spudich (1986).

Kinesin may also be used in an in vitro motility assay with microtubules (Vale et al., 1985). In such experiments, kinesin is purified (for example from bovine brain) and allowed to bind to the glass surfaces of a flow cell (Figure 21.1). Microtubules, repolymerized from purified tubulin monomers, are introduced and interact with the surface-bound kinesin. Under the dark-field microscope (or under fluorescence, if the tubulin is fluorescently labeled), one can observe or videotape the diffraction images of the microtubules as they move across the glass at speeds of about 1 μm per second.

21.2.2 Movement by single motors

The problem with actomyosin gliding assays is that dilute myosin preparations do not successfully move actin filaments. As one dilutes the concentration of motor molecules on the surface, there appears to be a minimum density at which filament movement stops, whereas if one motor molecule could move a filament, motion should be observed at arbitrarily low densities. One possible inference is that the cycle of interaction of myosin and actin includes a period of no binding, or of binding with less-than-thermal energy, that lasts long enough for escape to occur. This suggests that numerous myosin heads need to be operating on the same filament, which would be the case in skeletal muscle.

A crucial advantage of kinesin with regard to investigating molecular properties is that even a single dimer is capable of motor activity in such experiments (Block et al., 1990; Howard et al., 1989). This was deduced from two observations (Howard et al., 1989). First, as one reduces the concentration of kinesin on the surface, unlike the case with myosin, one finds that filament motion is still occurring, and the frequency of gliding events has the expected linear dependence on kinesin density. Second, one can observe moving microtubules pivoting about their point of attachment to the motor as they move.

Because single motors can move microtubules, there must be no prolonged detached period in the kinesin–microtubule cycle. This feature is consistent with kinesin's supposed role in axonal transport, because it would allow only a few kinesin molecules to effectively move a small object large distances along a microtubule.

21.2.3 Path of kinesin on the microtubule surface

The tubulin lattice that makes up the wall of a microtubule is not isotropic; tubulin dimers may be viewed as primarily attached head-to-tail into protofilaments, which are secondarily associated side by side to make up the hollow cylinder. In cells, most microtubules contain 13 protofilaments. An interesting aspect of microtubules repolymerized in vitro is that the number of protofilaments is variable, depending in large part on the conditions of polymerization, ranging from as few protofilaments as 9 to as many as 19. Furthermore, there is an offset between adjacent protofilaments that is (remarkably) quite commensurate with 13 protofilaments; when microtubules contain more or fewer than this number, the extra offset is accommodated by a gentle supertwist of the protofilaments, so that in electron micrographs one can see Moiré patterns caused by the changing protofilament overlap (Chrétien and Wade, 1991).

To determine the path along which kinesin moves across the surface lattice of a microtubule, Ray et al. (1993) have used this supertwist to advantage. One can preferentially make microtubules with particular protofilament numbers by choice of the buffer conditions. Because of their lack of supertwist, the 13-protofilament microtubules (13-mers) did not rotate when gliding across a kinesin-coated surface. Non-13-mers (which had supertwisted protofilaments) rotated, and the pitch and handedness of the rotation accorded with the supertwist measured by electron cryomicroscopy. The results showed that kinesin follows a path parallel to the protofilaments with high fidelity.

Incidentally, the fact that kinesin follows a single protofilament implies that the distance between consecutive kinesin-binding sites along the microtubule must be an integral multiple of 4 nm, the tubulin monomer spacing along the protofilament.

The work of Harrison et al. (1993) shows kinesin binding to saturate at one kinesin head per tubulin dimer; therefore, we conclude that the cycle step length should be 8 nm. Recently, using optical-trapping methods, Svoboda et al. (1993) have demonstrated discreteness in the motion of kinesin along microtubules that seems consistent with 8-nm steps.

21.2.4 Flexibility of the kinesin motor

Another important issue in interpreting motility is the question of orientation of the motor. One does not know whether the population of moving microtubules in a low-density in vitro assay has been restricted to cases where the single motor and the microtubule had some particular relative orientation so that the force-producing cycle could proceed (the pivoting during motion that was mentioned previously is fairly small).

To address this issue, Hunt and Howard (1993) have looked at the attachment of microtubules to a surface by single motors without ATP; in this situation (the "rigor" state, by analogy with myosin in muscle), one can watch the pivoting, mentioned previously, that occurs about the motor for an extended time. The result was that the kinesin attachment had a remarkable degree of rotational freedom. The tethered microtubules underwent as many as four complete rotations about their point of attachment over a period of several minutes. Furthermore, it is possible to establish an elastic restoring force for this motion, that is, a rotational compliance in the kinesin molecule, of about 1 kT per rotation.

21.2.5 Flexural rigidity of microtubules

To interpret some in vitro motility experiments and to study structural functions of microtubules in general, one needs to know the mechanical properties of microtubules. Because of the thinness of the microtubules (28 nm) compared to the scale of bending in these experiments (many μm), the elastic theory of flexible molecules or thin rods should be an excellent approximation. An important point in this regard is that in an elastic rod of diameter d, the bending energy is proportional to d^4, whereas the stretching energy is proportional to d^2. Therefore, stretching becomes negligible as d becomes small. One can thus take a thin rod to be inextensible when considering bends. The appropriate theory depends on only one parameter: the flexural rigidity.

The flexural rigidity, denoted by EI, is defined by the bending free energy g per unit length, which for two-dimensional bends is $g = (1/2)EI (d\theta/ds)^2$, where $\theta(s)$ is the tangent angle of the two-dimensional shape of a microtubule, and the curvature is $d\theta/ds$. We must also subtract off a "natural" curvature, due to structural defects. Because g is the first term in an expansion in the curvature $d\theta/ds$, its use requires only that the curvature is small enough that local deformation is small. Bending a microtubule into a 10-μm-diameter circle, for example, implies local stretching and compression of the polymer structure by about 0.3%.

The notation *EI* for the flexural rigidity arises in connection with elastic theory. If a rod is composed of an isotropic elastic substance, then one can separate the flexural rigidity into the product of Young's modulus, *E*, and the "moment of inertia" *I* of the cross section, which for a hollow cylinder of inner radius *a* and outer radius *b* is $I = (\pi/4)(b^4 - a^4)$. This expression is useful for interpreting the bending modulus; however, *EI* retains a precise meaning even when *E* and *I* separately do not.

We have measured (Gittes et al., 1993) the flexural rigidity of individual microtubules as well as of actin filaments. To do this, we devised a method of resolving the thermal shape fluctuations into independent configurational modes of bending. or microtubules (as opposed to actin filaments) the measurement was complicated by their relatively great stiffness, as well as by the fact that with in vitro–polymerized microtubules, such thermal shape changes are usually small compared to larger, "natural" structural bends due to defects present in the structure.

Microtubules between 25 and 65 μm in length were confined to a narrow region between a slide and a cover slip, and so were constrained to have an approximately two-dimensional shape. The microtubules were videotaped as they fluctuated slightly about their equilibrium shape. From the digitized image, we decomposed the shape θ(*s*) into cosines,

$$\theta(s) = \sum_{n=0}^{\infty} a_n \cos(\frac{n\pi s}{L}) \tag{21.1}$$

because in thermal fluctuations the a_n are independent, with variance

$$\text{var}(a_n) = \frac{kT}{EI}\left(\frac{L}{n\pi}\right)^2, \tag{21.2}$$

giving for each mode *n* an independent estimate of the flexural rigidity *EI*.

For taxol-stabilized microtubules we found the flexural rigidity to be 2.2×10^{-23} Nm^2. If one were to model the microtubule as a hollow cylinder, this stiffness would correspond to a Young's modulus of about 1.2 GPa, about that of Plexiglas and rigid plastics. We estimated the Young's modulus of actin filaments, in other experiments, to be similar.

21.3 Force measurements for single molecules

To understand the motions of cellular systems, it is important to understand the capabilities of these motor proteins. In the case of muscle contraction, macroscopic measurements relating force, time, and displacement have long been available (Bagshaw,

1982), and one focus of such measurements is the dependence of force on velocity. The velocity of contraction has long been known to decrease from its maximum at zero force (that is, unloaded), reaching zero velocity at its maximum isometric force.

If one wants to look at the general dependence of force on velocity, extrapolation from muscle fibers down to individual actomyosin interactions is difficult because many myosin motors act on each actin filament; in addition, because the actin and myosin spacings are different, motors are equally likely to be at all positions relative to possible binding sites along the actin filament. To interpret the total force and velocity in molecular terms, one must turn to complicated crossbridge theories, the classic treatment being that of Hill (1974). However, one can extrapolate downward from the maximum isometric force of muscle, using an estimate of the number of myosin heads acting in parallel in a cross section of the muscle. In this case, the maximum isometric force of a single myosin head is estimated to be about 1.4 pN (Lombardi et al., 1992).

The interest in force–velocity relations is very natural from a mechanical point of view. In engineering, for example, one can for many purposes summarize the characteristics of a motor by its torque versus its rpm (the rotational counterparts of force and velocity) and incorporate the motor into mechanical design on the basis of this information alone. In the same way, force–velocity information should be extremely important in understanding the "mechanical design" of the cell.

Force–velocity relations for individual motor molecules, if detailed enough, would complement macroscopic muscle data. When many motors act on one filament, the boundary condition on a single motor is constant velocity. The measured force (per motor) is time averaged; the actual force per motor will rise and fall as it goes through its cycle. In contrast, single-motor measurements might measure velocity at truly constant force. Comparison with macroscopic behavior might shed light on the mechanisms underlying force–velocity curves in muscle.

Estimates have been made in other laboratories of forces from a small number of motor molecules in vitro, using force fibers (Ishijima et al., 1991), optical traps (Block et al., 1990; Finer et al., 1994; Kuo and Sheetz, 1993; Svoboda and Block, 1994; Svoboda et al., 1993), and a centrifugal microscope that imposes an inertial load (Hall et al., 1993; Oiwa et al., 1990). The powerful method of optical trapping is progressing rapidly. Against an optical trap, Svoboda et al. report a maximum force of about 5 pN for kinesin, whereas Kuo and Sheetz report the maximum force as 1.95 ± 0.4 pN. However, questions remain about the nature of the trapping in these experiments and about the method of calibration.

We will describe several techniques we are using for force estimation: putting a viscous load on gliding microtubules in single-motor conditions, measuring the deflection of glass needles, and observing controlled deformations of the microtubules themselves.

21.3.1 Viscosity experiment

One approach to force measurement (Hunt et al., 1994) is to load individual kinesin molecules by increasing the viscosity of the solution in the low-density motility assay. When the viscosity is high enough, the longer microtubules, which experience a larger drag force from the fluid, move more slowly than the shorter ones.

The challenge of this approach is to provide sufficient viscous force. A particular mixture of polysaccharides and polypeptides increased the viscosity to about one hundred times that of water. In its appearance and texture, the viscous solution somewhat resembled rubber cement, and yet did not interfere with the biochemical cycle. This latter fact was checked by selectively cleaving long polymers in the mixture to reduce its viscosity drastically while leaving the chemical environment unchanged; this procedure restored full motor speed.

To calculate forces from the observed speeds of a microtubule, one needs to know its drag coefficient per unit length. Virtually all the drag comes from the localized shear gradient between the microtubule and the adjacent surface. Because the relevant nanometer-scale geometry was not directly observable, the height of the microtubule above the surface, about 10 nm, was deduced from the rotational diffusion times of microtubules pivoting about single kinesin molecules, as described earlier. As a further complication, the viscous solution exhibits strongly non-Newtonian behavior, and this had to be accounted for in calculating the viscous drag.

In general terms, the speed of movement of a microtubule depends approximately linearly on the drag force that loads the motor. At the lowest kinesin densities, where dilution experiments indicate that the movement is due to a single kinesin molecule, one can extrapolate a stalling force that is roughly 4 to 6 pN per motor (Figure 21.2).

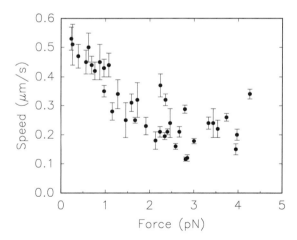

Figure 21.2. Speed versus calculated viscous loading of microtubules being moved by single kinesin motors in an in vitro experiment. Each data point corresponds to one microtubule whose length and speed, together with a non-Newtonian drag coefficient, were used to obtain a force value for the viscous load. The point at far right may result from more than one motor.

This experiment provides an interesting confirmation that a single motor is responsible for motion. When a motility assay is carried out at high enough viscosity that a single motor moves a microtubule more slowly than multiple motors, one can look at the variability of speed of microtubule motion as one lowers the number of kinesin motors on the surface. At high kinesin density, the microtubules move steadily. At intermediate density, one observes the speed to vary, as would be expected if the number of motors acting at any one time was a small but changing number. At the lowest densities, the speed of those microtubules that move once again becomes steady, consistent with the action of a single motor.

21.3.2 Measuring forces with glass needles

The forces generated by unknown but fairly small numbers of myosin molecules have been measured by measuring the deflection of calibrated glass needles to which actin filaments have adhered (Ishijima et al., 1991; Kishino and Yanagida, 1988). There have even been force–velocity curves (for many motors) measured with myosin-coated glass microneedles moved by actin cables in vitro (Chaen et al., 1989).

We have been repeating these experiments with kinesin, which is able to yield single-motor forces (Meyhöfer and Howard, 1995). In these experiments a microtubule, attached to the tip of a minute glass fiber that is maneuvered with a micromanipulator, is allowed to interact with kinesin motor molecules on the slide surface (Figure 21.3). Given a sufficiently low density, the microtubule will interact with only a single kinesin motor molecule that, in the presence of ATP, will displace the microtubule and deflect the tip of the glass needle. The deflection can be measured with high temporal and spatial precision using dark-field microscopy and a position detector; from the deflection, the motor force can be deduced.

The aim of these experiments is to observe the maximum force that a single kinesin molecule can exert, as well as to observe (if possible) the size and frequency of molecular steps, and perhaps the stiffness changes within the cycle of

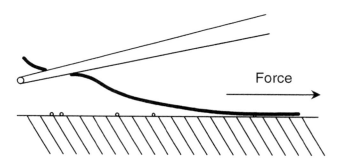

Figure 21.3. Schematic drawing of a microtubule (dark curve) attached to a glass needle roughly 0.5 μm in diameter. The microtubule interacts with kinesin molecules on the glass surface.

a single kinesin motor. The force observed when the kinesin motor can no longer translate the microtubule against the restoring force of the glass fiber represents the isometric, or stalling, force.

The glass needles are several centimeters in length but are drawn down to final diameters of roughly 0.5 μm. They have tip stiffnesses ranging from 10 to 1,000 pN/nm and, consequently, give rise to rms Brownian motions ranging from 20 to 2 nm, respectively. For each particular needle, the Brownian motion is used to calibrate its stiffness (Howard and Hudspeth, 1988). Furthermore, a reduced degree of thermal fluctuation should reveal the compliance of the kinesin itself (for myosin in muscle, the effective stiffness of an attached crossbridge is expected to be roughly 1 pN/nm (Bagshaw, 1982). In this situation, the resolution of force measurement is limited by the Brownian motion of the glass needle and not by resolution of the position detector. The actual attachment of microtubules to the needle is done by biotin labeling the tubulin in the microtubule and then coating the needle with streptavidin, which binds to the biotin.

Our glass needle experiments yield kinesin stalling forces of about 5 pN, together with an approximately linear dependence of motor speed on load force (Meyhöfer and Howard, 1995). Both of these results are consistent with the viscosity experiment described previously, as well as with optical trap experiments (Svoboda and Block, 1994; Svoboda et al., 1993).

21.3.3 Buckling of microtubules by kinesin motors

In low-density motility assays of kinesin with microtubules, a moving microtubule occasionally encounters an obstacle, and the kinesin motors buckle the microtubule. With regard to the unknown forces that are developed by the kinesin motors, such events are very interesting; in particular, a knowledge of the microtubule stiffness should allow an estimate of the force involved in these events. Furthermore, the time course of the buckling event should provide more detailed information about how the force depends on velocity, as well as on the angle through which the microtubule has pivoted about the fixed motor.

In a sense, the occurrence of buckling events in low-density kinesin preparations is a fortunate match between the bending stiffness of the microtubule and the force of the kinesin motors. In the classic problem of elastic stability (first considered by Euler), a rod becomes unstable to buckling and collapses when the longitudinal force on the end exceeds a critical buckling force F_B. Conversely, for a given force F_B, the rod buckles when the rod length exceeds a certain critical length L_B. The buckling point depends on the flexural rigidity EI, described earlier, and on the boundary conditions. For example, if the rod is clamped at one end and the force F is applied to the other, one has $F_B = \gamma EI/L_B^2$, where $\gamma \cong 4.49$. If the motor force F_B is on the order of piconewtons, L_B is on the order of 10 or 20 μm, which is a very convenient length of microtubule to work with in vitro. In contrast, for actin filaments, *EI* and L_B are hundreds of times smaller and buckling events cannot be distinguished from thermal fluctuations.

The rarity and the varying circumstances of these bending events were obstacles to their detailed analysis. For this reason we have found an experimental method (Gittes et al., 1996) in which bending events occur in a controlled and repeatable fashion, with precisely the "clamped" boundary conditions mentioned earlier. We specifically bind a short segment of each microtubule to the glass surface with biotin–streptavidin bonds, via a biotinylated albumin spacer molecule (Figure 21.4). The major part of the microtubule has no biotin label and does not bind to the surface, but sweeps back and forth in thermal motion. When the free portion with the proper polarity encounters a kinesin molecule, the kinesin binds and generates force, buckling the free segment of microtubule, provided the segment is long enough. Besides the well-defined boundary conditions, this experiment has the advantage that buckling events will usually repeat themselves a number of times.

The simplest way to use the clamped-buckling experiment for inferring motor force is to catalog as many buckling events as possible and to plot the frequency of events as a function of microtubule length. Video recordings were made of buckling events where the microtubules were moderately straight, and the onset of buckling was consistent with motion through a single point. Events with buckling lengths less than 3–4 μm were absent, reflecting the limited force of the kinesin motor. Dilution of kinesin had no significant effect on the distribution of buckling lengths found, consistent with the action of single motors. However, at longer L_B, one also sees a declining number of events at longer buckling lengths, due in part to a population bias and also a selection bias inherent in rejecting multiple-motor events.

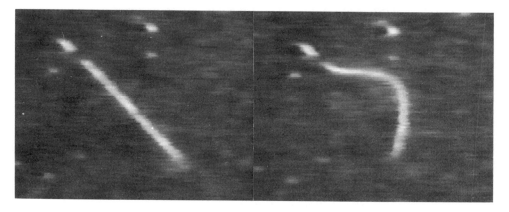

Figure 21.4. Images of a microtubule being buckled by a kinesin motor. The short dark portion is biotin labeled and, thereby, "clamped" to the glass surface. The longer of the two unclamped fluorescent segments, buckled by the motor, is about 10 μm in length. The duration of this event is about 4 seconds.

We can construct a distribution of putative single-motor buckling events in which the population and selection biases were removed. This distribution decreases for L_B below about 10 μm, suggesting that buckling at these lengths requires greater forces than the motors can typically develop. Using the microtubule's flexural rigidity *EI* found as described previously, an event with $L_B = 10$ μm would correspond to a buckling force of about 4 pN. Several results require further study, including understanding the number of events at much shorter lengths, which perhaps reflect defective microtubules, as well as understanding the distribution decreases *above* about 10 μm. This latter result probably arises because the microtubules curve away from the surface.

A more detailed mathematical shape analysis of the buckling microtubules reveals how the kinesin motor speed varies when the load force is no longer parallel to the microtubule axis (Gittes et al., 1996). Such nonparallel force effects give information about the direction of conformational change within the kinesin molecule.

21.4 Conclusion

We, and other laboratories, are arriving at estimates of elementary motor–protein forces probably on the order of 4 or 5 pN. This represents a considerable reserve of force for the intracellular transport of vesicles through a crowded cytoplasm. The available work derived from ATP hydrolysis under cellular conditions is about 80×10^{-21} J, which implies that, given steps of 8 nm per ATP, the efficiency of the kinesin motor is about 40%.

We are seeking a mechanical description of how kinesin molecules interact with microtubules in vitro, and we wish to observe elementary displacements of these molecules. The issue of mechanical mechanisms in cells is of basic importance in cell biology, and forces and mechanical quantities are fundamental properties of cellular motor proteins.

References

Bagshaw, C. R. (1993) *Muscle Contraction*, 2nd Ed. London: Chapman and Hall.

Block, S. M., Goldstein, L. S. B., and Schnapp, B. J. (1990) Bead movement by single kinesin molecules studied with optical tweezers. *Nature* 348:348–52.

Bloom, G. S. (1992) Motor proteins for cytoplasmic microtubules. *Curr. Opin. Cell Biol.* 4:66–73.

Brady, S. T. (1985) A novel brain ATPase with properties expected for the fast axonal transport motor. *Nature* 317:73–75.

Chaen, S., Oiwa, K., Shimmen, T. Iwamoto, H., and Sugi, H. (1989) Simultaneous recordings of force and sliding movement between a myosin-coated glass microneedle and actin cables in vitro. *Proc. Nat. Acad. Sci. USA* 86:1510–1514.

Chrétien, D. and Wade, R. H. (1991) New data on the microtubule surface lattice [see erratum (1991) *Biol. Cell.* 72:284]. *Biol. Cell.* 71:161–174.

Finer, J. T., Simmons, R. M., and Spudich, J. A. (1994) Single myosin molecule mechanics: piconewton forces and nanometre steps. *Nature* 368:113–119.

Gittes, F., Meyhöfer, E., Baek, S., and Howard, J. (1996) Directional loading of the kinesin motor molecule as it buckles a microtubule. *Biophys. J.* 70:418–429.

Gittes, F., Mickey, B., Nettleton, J., and Howard, J. (1993) Flexural rigidity of micro-tubules and actin filaments measured from thermal fluctuations in shape. *J. Cell Biol.* 120:923–934.

Goldstein, L. S. B. (1993) Functional redundancy in mitotic force generation. *J. Cell Biol.* 120:1–3.

Hall, K., Cole, D. G., Yeh, Y., Scholey, J. M., and Baskin, R. J. (1993). Force-velocity relationships in kinesin-driven motility. *Nature* 364:457–459.

Harrison, B. C., Marchese-Ragona, S. P., Gilbert, S. P., Cheng, N., Steven, A. C., and Johnson, K. A. (1993) Kinesin decoration of the microtubule surface: one kinesin head per tubulin heterodimer. *Nature* 362:73–75.

Hill, T. L. (1974) Theoretical formulation for the sliding-filament model of contraction of striated muscle. *Prog. Biophys. Mol. Biol.* 28:267–340.

Howard, J., and Hudspeth, A. J., (1988) Compliance of the hair bundle associated with gating of mechanoelectrical transduction channels in the bullfrog's saccular hair cell. *Neuron* 1:189–199.

Howard, J., Hudspeth, A. J. and Vale, R. D. (1989) Movements of microtubules by sin-gle kinesin molecules. *Nature* 342:154–158.

Hunt, A. J., Gittes, F., and Howard, J. (1994) The force exerted by kinesin against a viscous load. *Biophys. J.* 67:766–781.

Hunt, A. J. and Howard, J. (1993) Kinesin swivels to permit microtubule movement in any direction. *Proc. Nat. Acad. Sci. USA* 90:11653–11657.

Ishijima, A., Doi, T., Sakurada, K., and Yanagida, T. (1991) Sub-piconewton force fluc-tuations of actomyosin in vitro. *Nature* 352:301–306.

Kishino, A. and Yanagida, T. (1988) Force measurements by micromanipulation of a sin-gle actin filament by glass needles. *Nature* 334:74–76.

Kron, S. J. and Spudich, J. A. (1986) Fluorescent actin filaments move on myosin fixed to a glass surface. *Proc. Nat. Acad. Sci. USA* 83:6272–6.

Kuo, S. C. and Sheetz, M. P. (1993) Force of single kinesin molecules measured with optical tweezers. *Science* 260:232–234.

Lombardi, V., Piazzesi, G., and Linari, M. (1992) Rapid regeneration of the actin-myosin power stroke in contracting muscle. *Nature* 355:638–641.

Meyhöfer, E. and Howard, J. (1995) The force generated by a single kinesin molecule against an elastic load. *Proc. Natl. Acad. Sci. USA* 92:574–578.

Oiwa, K., Chaen, S., Kamitsubo, E., Shimmen, T., and Sugi, H. (1990) Steady-state force-velocity relation in the ATP-dependent sliding movement of myosin-coated beads on actin cables in vitro studied with a centrifuge microscope. *Proc. Nat. Acad. Sci. USA* 87:7893–7897.

Ray, S., Meyhöfer, E., Milligan, R. A., and Howard, J. (1993) Kinesin follows the microtubule's protofilament axis. *J. Cell Biol.* 121:1083–1093.

Rayment, I., Holden, H. M., Whittaker, M., Yohn, C. B., Lorenz, M., Holmes, K. C., and Milligan, R. A. (1993a) Structure of the actin-myosin complex and its implications for muscle contraction. *Science* 261:58–65.

Rayment, I., Rypniewski, W. R., Schmidt-Bäse, K., Smith, R., Tomchick, D. R., Benning, M. M., Winkelmann, D. A., Wesenberg, G., and Holden, H. M. (1993b) Three-dimensional structure of myosin subfragment-1: A molecular motor. *Science* 261:50–58.

Sheetz, M. P. and Spudich, J. A. (1983) Movement of myosin-coated fluorescent beads on actin cables in vitro. *Nature* 303:31–35.

Svoboda, K. and Block, S. M. (1994) Force and velocity measured for single kinesin molecules. *Cell* 77:773–784.

Svoboda, K., Schmidt, C. F., Schnapp, B. J., and Block, S. M. (1993). Direct observation of kinesin stepping by optical trapping interferometry. *Nature* 256:721–727.

Vale, R. D., Reese, T. S., and Sheetz, M. P. (1985) Identification of a novel force-generating protein, kinesin, involved in microtubule-based motility. *Cell* 42:39–50.

Yang, J. T., Saxton, W. M., Stewart, R. J., Raff, E. C., and Goldstein, L. S. B. (1990) Evidence that the head of kinesin is sufficient for force generation and motility in vitro. *Science* 249:42–47.

22

Contemporary problems in biology: contractile materials

GERALD H. POLLACK

Center for Bioengineering, Box 35-7962
University of Washington
Seattle, WA 98195-7962

22.1 Introduction

This chapter presents an overview of the field of biological motion, more specifically, of muscle contraction. The overview will depart from the traditional textbook treatment in that the current theory will be presented – but not necessarily as the final answer. Evidence that seems inconsistent with this theory will be put forth, and alternative paradigms that explain these observations, and that are beginning to draw attention, will be considered. Finally, because this field is one in which there is a natural application of nanofabricated materials, some potential uses of these devices in the field's research endeavors will be presented, along with predictions of how such approaches may lead to resolution of the central issue of contraction: the underlying molecular mechanism.

22.2 Crossbridge theory

For some 40 years, the field of muscle contraction has been dominated by a single paradigm (Huxley and Hanson, 1954; Huxley and Niedergerke, 1954). Commonly known as the swinging crossbridge theory, this paradigm posits a set of interdigitating filaments that are propelled to slide past one another as a result of the driving force imparted by crossbridges (Figure 22.1). Working independently and stochastically, these crossbridges are thought to function in an oarlike manner, propelling the thin filaments to slide toward the center of the sarcomere, thereby causing the sarcomere to shorten.

The theory prevails as strongly as it does for at least two good reasons. First, it is intuitive and elegant; it satisfies the widely held notion that nature's mechanism is likely to be simple. Second, the volume of evidence consistent with the theory has been appreciable. It includes x-ray evidence for underlying structure, biochemical and mechanical evidence for underlying function, and particularly the recent crystallographic evidence for the structure of actin (Holmes et al., 1990; Kabsch et al., 1990) and myosin (Rayment et al., 1993a, 1993b), which provides

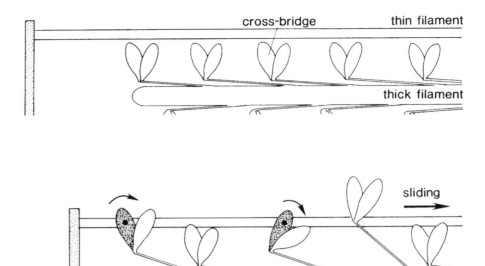

Figure 22.1. Schematic representation of the crossbridge theory. The sarcomere shortens (bottom) as crossbridges from the thick filament attach to the thin filament, rotate clockwise, and thereby drive the thin filament rightward. Each crossbridge may undergo many such strokes during contraction. (From Pollack, 1990.)

a basis for the realization of the crossbridge stroke. Details of the evidence favoring the swinging crossbridge theory can be found in recent reviews (see Cooke, 1986; Squire, 1990).

Although most of the field's investigators routinely invoke this theory as their working paradigm, several areas of recent evidence have provoked doubt. First, in spite of numerous attempts to detect bridge-angle changes anticipated to occur with the power stroke, the most recent experiments have proved consistently negative (Cooke, 1986; Hirose et al., 1993; Pollard et al., 1993; Suzuki et al., 1993; Thomas, 1987). The failure to find clear evidence of substantial angle change now appears to be of concern even to some long-term advocates of the theory (see Thomas, 1993).

The second area of doubt has been the "step size," that is, the amount of filament sliding produced by hydrolysis of one molecule of ATP. An early report implied that the size of the step was in accord with the theory's expectation, that is, 10–20 nm (Toyoshima et al., 1990), but other results imply steps of unexpectedly large magnitude – too large to be compatible with a realistic crossbridge stroke (Harada et al., 1990; Higuchi and Goldman, 1991; Kellermayer et al., 1995; Lombardi et al., 1992; Ohno and Kodama, 1991). This latter evidence implies that the energy derived from the hydrolysis of a single molecule of ATP is not released in a single power stroke; to explain the large size of the step, the energy must instead be assumed to be partitioned over a series of strokes. Partitioning of energy is typically invoked, but there has been concern whether the implied fractionation of energy of a single bond is theoretically sound (Schutt and Lindberg, 1993). Even if it is, some researchers see the concept of energy partitioning as detracting from the elegance of the original one-stroke–one-ATP model.

A third – and longer standing – concern is the shape of the length–tension relation. The theory predicts that tension varies directly with the number of cross-bridges available to make contact with the actin filament. Textbooks report the classical result (Gordon et al., 1966), which fits the theory's prediction, but they usually offer no hint of those results that are in conflict. The state of the conflict has been reviewed in detail at various stages (Pollack, 1983, 1990 [Chapter 12]) and has finally been resolved by experimental evidence published recently (Horowitz et al., 1992, 1993). This evidence demonstrates that the anticipated linear relation can be obtained, but it does not reflect the sarcomere's full tension-generating capacity. With full steam, tension is relatively insensitive to the number of overlapping crossbridges (Horowitz et al., 1992, 1993), a result that does not square with the theory's prediction.

These are not the only conflicts. Issues of structure, energetics, biochemistry, and mechanics raise additional concerns, which are treated in depth in a recent book (Pollack, 1990), to which readers interested in further explanation are referred.

For most researchers, these difficulties have not proved reason enough to reject the theory, and the majority of the field's workers continue to invoke the theory to explain their experimental observations. For a small but growing number of researchers, however, the theory simply does not work; the oarlike mechanism is convincingly at odds with enough evidence to prompt a serious search for alternatives.

22.3 An actin-based motor?

The actin-motor hypothesis proposes that some action within the actin filament drives motility and contraction. Several plausible types of action have recently been proposed. Among these is the "thermal ratchet" mechanism, which posits that thermal energy can be harnessed by the actin filament. Harnessing of this energy, through splitting of ATP, constrains the thermal motion to be unidirectional (Nakata et al., 1993; Vale and Oosawa, 1990). In another proposal, the actin filament (or regions thereof) undergoes transition from a "ribbon" configuration to a helix configuration (Schutt and Lindberg, 1992, 1993). The rationale for such a transition is based on crystallographic evidence for two configurations of actin packing, with one longer and the other shorter. Thus, segmental shortening and relengthening of the actin filament could drive the actin filament past the myosin filament – much like the action of a caterpillar.

These theories are based on a growing body of evidence – much of it gathered only during the past several years – that implicates the actin filament in a more central role. Briefly, this evidence falls naturally into three categories:

(1) *Global transitions in the actin filament.* Several lines of evidence point to the possibility that the actin filament has two interconvertible states. These lines of evidence include electron microscopy (Janmey et al., 1990; Jonas et al., 1993; Trombitás et al., 1988a, 1988b), crystallography (Schutt and Lindberg, 1992; Schutt et al., 1989), optics (Kobayashi, 1964; Prochniewicz-Nakayama et al., 1983; Yanagida and Oosawa, 1978; Yanagida et al., 1974), and x-ray diffraction (Wakabayashi et al., 1988). Collectively, the evidence implies that the actin filament is not at all static; some major transition apparently occurs along the filament during contraction.

(2) *Propagation.* Several additional lines of evidence imply that the transitions previously mentioned occur locally and propagate along the filament. Described variously as snakelike motion, wiggling, undulation, and reptation, such propagated motion is detectable not only in time-resolved electron microscopy (Ménétret et al., 1991) but also in real-time optical microscopy of fluorescently labeled actin filaments (Toyoshima et al., 1987; Yanagida and Oosawa, 1978, 1980; Yanagida et al., 1984). The wiggling motion is especially prominent when ionic strength is edged just above the critical level required to stop translational motion; at that point, wiggling is fully unmasked (Kellermayer and Pollack, 1995). Careful measurements of actin-filament-translation velocity in the in vitro motility assay support such propagated motion: the velocity has a strong periodic component (de Beer et al., 1995). Thus, the transitions that occur in actin packing appear to propagate along the filament in periodic waves.

(3) *Other observations.* Additional experimental evidence points to the possibility that these phenomena are not merely incidental, but are centrally relevant to the contractile process. First, actin splits ATP; that is, it is an ATPase (Szent-Györgyi and Prior, 1966). The ATPase has recently been presumed related to polymerization rather than to contraction, because there is increasingly clear documentation of ATP's involvement in actin polymerization. But ATP continues to be split following polymerization (Pollard, 1986), implying that the once-thought central role of actin ATPase in contraction may indeed hold. A second piece of implicative evidence is that actin residues situated far from the myosin-docking site show unexpectedly high functional relevance. Mutation of the single actin residue E316 K, for example, decisively impacts mechanics (Drummond et al., 1990). This would not be expected if the role of the actin filament was merely passive.

A few general conclusions seem to emerge from this collection of evidence. First, the actin filament appears to have two distinct states. These states have different physical properties, and possibly different packing arrangements. Second, whatever transition takes place in the filament – possibly a conversion from one state to another – may propagate cooperatively along the length filament. Some of the experiments imply a wavelike process, not unlike that proposed by Schutt and colleagues (Schutt and Lindberg, 1992; Schutt et al., 1989).

The actin-motor concept is not new. In his classic 1969 *Science* article, Huxley (1969) described as an alternative the possibility that "the cross bridges might remain rigidly fixed in position while repetitive internal changes in the actin filaments enabled them to crawl along the series of fixed points so provided." With new evidence for conformational change in actin, and with great interest in actin from those in the nonmuscle motility field, Huxley's proposed alternative seems finally to be gaining support.

22.4 A myosin-rod–based motor?

At about the same time that Huxley put forth the actin-crawling hypothesis, a mechanism involving the myosin rod was advanced by Harrington (1971). In that hypothesis, it was not the myosin head (or the actin filament) that produced the power stroke, but a segment of the myosin rod. A helix-coil transition in the rod's so-called "hinge" region generates the force that drives contraction (Figure 22.2). Like the shrinkage of wool, the helix-coil transition, or melting, of the rod was theorized to underlie the power stroke. The proposed mechanism was based on physicochemical and electron micrographic observations (Burke et al., 1973; Segal and Harrington, 1967; Ueno and Harrington, 1984; Walker et al., 1985); it could account for elementary ATPase properties of myosin. It was also shown to explain basic mechanical features of contraction (Harrington, 1979).

The motor, in Harrington's scheme, lies within the hinge. As the crossbridge swings out and attaches to actin, the hinge region melts. Melting causes shortening of the polypeptide chain, or, if the chain is held at constant length, tension develops (Flory, 1956). Thus, the helix-coil transition in the hinge region is viewed as the source of the power stroke.

The proposal of hinge-region involvement was not supported by the advent of the in vitro motility assay (Harada et al., 1987; Kron and Spudich, 1986). It became clear that relative motion of actin and myosin could occur with the hinge region absent from the preparation. To many, this observation has seemed straightforward and compelling: the hinge region could not be the central element of contraction, irrespective of any evidence lending support to such a notion.

A potential solution to this dilemma was offered by Morel (1991). Morel noted that the force measured in the in vitro assay was sharply lower than the force (per molecule) measured in intact specimens. This was not true of velocity, however, which was comparable. The disparity in force implied the possibility of

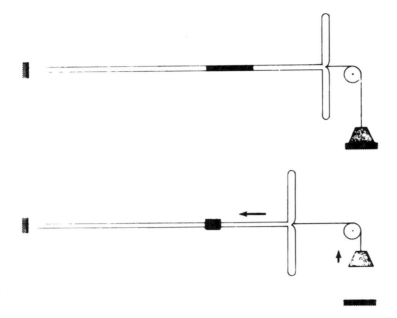

Figure 22.2. Helix-coil melting mechanism. A helix-coil transition in the rod of the myosin molecule generates force and shortening. (From Pollack, 1990.)

a force-generating mechanism present in the intact specimen but absent in the motility assay. If this mechanism was responsible for the production of high tensions, but played a lesser role in translational movement, the disparity between intact and in vitro measurements might be reconciled.

In fact, hinge-region melting produces the correct order of magnitude of force. The helix-coil transition is a physicochemical process that is reasonably well understood (Flory, 1956, 1961), from which force calculations are straightforward. Harrington (1979) initially showed that the correct magnitude of force is produced by this process; later, this was confirmed in a different context by Pollack (1990, p. 210). Thus, the mechanism "missing" in the motility assay could well be the helix-coil transition in the myosin rod.

Evidence in support of this mechanism comes from several areas. Direct evidence for myosin-rod shortening (expected from melting) has been obtained from measurements on electron micrographic (EM) images (Walzthöny et al., 1986a, 1986b). Up to 20 nm of hinge shortening has been detected at physiological temperatures (Walker and Trinick, 1986). More recently, a number of investigators have used hinge-directed antibodies as probes of hinge function (Harrington et al., 1990; Margossian et al., 1991; Sugi et al., 1992). The antibodies brought about an inhibition of contraction, and these results were interpreted to imply a central role of the hinge region in contraction.

Molecular biological evidence from *Drosophila* muscles also lends support to such a proposal (Collier et al., 1990; Hastings and Emerson, 1991). Exons 15a and 15b encode the central region of the hinge. In muscles that contract slowly (larval and adult body wall), only exon 15b is used. In muscles that contract rapidly (indirect flight muscle) or generate high levels of force (jump muscles), only exon 15a is used. In muscles that contract at intermediate rates (leg, proboscis), some transcripts contain exon 15a, whereas others contain 15b. Other regions along the rod remain invariant (Collier et al., 1990; George et al., 1989). Thus, the *Drosophila* evidence implies a central role of the hinge in the contractile process.

If hinge-region melting does take place during contraction, as much of this evidence implies, there are (at least) two ways in which it could be manifested. The distinction depends on whether the head–rod junction is bound to the thick filament backbone during melting. If the head–rod junction is not bound (as in the Harrington scheme), the melt force is exerted directly on the thin filament. The resulting scheme is much like the swinging crossbridge mechanism, except that power comes from head translation, not rotation.

If the head is bound to the filament backbone during the melt, the consequence is entirely different. In this case, melting force is exerted on another myosin in the backbone, and the rod of that molecule is driven past the one that melts. The net effect is a small amount of localized thick filament shortening. If additional molecules melt, the thick filament shortens more (Pollack, 1990). A paradigm built on this mechanism can account for a large body of data, including the fact that the thick filament is sometimes observed to shorten during contraction (for review, see Pollack, 1983).

Thus, there are at least two variations on the helix-coil–melting scheme: one in which the thick filament backbone remains stable, and another in which the thick filament can shorten.

22.5 Modern experimental approaches in search of the answer

Two near-molecular approaches that may help resolve the issue of the contractile mechanism are the single myofibril and the in vitro motility assay.

The single myofibril is the smallest functional unit that retains muscle's natural structural lattice. It is a bundle of several hundred parallel filaments, typically 1 μm in diameter (Figure 22.3). Its length may run from a few sarcomeres to many hundreds of sarcomeres.

Because the single myofibril is a unitary structure, standing alone, it bears all the tension measured at the end of the specimen. In larger preparations, tension is borne by many myofibrils in parallel; only if tension is assumed to be shared equally among all myofibrils can functional relations be drawn between tension and sarcomere dynamics. In the single myofibril, no such assumption need be made; measured tension can be related directly to each of the observed sarcomeres in the optical field.

The challenge is to devise transducers whose sensitivity is sufficient to detect the microgram-level forces generated by these specimens, and to develop means to cope with the specimens' delicacy. Pioneering work in these areas was carried out by Iwazumi, who spent some years developing an apparatus for these measurements (Iwazumi, 1987). A few experimental results have been published in the interim (Iwazumi, 1988).

We began designing a single myofibril setup about six years ago. In order to facilitate rapid data collection, we opted for a device simpler than Iwazumi's. The apparatus is built on a Zeiss inverted microscope equipped with phase-contrast optics. On a specially designed stage (Bartoo et al., 1993), we constructed a mechanical system to hold the myofibril. At each end of the specimen is a piggy-back combination of a piezoelectric motor mounted on a hydraulic micromanipulator. Thus, length changes can be imposed on either end of the specimen.

The striation pattern is viewable through the binocular eyepieces. The pattern is also projected through a port, onto a sensor, from which sarcomere length may be computed by two distinct methods, one based on a photodiode array, the other on a vibrating mirror and slit (Bartoo et al., 1993; Linke et al., 1993). Each scan produces an intensity profile of the striation pattern, which is then analyzed using a series of algorithms.

Tension is measurable by means of a specially designed force transducer, details of which were recently published (Fearn et al., 1993). In essence, the transducer is a bendable optical fiber beam through which light passes. Emerging light falls on a pair of optical fibers. As the beam deflects, the fraction of light falling differentially upon the two optical fibers changes, and thus tension is detected. The method works reasonably well, except for drift and limited bandwidth. A better transducer can be nanofabricated by optical or electron-beam lithography, and a preliminary version based on a deflectable silicon-nitride beam has proven robust and up to the task.

Because tension and sarcomere length are directly and unambiguously measured, the single myofibril preparation offers an excellent vehicle by which molecular theories can be tested. For example, the crossbridge theory predicts that tension should vary inversely with sarcomere length: longer sarcomeres have fewer working crossbridges, and therefore commensurately lower force-generating ability. On the contrary, both the actin-motor hypothesis and the hinge-melting hypothesis imply little falloff of tension with stretch (Pollack, 1990). In attempting to distinguish these possibilities in stretched, activated specimens, we have not been as successful as hoped in maintaining the uniformity of sarcomere length (Bartoo et al., 1993). Generally, sarcomeres lengthen or shorten upon activation, and then tend to maintain the new length for an extended period. The very

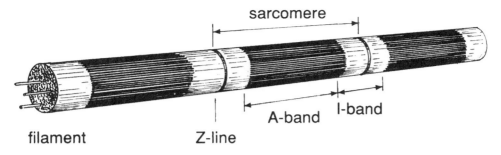

Figure 22.3. Schematic drawing of a single myofibril. The myofibril may extend for hundreds, even thousands, of sarcomeres. The Z-line denotes the sarcomere's boundary. (From Pollack, 1990.)

fact, however, that sarcomeres of varying length support the same tension implies that length does not matter; the same tension is generated by sarcomeres of different lengths. Thus, the length–tension curve may well be flat, as predicted by the newer theories. Additional evidence is required.

Another topic of interest is the comparative amount of tension generated by different types of muscle. By examining differences in molecular structure in the respective muscle types, it may be possible to elucidate the mechanism. For example, we have found that skeletal muscle myofibrils (rabbit psoas) generate four times more tension than cardiac muscle myofibrils (rabbit ventricle) under identical experimental conditions (Bartoo et al., 1993; Linke et al., 1993). There are differences in amino-acid sequence in a critical region of the hinge, and we are currently examining whether such differences may account for the difference of tension-generating ability.

The second preparation of interest is the in vitro motility assay. From pioneering work in the laboratories of Spudich (Sheetz and Spudich, 1983) and Yanagida (Yanagida et al., 1984), it has become possible to study directly the dynamics of interaction of the molecules responsible for contraction. In the in vitro motility assay, myosin molecules are planted on a substrate of silicon or nitrocellulose. Fluorescently labeled actin filaments are then added, and they bind to the lawn of myosins. When ATP is added, the actin filaments can be seen to translate over the myosins. Because of the potential of this approach in ferreting out molecular transduction mechanisms, we began setting up our own apparatus several years ago and have used it to complete a series of studies relevant to the contractile mechanism.

In setting up the assay, we came upon some interesting findings concerning motility at low ATP concentrations. Briefly, we found that, following exposure to millimolar ATP, we could reduce the ATP level to nanomolar and still not abolish motility (Kellermayer and Pollack, 1995; Kellermayer et al., 1995a, 1995b).

These studies imply that the assay system contains a mode of energy storage. Contractile motion can occur with very little ATP in solution, and so it appears that the pre-exposure to high ATP confers potential energy on the system and that this potential energy can then be tapped. In the crossbridge theory, there is no provision for energy storage; once a bridge cycles, its energy store is depleted, and additional ATP is required for the next cycle. In the actin-motor mechanism, ATP is split by actin. Because each actin subunit of the filament contains at least one ATP, a large number of ATP molecules are potentially available for energy supply, even at low-solution ATP.

The main reason for setting up the assay, however, was to probe filament dynamics and thereby infer something of the underlying mechanism. We, and others, had noticed that thin filament translation over myosin is not regular; the filament tends to reptate in snakelike fashion as it translates. To quantitate what was otherwise merely a visual impression, we carried out detailed measurements of filament-translation dynamics. We tracked the position of the leading tip, centroid, and trailing end of the actin filament. The velocity of translation of each of these points was found indeed to be periodic. Velocity oscillated between high and low values. Cross-correlation and autocorrelation functions revealed a periodicity of 2–4 Hz. Persistence time was several seconds, implying long-term memory of past events, and thus implying a deterministic process. Because phase differences were found between motions of leading and trailing ends of the filament, the deterministic process may involve a traveling wave, the visual manifestation of which is reptation. To check this hypothesis, we measured the frequency of the reptational wiggles and compared it with the frequency of periodic translation velocity. We confirmed that the frequencies were the same (controls, in rigor, showed no wiggling). Thus, the traveling-wave hypothesis along the actin filament is supported. Reptational wiggling had been previously thought to be a random process, arising out of Brownian motion, but the finding of a distinct wiggling frequency implies that the process is not random. It is apparently a manifestation of the actin-motor mechanism (de Beer et al., 1995).

A recent complementary study was carried out using a pair of optical traps to hold either end of the filament (Finer et al., 1994). A simpler and more natural approach is to attach a deflectable lever to the rear end of the filament. This has been accomplished by Yanagida and colleagues, who used a glass micropipette as the deflectable lever (Kishino and Yanagida, 1988).

Here again, nanofabrication offers considerable promise. First, nanofabricated levers of suitably thin scale can be used as the deflectable lever element. Such levers are precision made, of known material, and hence of known compliance. Thus, they are in essence precalibrated. Our limited experience with such levers is that they are remarkably robust. A second interesting application is in the design of surface tracks. At present, myosin is laid down randomly on the chamber surface. By creating tracks of nanometer width, myosin can be laid down in predesigned patterns. It would be interesting, for example, to determine how easily a filament can turn a right-angle bend.

22.6 Conclusion

At present, the mechanisms of contraction and motility are unknown. The crossbridge theory has been the central paradigm for some 40 years, but increasing evidence in conflict with predictions of the theory raise doubts as to its adequacy. Two alternative theories are under active consideration by a number of groups: the actin-motor hypothesis and the myosin-hinge–melting hypothesis. Each is supported by an appreciable body of evidence. The two mechanisms could work in a complementary way, with the first generating large-scale shortening and the second being responsible for the high tensions characteristic of muscle.

The most promising approaches to testing these paradigms involve experiments on preparations from which molecular-scale information can be extracted. In particular, the single myofibril offers information near the molecular scale in a preparation that retains the normal myofilament lattice in its pristine condition. The in vitro motility assay offers true molecular-scale information, albeit in a preparation in which the filament lattice has been removed. In combination, and especially with the help of nanofabricated devices, the two approaches promise new revelations – and perhaps final resolution of the question of how muscles contract.

References

Bartoo, M. L., Popov, V. I., Fearn, L., and Pollack, G. H. (1993) Active tension generation in isolated skeletal myofibrils. *J. Mus. Res. Cell Motil.* 14:498–510.

Burke, M., Himmelfarb, S., and Harrington, W. F. (1973) Studies on the "hinge" region of myosin. *Biochemistry* 12:701–710.

Collier, V. L., Kronert, W. A., O'Donnell, P. T., Edwards, K. A., and Bernstein, S. I. (1990) Alternative myosin hinge regions are utilized in a tissue-specific fashion that correlates with muscle contraction speed. *Gene Dev.* 4:885–895.

Cooke, R. (1986) The mechanism of muscle contraction. *CRC Crit. Rev. Biochem.* 21:53–118.

de Beer, E. L., Sontrop, A. M. A. T. A., Kellermayer, M. S. Z., and Pollack, G. H. (1995) Actin–filament motion in the *in vitro* motility assay is periodic. *Submitted.*

Drummond, D. R., Peckham, M., Sparrow, J. C., and White, D. C. S. (1990) Alteration in crossbridge kinetics caused by mutations in actin. *Nature* 348:440–442.

Fearn, L. A., Bartoo, M. L., Myers, J. A., and Pollack, G. H. (1993) An optical fiber transducer for single myofibril force measurement. *IEEE Trans. Biomed. Eng.* 40:1127–1132.

Finer, J. T., Simmons, R. M., and Spudich, J. A. (1994) Single myosin molecule mechanics: piconewton forces and nanometre steps. *Nature* 368:113–119.

Flory, P. J. (1956) Role of crystallization in polymers and proteins. *Science* 124:53–60.

Flory, P. J. (1961) Phase changes in proteins and polypeptides. *J. Polym. Sci.*
49:105–128.

George, E. L., Ober, M. B., and Emerson, C. P., Jr. (1989) Functional domains of the
Drosophila melanogaster muscle myosin heavy-chain gene are encoded by alternatively spliced exons. *Mol. Cell Biol.* 9:2957–2974.

Gordon, A. M., Huxley, A. F., and Julian, F. J. (1966) The variation in isometric tension
with sarcomere length in vertebrate muscle fibres. *J. Physiol.* (London)
184:170–192.

Harada, Y., Noguchi, A., Kishino, A., and Yanagida, T. (1987) Sliding movement of single actin filaments on one-headed myosin filaments. *Nature* 326:805–808.

Harada, Y., Sakurada, K., Aoki, T., Thomas, D. D., and Yanagida, T. (1990)
Mechanochemical coupling in actomyosin energy transduction studied by *in vitro*
movement assay. *J. Mol. Biol.* 216:49–68.

Harrington, W. F. (1971) A mechanochemical mechanism for muscle contraction. *Proc.
Natl. Acad. Sci. USA* 68:685–689.

Harrington, W. F. (1979) On the origin of the contractile force in skeletal muscle. *Proc.
Natl. Acad. Sci. USA* 76:5066–5070.

Harrington, W. F., Karr, T., Busa, W. B., and Lovell, S. J. (1990) Contraction of myofibrils in the presence of antibodies to myosin subfragment 2. *Proc. Natl. Acad. Sci.
USA* 87:7453–7456.

Hastings, G. A. and Emerson, C. P., Jr. (1991) Myosin functional domains encoded by
alternative exons are expressed in specific thoracic muscles of *Drosophila. J. Cell
Biol.* 114:263–276.

Higuchi, H. and Goldman, Y. (1991) Sliding distance between actin and myosin filaments
per ATP molecule hydrolysed in skinned muscle fibres. *Nature* 352:352–354.

Hirose, K., Lenart, T. D., Murray, J. M., Franzini-Armstrong, C., and Goldman, Y. E.
(1993) Flash and smash: Rapid freezing of muscle fibers activated by photolysis
of caged ATP. *Biophys. J.* 65:397–408.

Holmes, K. C., Popp, D., Gebhard, W., and Kabsch, W. (1990) Atomic model of the
actin filament. *Nature* 347:44–49.

Horowitz, A. and Pollack, G. H. (1993) Force-length relation of isometric sarcomeres in
fixed-end tetani. *Am. J. Physiol.* 264 (*Cell Physiol.* 33):C19–C26.

Horowitz, A., Wussling, H. P. M., and Pollack, G. H. (1992) Effect of small release on
force during sarcomere-isometric tetani in frog muscle fibers. *Biophys. J.* 63:3–17.

Huxley, H. E. (1969) The mechanism of muscular contraction. *Science* 164:1356–1366.

Huxley, A. F. and Hanson, J. (1954) Changes in the cross-striations of muscle during
contraction and stretch and their structural interpretation. *Nature* 173:973–976.

Huxley, A. F. and Niedergerke, R. (1954) Structural changes in muscle during contraction: Interference microscopy of living muscle fibres. *Nature* 173:971–973.

Iwazumi, T. (1987) High-speed ultrasensitive instrumentation for myofibril mechanics
measurements. *Am. J. Physiol* 252 (*Cell Physiol.*):C253–C262.

Iwazumi, T. (1988) Myofibril tension fluctuations and molecular mechanism of contraction. *Adv. Exp. Med. Biol.* 226:595–608.

Janmey, P. A., Hvidt, S., Oster, F. F., Lamb, J., Stossel, T. P., and Hartwig, J. H. (1990)
Effect of ATP on actin filament stiffness. *Nature* 347:95–99.

Jonas, M., Fearn, L. A., and Pollack, G. H. (1993) I-band periodicity in rabbit psoas
fibers measured using image analysis techniques. *J. Electr. Micr.* 42:285–293.

Kabsch, W., Mannherz, D., Suck, E., Pai, E. F., and Holmes, K. C. (1990) The structure
of the actin-DNase I complex. *Nature (Lond.)* 347:37–44.

Kellermayer, M. S. Z., Hinds, T. R., and Pollack, G. H. (1995a) Persisting *in vitro* motility of actin filaments at nanomolar ATP concentrations after ATP pretreatment.
Biochim. Biophys. Acta 1229:89–95.

Kellermayer, M. S. Z., Hinds, T. R., and Pollack, G. H. (1995b) Persisting *in vitro* motility at nanomolar adenosine triphosphate levels: Comparison of skeletal and cardiac myosins. *Submitted.*

Kellermayer, M. S. Z. and Pollack, G. H. (1995) Combined effect of adenosine triphosphate and ionic strength on actomyosin *in vitro* motility. *Submitted.*

Kishino, A. and Yanagida, T. (1988) Force measurements by micromanipulation of a single actin filament by glass microneedles. *Nature* 334:74–76.

Kobayashi, S. (1964) Effect of electric field on F-actin oriented by flow. *Biochim. et Biophys. Acta* 88:541–552.

Kron, S. J. and Spudich, J. A. (1986) Fluorescent actin filaments move on myosin fixed to a glass surface. *Proc. Natl. Acad. Sci. USA* 83:6272–6276.

Linke, W., Bartoo, M. L., and Pollack, G. H. (1993) Spontaneous sarcomeric oscillations at intermediate activation levels in single isolated cardiac muscle fibers. *Circ. Res.* 73:724–734.

Lombardi, V., Piazzesi, G., and Linari, M. (1992) Rapid regeneration of the actin–myosin power stroke in contracting muscle. *Nature* 355:638–641.

Margossian, S. S., Kreuger, J. W., Sellers, J. R., Cuda, G., Caulfield, J. B., Norton, P., and Slayter, H. S. (1991) Influence of the cardiac myosin hinge region on contractile activity. *Proc. Natl. Acad. Sci. USA* 88:4941–4945.

Ménétret, J.-F., Hofmann, W., Schröder, R. R., Rapp, G., and Goody, R. S. (1991) Time-resolved cryo-electron microscopic study of the dissociation of actomyosin induced by photolysis of photolabile nucleotides. *J. Mol. Biol.* 219:139–144.

Morel, J. E. (1991) The isometric force exerted per myosin head in a muscle fibre is 8 pN. Consequence on the validity of the traditional concepts of force generation. *J. Theor. Biol.* 151:285–288.

Nakata, T., Sato-Yoshitake, R., Okada, Y., Noda, Y., and Hirokawa, N. (1993) Thermal drift is enough to drive a single microtubule along its axis even in the absence of motor proteins. *Biophys. J.* 65:2504–2510.

Ohno, T. and Kodama, T. (1991) Kinetics of adenosine triphosphate hydrolysis by shortening myofibrils from rabbit psoas muscle. *J. Physiol. (Lond.)* 441:685–702.

Pollack, G. H. (1983) The sliding filament/cross-bridge theory. *Physiol. Reviews* 63:1049–1113.

Pollack, G. H. (1990) *Muscles and Molecules: Uncovering the Principles of Biological Motion.* Seattle: Ebner & Sons.

Pollard, T. D. (1986) Actin and actin-binding proteins. A critical evaluation of mechanisms and functions. *Am. Rev. Biochem.* 55:987–1035.

Pollard, T. D., Bhandari, D., Maupin, P., Wachsstock, D., Weeds, A. G., and Zot, H. G. (1993) Direct visualization by electron microscopy of the weakly bound intermediates in the actomyosin adenosine triphosphatase cycle. *Biophys. J.* 64:454–471.

Prochniewicz-Nakayama, E., Yanagida, T., and Oosawa, F. (1983) Studies on conformation of F-actin in muscle fibers in the relaxed state, rigor, and curing contraction using fluorescent phalloidin. *J. Cell Biol.* 97:1163–1667.

Rayment, I., Holden, H. M., Whittaker, M., Yohn, C. B., Lorenz, M., Holmes, K. C., and Milligan, R. A. (1993a) Structure of the actin-myosin complex and its implications for muscle contraction. *Science* 261:58–65.

Rayment, I., Rypniewski, W. R., Schmidt-Base, K., Smith, R., Tomchick, D. R., Benning, M. M., Winkelmann, D. A., Wesenberg, G., and Holden, H. M. (1993b) Three-dimensional structure of myosin subfragment-1: A molecular motor. *Science* 261:50–58.

Schutt, C. E. and Lindberg, U. (1992) Actin as the generator of tension during muscle contraction. *Proc. Natl. Acad. Sci. USA* 89:319–323.

Schutt, C. E. and Lindberg, U. (1993) A new perspective on muscle contraction. *FEBS* 325:59–62.

Schutt, C. E., Lindberg, U., Myslik, J., and Strauss, N. (1989) Molecular packing in profilin: Actin crystals and its implications. *J. Mol. Biol.* 709:735–746.

Segal, D. M. and Harrington, W. F. (1967) The tritium-hydrogen exchange of myosin and its proteolytic fragments. *Biochemistry* 6:768–787.

Sheetz, M. P. and Spudich, J. A. (1983) Movement of myosin-coated fluorescent beads on actin cables *in vitro*. *Nature 303,* 31–35.

Squire, J. M. (1990) *Molecular Mechanisms in Muscular Contraction.* Boca Raton, FL: CRC Press.

Sugi, H., Kobayashi, T., Gross, T., Noguchi, K., Karr, T., and Harrington, W. F. (1992) Contraction characteristics and ATPase activity of skeletal muscle fibers in the presence of antibody to myosin subfragment 2. *Proc. Natl. Acad Sci USA* 89:6134–6137.

Suzuki, S., Oshimi, Y., and Sugi, H. (1993) Freeze-fracture studies on the cross-bridge angle distribution at various states and the thin filament stiffness in single skinned frog muscle fibers. *J. Electron Microsc.* 42:107–116.

Szent-Györgyi, A. G. and Prior, G. (1966) Exchange of adenosine diphosphate bound to actin in superprecipitated actomyosin and contracted myofibrils. *J. Mol. Biol.* 15:515–538.

Thomas, D. D. (1987) Spectroscopic probes of muscle crossbridge rotation. *Ann. Rev. Physiol.* 49:891–909.

Thomas, D. D. (1993) Pollard to actomyosin: "Freeze! Don't even move your head!" *Biophys. J.* 64:297–298.

Toyoshima, Y. Y., Kron, S. J., McNally, E. M., Niebling, Toyoshima, C., and Spudich, J. A. (1987) Myosin subfragment-1 is sufficient to move actin filaments *in vitro*. *Nature* 328:536–538.

Toyoshima, Y. Y., Kron, S. J., and Spudich, J. A. (1990) The myosin step size: measurement of the unit displacement per ATP hydrolyzed in an in vitro assay. *Proc. Natl. Acad. Sci. USA* 87:7130–7134.

Trombitás, K., Baatsen, P. H. W. W., and Pollack, G. H. (1988a) Effect of tension on the rigor cross-bridge angle. In *Molecular Mechanism of Muscle Contraction*, ed. H. Sugi and G. H. Pollack, pp. 17–30. New York: Plenum Press.

Trombitás, K., Baatsen, P. H. W. W., and Pollack, G. H. (1988b) Rigor cross-bridge angle: Effect of stress and preparative procedure. *Ultrastruct. Molec. Struct. Res.* 97:39–49.

Ueno, H. and Harrington, W. F. (1984) An enzyme-probe study of motile domains in the subfragment-2 region of myosin. *J. Mol. Biol.* 180:667–701.

Vale, R. D. and Oosawa, F. (1990) Protein motors and Maxwell's demons: Does mechanochemical transduction involve a thermal ratchet? *Adv. Biophys.* 26:97–134.

Wakabayashi, K., Ueno, Y., Amemiya, Y., and Tanaka, H. (1988) Intensity of actin-based layer lines from frog skeletal muscles during an isometric contraction. *AEMB* 226:353–367.

Walker, M., Knight, P., and Trinick, J. (1985) Negative staining of myosin molecules. *J. Mol. Biol.* 184:535–542.

Walker, M. and Trinick, J. (1986) Electron microscope study of the effect of temperature on the length of the tail of the myosin molecule. *J. Mol. Biol.* 192:661–667.

Walzthöny, D., Eppenberger, H. M., and Wallimann, T. (1986a) Melting of myosin rod as revealed by electron microscopy. I. Effects of glycerol and anions on length and stability of myosin rod. *Europ. J. Cell Biol.* 41:33–37.

Walzthöny, D., Eppenberger, H. M., and Wallimann, T. (1986b) Melting of myosin rod as revealed by electron microscopy. II. Effects of temperature and pH on length and stability of myosin rod and its fragments. *Europ. J. Cell Biol.* 41:38–43.

Yanagida, T., Nakase, M., Nishiyama, K., and Oosawa, F. (1984) Direct observation of single F-actin filaments in the presence of myosin. *Nature* 307:58–60.

Yanagida, T. and Oosawa, F. (1978) Polarized fluorescence from e-ADP incorporated into F-actin in a myosin-free single fiber: Conformation of F-actin and changes induced in it by heavy meromyosin. *J. Mol. Biol.* 126:507–524.

Yanagida, T. and Oosawa, F. (1980) Conformational changes of F-actin–e-ADP in thin filaments in myosin-free muscle fibers induced by Ca^{2+}. *J. Mol. Biol.* 140:313–320.

Yanagida, T., Taniguchi, M., and Oosawa, F. (1974) Conformational changes of F-actin in the thin filaments of muscle induced *in vivo* and *in vitro* by calcium ions. *J. Mol. Biol.* 99:509–522.

23

Technology needs for the Human Genome Project

DAVID T. BURKE

Department of Human Genetics, Human Genome Center
University of Michigan
Ann Arbor, MI 48104

23.1 Introduction

The Human Genome Project is a challenge to the international science community to develop an informational and technological infrastructure for medicine and the biological sciences. In the United States, two Federal grant-funding agencies – the National Institutes of Health and the Department of Energy – have taken the lead in promoting this work. The current annual allocation is over $160 million, with funding for both technology development and data acquisition. Although the Genome Project has defined as its goal the assembly of the three billion basepairs of the human genome, the demand for additional DNA sequence information resulting from that knowledge will be unlimited. Future progress in biological science will require improvements in all aspects of DNA biochemistry, as well as in sample handling and data analysis. The Genome Project is intended to be the leading edge for the development of new ideas in biology.

Future genome-related technologies should attempt to combine the various aspects of DNA isolation, purification, and sequencing biochemistry within a single, integrated platform. In the ideal case, data detection, analysis, and decision-making would also reside in the same system. The use of micro- or nanoscale technology is an obvious choice for this platform. Microscale biochemistry will reduce costs associated with expensive biological reagents, and lithographic fabrication techniques can provide inexpensive identical components. Additionally, the biochemical systems may be directly controlled by electronic circuitry fabricated within the same substrate. The integration of information processing and sample processing on a single platform will reduce human interaction in the routine acquisition of data, allowing a greater emphasis on biological interpretation.

This chapter will provide brief definitions of genes, sequences, clones, and genomes. Then the current status of the Human Genome Project is outlined. Finally, a list of challenges is set forth regarding the technology. It is essential to note that the current 10- and 15-year goals of the Project will not be attainable with current technology.

23.1.1 Genes

Genes, simply put, are packets of information. Each gene represents an instruction for making one component of a complete organism. It is estimated that approximately 100,000 to 300,000 genes are required to construct and maintain an organism as complex as a human being. A gene can be observed by its effects on the whole organism, even without knowing what precise component is made from the gene. As an example, in laboratory mice, gray or black fur color is regarded as the end product of two varieties of the same gene, even though the molecules involved in the color formation process are unknown. Similarly, defective genes can be observed as either a single defective component or as an incorrectly functioning complex system.

Genetic information is known to be stored in the molecule DNA; consequently, the two have become inextricably bound together in both scientific and popular understanding. The genetic information is encoded in the DNA molecule. A complex set of biochemical machinery is required to decode the information and produce the final, functioning biological components. Proteins form the major class of gene products and are the working parts of the cell. However, DNA molecules are nearly all that is passed from generation to generation. To a first approximation, a full complement of DNA-encoded genetic information is enough to construct a new individual. Biological science is just beginning to understand how the elaboration of the encoded information occurs during the complex process of development.

In human genetics there is no gene for a particular disease. A full set of genes is found in every human. Typically, there exist variants of a gene, and certain variants can lead to disease. One class of variants results in the complete lack of a particular gene product; these are termed "null" variants. Often this is the result of a gene version whose information has been scrambled so badly that it can no longer be decoded by the organism. Null variants, as a class, are usually lethal in the developing individual. Some null variants can be born, however, with the full effect's being delayed. Examples are the childhood-onset human diseases Duchenne muscular dystrophy and Tay–Sachs disease.

Other gene variants are more subtle, providing sufficient function in the young individual and proving inadequate only at a later age. Known genes of this class include Huntington disease, and some rare forms of cancer and heart disease. The future of human genetics, and one of the driving forces behind the Human Genome Project, is the uncovering of the genetic influences in common diseases such as diabetes, high blood pressure, and cancer. Common diseases typically appear later in life and have both a complex, multigene inherited component and a significant environmental influence.

23.1.2 Sequences

In any discussion about genes at the physical level it becomes important to understand the chemistry of DNA (Stryer, 1988). There are, of course, plenty of good books for detailed information. Essentially, DNA is a long linear polymer molecule. The long chain polymer is made of four possible monomer units, called bases (abbreviated as A, C, G, and T). The genetic information is contained in the pattern of the four monomer varieties along the chain. The unique biological characteristic of DNA is its double-stranded nature. Each of the four monomer units can only "pair" with one of the others on the opposite strand; A pairs with T, and G pairs with C. Consequently, the encoded information exists on each strand, with each using the complementary set of monomer bases. When a cell divides, it makes use of the double-stranded property by sending one strand to each new cell. The missing strand is then synthesized from a pool of available monomers, using the intact strand as a template.

Determining the pattern of monomer units along a strand is called DNA sequencing. Several biochemical methods exist for determining the monomer pattern from a portion of DNA. At the current state of technology, the length of DNA that can be sequenced in a single biochemical reaction is between 300 and 1,000 monomers long. The typical human "gene" – a single instruction to perform a single task – uses about 30,000 units or basepairs of information. The largest human gene known uses about 2.5 million units; the smallest gene uses about 80 units. The information is truly "encoded"; it is not clear how to decode the sequence into human-readable information. One of the major challenges of the Human Genome Project will be to perform the necessary biological experiments and develop algorithms for identifying the important information from a long string of monomer units.

23.1.3 Clones

One of the few molecular biology terms to make it into the vernacular, *clone* predates molecular biology by several decades. Its original definition still provides the most accurate description. A clone was originally described as the set of propagated cuttings derived from a single parent plant. All the new individuals are direct progeny of the original. The important feature of clonal reproduction is that the DNA information of the progeny is identical to that of the original plant. Cloning is very unlike standard sexual reproduction, where the DNA information from both parents is "reshuffled" prior to combining in the offspring.

Any modifications to the DNA of the single, parental organism are propagated intact to its clonal offspring. In particular, if a new, foreign piece of DNA is inserted into the parent, that fragment of DNA will be propagated in all of the offspring. The single piece of DNA originally captured by the parent is now reproduced thousands or millions of time over, once in each descendant. Thus, DNA "cloning" is essentially the amplification of a small initial signal. Because the DNA replication machinery of a cell is very accurate (it needs to be, to keep from introducing errors into its own essential information), the DNA that has gone "along for the ride" is replicated accurately as well.

23.1.4 Genomes

The word *genome* is used to describe the complete set of genes required to make an organism (Alberts et al., 1994). It implies knowledge of all of the DNA found in the cells of a particular organism. Consequently, there is no "Human Genome." Each of us is a unique collection of gene variants; therefore, we each have our own genomes. The term *human genome* is used to define the collection of genes required to make an "archetypal" human being. When the sequence of the first human genome is complete, it will be a reference genome used for further detailed characterization of the basic set of information. The basic set of information will be identical in each human individual. However, each individual contains a unique set of variants, with changes on the order of one per thousand bases.

The 6 billion basepairs of DNA in each human cell can be thought of as two immensely long DNA molecules of 3 billion units each. Each 3-billion-unit molecule encodes a complete set of "human-specific" information, with one set donated from each parent. (This redundancy is important, because it provides a functioning "back-up copy" in cases where a defective instruction has been inherited from one parent.) In reality, the 3 billion basepairs of DNA are divided into 23 rather arbitrary units or chromosomes, ranging from about 200 million to 50 million basepairs in length.

The sequence-pattern structure of the human genome is complex. The rules used by the cell to recognize encoded information involve both specific and variable basepair orderings. Also, many sequence patterns are repeated in the genome, from stretches of small repeats (two-, three-, and four-member sequences; such as GTGTGTGTGTG . . .), to massive repeated segments of 100,000 basepairs or larger. One of the most confounding sequence structures – for scientists actively sequencing large stretches of human DNA – is an element of about 280 basepairs that occurs randomly over 100,000 times in the genome. This element is found on average several times every 10,000 basepairs. Finally, many individuals are born that are missing fragments of their genomes (from single basepairs to 105 basepairs). From this observation it is clear that not all of the DNA-encoded information is essential. The basepairs of the genome that are essential, or superfluous, are not understood by simple examination of their sequence. Consequently, the scientific challenges of DNA-sequence analysis are of encoding or encryption, rather than of structural understanding. The placement of biological or sequence-based information at specific locations along the genome constitutes mapping.

23.2 The Human Genome Projects

23.2.1 Defined goals of the NIH/DOE Project

The Human Genome Project is a large, directed effort to establish a common data infrastructure for human genetics and the biological sciences. The scope and time scale of the project were proposed in a 1988 National Research Council (NRC) report *Mapping and Sequencing the Human Genome* (National Research Council,

1988). The 15-year project is jointly administered by the National Institutes of Health (NIH) and the Department of Energy (DOE), and has a current-year 1994 budget of $160 million. The explicit goals are the determination of the complete DNA sequence of the human genome and of the genomes of several experimental organisms. The experimental organisms have been chosen based on their historical use as genetic systems. They include the fruit fly *Drosophila melanogaster*, brewer's yeast *Saccharomyces cerevisiae*, the roundworm *Caenorhabditis elegans*, the bacterium *Escherichia coli*, and the laboratory mouse *Mus musculus*.

The NRC report anticipated three 5-year periods, each defined by milestones unattainable by the then-available level of technology (U.S. Dept. of Health and Human Services, 1990). To date, those expectations have been fulfilled, with technical developments occurring at a rapid pace. Most of the developments have been at the level of basic biochemical manipulation of DNA, including new methods for DNA isolation, large-DNA-fragment cloning, DNA amplification (polymerase chain reaction, PCR), and DNA chemical synthesis. These have led directly to a complete moderate-resolution gene map and the first low-resolution DNA fragment map of the human genome. Detailed descriptions of several experimental organisms have also been completed. The second and third 5-year periods are expected to refine the crude maps, annotate them with important biological information including genes, and finally, complete the full DNA sequence of all of the genomes.

However, the 10- and 15-year goals of the Project are clearly not attainable with the current level of technology. Technical advances that increase sequencing abilities by over 100-fold must occur, not only to acquire the initial information but to implement the sequence information in solving real-world biological problems.

23.2.2 *The implied goals of genome science*

Complete sequencing implies more than the simple listing of the pattern of 3 billion basepair units. The DNA substrate for the sequencing reactions must be isolated (purified) as smaller cloned fragments. The fragments must be correctly ordered to reflect the true organization of the genome as a linear genome clone map, and they must be permanently stored for future use by the research community. Subsequently, the clones can be sequenced based on their map location to provide complete genomic coverage. Finally, a continuing process of biological analysis will reveal the encoded information and provide an annotated map with the location of individual genes.

Mapping and sequencing the human genome is not an end in itself (Olson, 1993). It will not provide complete understanding of human biology and disease. The genome project is, instead, a directed attempt to obtain the raw materials for decades' worth of detailed biological work. The use of the term *maps* is appropriate. Maps provide guides for understanding and serve to display the accumulated set of knowledge from those that have gone before. The human genome sequence will provide the foundation for this map, with further clinical and basic science adding more layers of information to it in a continuing process.

Although the archetypal reference map needs to be completed only once, the end of the Human Genome Project is by no means the end of large-scale sequence analysis. The archetypal genome is only the first level. Like digital computing, DNA sequencing is a technology with potentially unlimited demand. In particular, the reference genome is not satisfactory from the clinical view. Clinical science is concerned with what is wrong with a particular human, and is less interested in humans as a general class. The study of human variation is the key to clinic-level genetic analysis. All humans are approximately 99.9% identical. However, an understanding of how humans are different as individuals is essential to clarifying why certain patients develop disease.

Information on human variability will have an enormous impact on health care of the future. It will, at least in part, form the basis of predictive health care. Most of the predictions will be conditional and will provide options for preventive health management. The information will emphasize individually tailored responses to specific drugs or individual predictions of prognosis and testing. The technologies developed for sequencing the reference human genome will be used in the clinic to identify thousands of DNA basepairs from millions of individuals.

Numerous biological tests can be made more quickly or more accurately by DNA-sequence analysis (Edwards et al., 1991; Melis et al., 1993; Pena and Chakraboty, 1994). DNA-based comparisons provide the highest level of discrimination between individual humans. Organ transplantation using accurate matches between the prospective donor and recipient can greatly reduce the rejection rate between unrelated individuals. The ability to distinguish people by their genetic makeup has made DNA analysis an indispensable tool in forensic police investigation and in relationship (paternity) testing. DNA sequence information is also useful for characterization of bacterial or viral infections in clinical patients. DNA identification of pathogens can be accomplished in a few hours, in contrast to conventional culturing techniques, which take several days. Rapid, inexpensive organism typing will be extremely useful for identifying the source of communicable disease outbreaks or for distinguishing between closely related disease organisms.

From a basic biological science standpoint, the genome projects will provide essential information. The emphasis on parallel sequencing of model organisms is crucial, both as a forum for testing hypotheses and for comparative analysis. Comparisons between organisms are the fundamental strength of biological science. In the same way that comparative anatomy – looking at frogs and rats – has led to a clearer understanding of human anatomy, comparative anatomy of the underlying information in DNA will be essential to human genetics.

23.2.3 An infrastructure for bioindustry

Both the biological information and the new technologies created by the Human Genome Project will provide the basis for many new branches of the biotechnology industry. DNA analysis is universal for the understanding of living things. Technologies developed for human DNA analysis are equally useful for the analysis of any other organism. Already, several multimillion-dollar companies have been formed to take advantage of human DNA sequence information from the Genome Project, with anticipated payoffs in pharmaceutical development, gene therapy, and bioengineering.

A prime example of a future DNA-based industry is disease diagnostics. It is not sufficient simply to discover the DNA sequence of a gene involved in a disease; that sequence must also be re-identified in thousands of individuals being tested for the disease. Population-based genetic studies will be essential to understanding the heritable components of these diseases; such studies will involve the genetic typing of hundreds to thousands of individuals. Localization of the genes involved in multigenic disorders will be even more complex and may involve broad surveys of the entire genome. Genotyping assays in such studies may easily approach several million.

Basic science laboratories will also be intensive users of new technologies. Technologies established for studying the human genome and the classical experimental organisms will enable researchers on other organisms to operate at a reduced cost. Many pathological organisms present an enormous health burden on the human species, including protozoans (malaria), parasites (schistomatosis), bacteria (cholera), and viruses (AIDS). Inexpensive genomic-analysis technology will permit rapid diagnosis, tracking of populations, and a clearer understanding of the underlying biological basis of these diseases.

Agricultural science of animals and plants is a traditional user of genetic technology, with annual expenditures of over a billion dollars per year. DNA technologies can improve the speed and effectiveness of selective breeding for pathogen resistance, improved crop yields, and improved nutritional value. The implementation of modern DNA techniques to crops important in the developing world has often been held back by high costs. Inexpensive DNA analysis will be instrumental to expanding genetic analysis to indigenous crop species. Finally, preservation programs for endangered species are rapidly expanding their use of DNA analysis to improve the understanding of breeding schemes and genetic variation.

23.3 The technology needs of the Human Genome Project

The cost–benefit analysis of individual genetics projects is often different from the optimal benefit for the community. Individual projects usually proceed faster by the intensive application of current methods rather than by taking time to develop newer, more efficient techniques. This is sometimes called the "undergraduate labor equation": Will results be obtained faster by hiring 10 undergraduate students (at $2,000 plus $3,000 in supplies per year, each) or by investing $50,000 in engineering costs? Each student will be performing proven, low-technology methods. The repetitive nature of the tasks is not important, since the labor source is replicable. In contrast, any engineering project entails significant risk. Almost without exception, successful projects to identify human genetic diseases have relied on the application of brute force rather than risky technology. This has led to a strong sense of "John Henryism" among the geneticists, and a less-than-enthusiastic response to new technologies.

This focus on the immediate needs of individual research projects has been successful to date – and funding for further research usually follows success. It is clear, however, to many genetic researchers that fundamental limits to the application of brute force are being approached. The current best level of DNA sequencing is being performed by two groups operating in parallel (one in the United States and the other in the United Kingdom) (Sulston et al., 1992). Both groups are able to generate finished sequence to an accuracy of 99.9% at the rate of 1.5 million bases per year. Each group maintains a full-time staff of about 10 scientists and 20 technicians and spends close to $6 million per year. Even a 10-fold improvement in current methodologies will not meet the defined goals of the Human Genome Project. Consequently, the NIH and DOE are providing a sizable portion of their budgets (approximately $30 million) for the next five years for development of new DNA-analysis technologies that can provide the 100- to 1,000-fold increase necessary.

Although a wide range of technologies is being examined at the pilot-project level, only methods that deliver over the next five years will be considered successful. The delivery of the technology must occur in the laboratory of human molecular geneticists. The payoff of success is large, with hundreds of basic science laboratories and thousands of clinical testing centers interested in the technology.

23.3.1 *Short-term needs*

The short-term needs of the genetics community are simple engineering problems defined by limited goals, time frames, budgets, and error parameters. The primary focus will be laboratory devices for reducing repetitiveness and operator error. Almost certainly, all the short-term developments will be made obsolete within 10 years. However, if planned properly, they can be enormously valuable first steps in the construction of more demanding technologies. As expected, the short-term solutions must be grounded in the current technology and must have a high likelihood of working.

Current sequencing technology was developed to look at small regions of DNA in high detail. Sequencing of 40,000 basepair units per day, as 100 sets of 400 basepairs each, is readily accomplished with the current generation of semi-automated sequencers. These machines are simple DNA-sequence readers and do not perform *any* of the difficult sample-preparation tasks. As a consequence, the human effort has been transferred from the actual DNA-sequence separation and reading to DNA-sample preparation. The machines are not robust and, in order to keep the machine "fed," the incoming samples must be completely uniform. Uniform, small DNA fragments are derived from the source DNA by a process known as "shotgun" cloning. The shotgun sequencing technique works by breaking up a large DNA fragment into little pieces, sequencing the small fragments, and reconstructing the original sequence pattern with specially designed software (Ansorge et al., 1992; Hunkapillar et al., 1991; Wilson et al., 1988).

A rough analogy is moving a house that is too big to lift and move in one piece. The first step is to break the house into many tiny fragments that can be easily lifted. The small fragments are moved and are used to reconstruct the house at another location. The process demands that no pieces are lost during transport. Because losses are inevitable, however, a group of one hundred identical houses is the starting material, and the houses are all broken into pieces. The fragments from each house will be slightly different, and the overlap between fragments from different houses will guide the reconstruction. The entire process is heavily redundant and computationally intensive.

Shotgun sequencing is a very similar brute force method. Many identical copies of a DNA fragment are available from a cloned source. The population of molecules is broken up randomly and the random portions are sequenced. The standard redundancy level is about six- to eight-fold, resulting in more than 600,000 bases of "raw sequence" being processed to obtain 100,000. Technology that reduces the human effort, or simplifies individual steps in the process, is of immediate importance. Examples include automated labeling systems, robotic liquid-handling devices, and simplified gel electrophoresis systems.

Alternative "bootstrap" methods make better use of the continuity of the starting DNA and require a lower level of redundancy. From information derived at the end of a completed sequence, new reagents (oligonucleotide primers) are prepared that can sequence the next several hundred bases. However, not all sequences are equally suited to the development of the new reagents, and small errors in the initial sequence determination can propagate into the new sequence. Additionally, the synthesis of new oligonucleotide reagents at each step is expensive. Novel technologies that can anticipate the need for oligonucleotide primers, either by rapid synthesis or by prepared arrays, would increase the utility of bootstrapping. In addition, there should be other strategies that combine the continuity of bootstrap methods with the uniform sample preparation of shotgun methods. These would break the initial large DNA fragment into pieces that retain some aspect of their original linearity, rather than having complete randomization. The redundancy of initial data collection and the associated effort of reassembly would be reduced.

23.3.2 Long-range needs

Ideally, new technologies would replace all of the separate processing steps currently found in the laboratory. The replacing technology should provide equal experimental value for comparable cost. The system must be modifiable, compatible with current technology, and must reduce the labor-intensive nature of the work. The future technology either must be compatible with the current hierarchical methods of DNA analysis or must eliminate them completely. No clear alternative yet exists that can take a chromosome-sized natural DNA molecule and complete the sequence analysis in a single step.

The large size of human genomic DNA reinforces the point that these are biological molecules. The machinery in human cells is able to copy DNA fragments on the order of 100 million bases (Mb) to 200 Mb long. The only known way to maintain and replicate large stretches of DNA accurately is in vivo. Consequently, biologists have focused on efficient methods for disassembling genomic DNA into cloned libraries that represent the total genome. Other types of microbial cells, such as yeast or bacteria, are used as the workhorses for maintaining smaller fragments of human DNA.

Currently the hierarchy is represented in four steps (Burke, 1991). First, from total genomic DNA (3 billion basepairs), large cloned DNA fragments of 1 million to 300,000 basepair, are maintained in brewer's yeast as yeast artificial chromosomes (YACs). A complete set of YAC clones for the human genome is currently being ordered and will be available within two years. At this stage, further detailed analysis can be performed by starting with YAC DNA, rather than with total human DNA. The complexity of the analysis is reduced by a factor of 1,000, but the number of analyses increases by several thousand. Analysis at each step requires logical methods for reassembling the fragments.

Subfragments of YACs on the size of 25,000 to 50,000 basepairs can be easily maintained in bacteria as either cosmid clones or bacteriophage. Each 1-million-basepair YAC is then represented by an overlapping set of 20 or so cosmid-containing clones. At the next level, the cosmids are further subcloned into M13 filamentous bacteriophage, which stably maintain DNA of less than 5,000 basepairs. Again, these must be arranged into an overlapping set of fragments that represent the source cosmid clone.

Finally, complete in vitro methods for DNA duplication become available at lengths below 5,000 basepairs. In the main, these are polymerase chain reaction (PCR) methods. Because it is completely ex vivo, PCR "cloning" is highly controllable, and it is the current favorite among molecular biologists. It still, however, is not a nonbiological chemical reaction, because it requires an enzyme catalyst and is prone to error not found in intact cellular replication. The ultimate sequencing reactions, as they exist today, are only effective at less than 1,000 bp.

Consequently, the genome biologist must transverse levels in this hierarchy by handling cloned DNA in living organisms. In the absence of a completely in vitro system for dealing with large complex samples of DNA, any complete process must still deal with cloning and subcloning to reduce the individual pieces of DNA to manageable size. Each organism or system for cloning has its own quirks and is not as definable as an in vitro system. Traversing the hierarchy involves constant growing of organisms to make DNA, harvesting the material, modifying it or reducing it to smaller fragments, and reintroducing the DNA into the next organism in the series. At each stage decisions are being made, and cells are being sorted by their biological response to the introduced DNA fragments. The number of individual samples that must be dealt with expands enormously.

In general, biological laboratory analysis uses natural "bioreactors" (cells) or in vitro biochemical reactions, whichever is most efficient. In vitro biochemistry has more control of inputs and outputs. However, in vitro systems have finely tuned automatic control mechanisms, including feedback regulation and error correction. Proposed systems for use in the near future may need to include both.

23.3.3 *Integrated microfluidic analysis*

One proposed solution is to develop a suite of components that work together and are interactive. The initial implementation can be rather crude and decidedly "large scale" for micro- or nanofabrication researchers. The goal is to develop a novel integrated format for DNA molecular analysis. All the processing steps for extracting genotype information from a DNA sample occur without the intervention of a human or robotic operator. The individual processing components are fabricated on a substrate that is readily miniaturized and is amenable to mass production technology. Integration requires that each step in the process be modified for direct compatibility with the subsequent step, with the minimum amount of operator involvement. Having a working system available and in the laboratory quickly will be important. The feedback from DNA researchers will assist in driving incremental improvements.

The silicon integrated computer circuit chip is a useful example of an integrated system. Each of the components of the desired circuit – whether transistors, capacitors, or resistors – is fabricated directly from the same silicon substrate. Once the characteristics of each type of silicon component are established, various components can be linked in a logical pattern to accomplish a task. Incremental improvements in each component can be developed by independent research teams, with the eventual goal of reintegration into the complete system.

If all the components for DNA analysis are to be fashioned from a single substrate, silicon is a natural choice (Petersen, 1982). The initial implementation should begin at a scale where biologists are confident of the parameters and variables. Potential new methods of DNA biochemistry, as well as of protein biochemistry, should remain compatible with the integrated microfluidic system. The enormously flexible design characteristics of silicon will allow improved versions to be developed rapidly as the basic biochemical methods advance.

At a technical level, several components need to be developed to complete an integrated DNA-analysis system. All components must be tested for biological and fabrication compatibility. As the overall system improves, it is reasonable to anticipate that the amount of microbial handling will decrease and the variety of biochemical handling steps will increase. Several research groups around the world are working on subsets of these components.

(1) *Sample injectors.* Biological samples are usually complex mixtures, available as volumes 10- to 100-fold larger than required for a microfabricated system. Clinical samples, in particular, are extremely heterogeneous and will require simple, robust methods for application to the microfabricated substrate. Reagents can be loaded and maintained in integral reservoirs.

(2) *Microbial sorting.* Individual cells can be segregated into chambers and expanded to form clonal colonies. An ideal sorter would also detect and make sorting decisions based on cell-surface markers.

(3) *Microbial growth chambers.* The chamber allows expansion of a single cell to isolate sufficient quantities of DNA. Nutrients must be transferred into the chamber and wastes flushed out. Many of the relevant biological molecules can be passed across a membrane with a molecular-weight cutoff of 50,000 daltons.

(4) *DNA purification.* Because the majority of the mass of any microbe is *not* DNA, the components must be chemically separated. Standard chemical purification requires detergents and relatively high pH, all of which must be neutralized before any subsequent handling steps. Filamentous bacteriophage, such as M13, are a promising alternative, since they are simply protein-coated DNA molecules. Filamentous viruses do not kill their hosts; instead, the virus particles are extruded through the cell membrane.

(5) *Transport channels.* Very small drops of liquid must be moved between components in the substrate. By analogy to microelectronics, the transport channels are the "wires" linking the functional components. The channels should be completely enclosed, to reduce evaporation, and may require several size or cross-sectional configurations. The motive force can be external or integral peristaltic pumps (Esashi et al., 1989; Pfahler et al., 1990; Smits, 1990; Zengerle et al., 1992).

(6) *Reaction chambers.* Biochemical reactions must occur under specific temperature and solution conditions. Chambers should be developed to mix and hold reaction components and to connect with transport channels. The reaction chambers may

require active heating elements and thermosensors to control temperature. Surface properties of the chambers also may be chemically modified to control adsorption and or to covalently bind enzymes.

(7) *Electrophoresis.* Currently, size separation of DNA molecules is performed most reproducible by electrophoresis and can work on the micron size-scale. Silicon channel electrophoresis may have several advantages over conventional systems, including greater mechanical stability and the possibility of direct integration with detector components (Brumley and Smith, 1991; Cox et al., 1982; Drossman et al., 1990; Harrison et al., 1992).

(8) *Detectors.* Integrated detectors will be essential at multiple steps in the processing system. The presence of liquid drops in a channel can be monitored by surface electrodes or capacitors. Temperatures can be detected by changes in the conductivity of metal components or by thermosensors. The electrophoresed DNA molecules can be observed by fluorescent-tag modification or by radioactive labeling.

(9) *Electronic controls and circuitry.* Electronic controls are available both as external circuitry or as "on-chip" silicon circuitry. At the development stage, a separate computer-controlled system will allow rapid testing and modification. However, the final working system would ideally have a user interface as the sole external equipment and extensive on-chip circuitry.

There is a reasonable expectation that the last two items will be the most important. First, the sensitivity of the detectors will ultimately determine the limits of scale. Increases in detector sensitivity will reduce the amount of reagents required and improve data accuracy. Improved detectors can also increase the speed at which the components operate and can increase sample throughput. Second, rapid control circuitry will allow decision making "on the fly" and will replace the current level of extensive human interaction.

The first steps in developing new technologies for DNA analysis are already being taken in many research laboratories. There is little doubt that the powerful incentive of the new genetics, together with the stimulus of the Human Genome Project, will continue to drive these developments. Only a determined effort to combine sets of components into an integrated system will push microfabricated systems into the laboratory, where real-world testing can prove their usefulness.

Acknowledgments

This work has been supported by grants from the National Institutes of Health, National Center for Human Genome Research. Additional support has been received from the Searle Foundation and the March of Dimes Foundation.

References

Alberts, B., Bray, D., Lewis, J., Raff, M., Roberts, K., and Watson, J. D. (1994) *Molecular Biology of the Cell,* pp. 291–336. New York: Garland Publishing Inc.

Ansorge, W., Voss, H., Wiemann, S., Schwager, C., Sproat, B., Zimmerman, J., Stegeman, J., Erfle, H., Hewitt, N., and Rupp, T. (1992) High-throughput automated DNA sequencing facility with fluorescent labels at the European Molecular Biology Laboratory. *Electrophoresis* 13:616–619.

Brumley R. L. and Smith L. M. (1991) Rapid DNA sequencing by horizontal ultrathin gel electrophoresis. *Nucleic Acids Res.* 19:4121–4126.

Burke, D. T. (1991) The role of yeast artificial chromosome clones in generating genome maps. *Curr. Opin. Genet. Dev.* 1:69–74.

Collins, F. S. and Galas, D. (1993) A new five-year plan for the U.S. Human Genome Project. *Science* 262:43–46.

Cox, I. J., Sheppard, C. J. R., and Wilson, T. (1982) Super resolution by confocal fluorescent microscopy. *Optik* 60:391–396.

Drossman, H., Luckey, J. A., Kostichka, A. J., D'Cunha, J., and Smith, L. M. (1990) High-speed separations of DNA sequencing reactions by capillary electrophoresis. *Anal. Chem.* 62:900–903.

Edwards, A., Civitello, A., Hammond, H. A., and Caskey, C. T. (1991) DNA typing and genetic mapping with trimeric and tetrameric tandem repeats. *Am. J. Hum. Genet.* 49:746–756.

Esashi, M., Shoji, S., and Nakano, A. (1989) Normally closed microvalve and micropump fabricated on a silicon wafer. In *International Workshop on Micro Electromechanical Systems* (MEMS 89), pp. 29–34.

Harrison, D. J., Seiler, K., Manz, A., and Fan, Z. (1992) Chemical analysis and electrophoresis systems integrated on glass and silicon chips. In *International Workshop on Solid-State Sensors and Actuators* (Hilton Head 92), pp. 110–113.

Hunkapillar T., Kaiser R. J., Koop, B. F., and Hood, L. (1991) Large-scale and automated DNA sequence determination. *Science* 254:59–67.

Melis, R., Bradley, P., Elsner, T., Robertson, M., Lawrence, E., Gerken, S., Albertsen, H., and White, R. (1993) Polymorphic simple-sequence-repeat markers for chromosome 21. *Geonomics* 16:56–62.

National Research Council Committee on Mapping and Sequencing the Human Genome (1988) *Mapping and Sequencing the Human Genome.* Washington, D.C.: National Academy Press.

Olson, M. V. (1993) The human genome project. *Proc. Natl. Acad. Sci. USA* 90:4338–4344.

Pena, S. D. J. and Chakraborty, R. (1994) Paternity testing in the DNA era. *Trends Genet.* 10:204–209.

Petersen, K. E. (1982) Silicon as a mechanical material. *IEEE Proc.* 70:420–457.

Pfahler, J., Harley, J., Bau, H., and Zemel, J. N. (1990) Liquid transport in micron and submicron channels. *Sensors Actuators* A21(23):431–434.

Smits, J. G. (1990) Piezoelectric micropump with three valves working peristaltically. *Sensors Actuators* A21(23):203–206.

Stryer, L. (1988) *Biochemistry,* pp. 120-123. New York: W. H. Freeman and Company.

Sulston, J., Du, Z., et al. (1992) The *C. elegans* genome sequencing project: A beginning. *Nature* 356:37–41.

U.S. Department of Health and Human Services and Department of Energy (1990) Understanding our genetic inheritance. *The U.S. Human Genome Project: The First Five Years.* Washington, D.C.: DHHS.

Wilson, R. K., Yuen, A. S., Clark, S. M., Spence, P., Arakelian, P., and Hood L. E. (1988) Automation of dideoxynucleotide DNA sequencing reactions using a robotic workstation. *BioTechniques* 6:776–777.

Index